NASA/SP—2016-627

# Bringing the Future Within Reach—

## Celebrating 75 Years of the NASA John H. Glenn Research Center

By: Robert S. Arrighi

National Aeronautics and
Space Administration

Glenn Research Center
Cleveland, Ohio 44135

February 2016

Available from

NASA STI Program
Mail Stop 148
NASA Langley Research Center
Hampton, VA 23681-2199

National Technical Information Service
5285 Port Royal Road
Springfield, VA 22161
703-605-6000

This report is available in electronic form at http://www.sti.nasa.gov/ and http://ntrs.nasa.gov/

ISBN 978-0-16-093210-6

9 780160 932106     90000

ISBN 978-0-16-093210-6

# Table of Contents

# Foreword

As the ground was broken in 1941 for the Aircraft Engine Research Laboratory (which would come to be known as the NASA Glenn Research Center), our predecessors had a solitary goal—to improve the state of aircraft engines. In the 75 years that followed, our goals evolved and multiplied as generations of researchers pushed the state of the art. Today we are experts in fields of research that had not even been dreamed of in 1941. We have done much more than just "improve the state of aircraft engines."

The engine components, fuels, and materials that we tested to make aircraft fly faster, higher, and more efficiently paved the way for the success of jet engines. Our fuels research was the proving ground for liquid hydrogen, which we managed through the Centaur program for journeys to the Moon and beyond. Our expertise in how fluids and combustion work in space became the foundation for our leading work in microgravity science. We find similar connections from the past to the present in all the work that we do.

We continue to lead NASA in propulsion, energy storage and conversion, materials research, and communications technology. For our work in these areas and others, we have been awarded with well over one hundred R&D 100 Awards (the "Oscars of Innovation" from R&D Magazine), two Robert J. Collier Trophies, an Emmy for technical achievement, and numerous other recognitions. As our namesake John Glenn stated, "Research and development has been mankind's most fundamental tool for meeting and shaping the challenges of the future." Today we provide key support to NASA's missions, and the impact of our research is felt the world over.

*Image 1: Glenn Center Director Jim Free (GRC–2013–C–00197).*

As our areas of research have evolved, one thing has remained constant—the spirit of service and humility that characterizes Glenn. For that spirit and for hard work, dedication, and innovation, I thank our past employees who have brought us here, our current employees who bring pride to Glenn through their work today, and employees yet to come, who will guide us through the next 75 years.

*Jim Free*
Director, Glenn Research Center

Image 2: Two women prepare instruments for data recording. The Aircraft Engine Research Laboratory (AERL) significantly increased the number of female employees during World War II (GRC–1944–C–05395).

NACA
C-5395
6-22-44

# Preface

This book seeks to spotlight the NASA Glenn Research Center's accomplishments, people, and research tools. It also aspires to elucidate the esoteric world of research laboratories, exhibit a large number of photographs and historical documents, and complement previous efforts to document the center's history.

The history of science and technology is a well-established area in the overall field of history, and the National Aeronautics and Space Administration (NASA) has carved out a significant place in that realm. This book unavoidably discusses technological topics, but it also seeks to present Glenn's role in the larger national advancement of technology, the effects of outside influences on research, and the evolution of technology over the years.

The advancement of research and technology is extremely difficult. For every success, there are failed concepts or successfully developed technologies that are shelved because of political, social, or budgetary reasons. Glenn's history includes examples of all of these categories. The advancement of the early turbojet engines, the demonstration of liquid hydrogen as a rocket propellant, the introduction of modern wind energy concepts, and the development of noise- and emission-reducing technologies for aircraft are just a few of Glenn's noteworthy achievements that have benefited the nation.

*Image 3: Test engineers in the Prop House control room in 1942 (GRC–1942–C–01072).*

Image 4: *Frontiers of Flight documented Lewis's achievements during the 1940s.*

Image 5: *Engines and Innovation provided the first in-depth history of the center (NASA SP–4306).*

Photographs play an essential role in this publication. There was a conscious effort to mix seminal images that have appeared in other center publications with newly discovered and lesser known examples. Photography has had a vital function at the center since its beginning, with photographers documenting the construction of facilities, capturing test footage for research, creating well-composed images for publicity and recruiting, and capturing visitors and staff. This book contains a mixture of these photographs, as well as the occasional candid photograph.

It also was important to include some of the actual documents from the Glenn History Collection that were used to trace the center's history. In 1999 Glenn established a History Office to collect and preserve historical materials and promote the center's history through publications, websites, and other media. (A list of these publications can be found in the appendixes.) Materials in the History Collection include

copies of official records sent to the National Archives, personal papers donated by some of the center's luminaries, the complete run of the center's newspaper, and various program and facility files.

Despite the demonstrated appreciation for history at the center and NASA's larger responsibility to share its history and research with the public, the effort to document Glenn's history was sporadic until the 1990s. Early efforts include Helen Ford's notes and timelines that document the construction and staffing during World War II (Helen Ford, "From a Historical Viewpoint," c1944, NASA Glenn History Collection, Directors Collection, Cleveland, OH); Associate Director Eugene Manganiello's lists highlighting the technical accomplishments during the 1940s and 1950s (Eugene Manganiello to Eugene Emme, 18 October 1960, "Additional Information for Draft Chronology" NASA Glenn History Collection, Directors Collection, Cleveland, OH); and Ronald Blaha's compilation of almost all the test schedules for the major facilities through the 1980s (Ronald Blaha, "Completed Schedules of NASA Lewis Wind Tunnels, Facilities and Aircraft 1944–1986," February 1987, NASA Glenn History Collection, Test Facilities Collection, Cleveland, OH).

Although the center newspaper has regularly written small pieces on different aspects of the center's history, there have been relatively few publications specifically about Glenn. The short list includes George W. Gray's *Frontiers of Flight: The Story of NACA Research* (New York: Alfred A. Knopf, 1948), which describes the early technical accomplishments at the center and its sister National Advisory Commitee for Aeronautics (NACA) laboratories; John Holmfeld's thesis "The Site Selection for the NACA Engine Research Laboratory: A Meeting of Science and Politics"

(Master's Essay, Case Institute of Technology, 1967) on the selection of Cleveland as the site for the NACA engine laboratory; and James Hawker and Richard Dali's documentation of the center's rebirth in 1980—"Anatomy of an Organizational Change Effort at the Lewis Research Center" (NASA–CR–4146).

The most notable effort, by far, is Virginia Dawson's *Engines and Innovation: Lewis Laboratory and American Propulsion Technology* (NASA SP–4306). This scholarly history, published in 1991, provides remarkable in-depth analysis of the center's first 50 years. Twenty-five years later, I hope that this publication will complement the previous efforts, provide new stories and perspectives, and share unique photographs and documents spanning Glenn's first 75 years.

*Image 6: Zella Morowitz views an NACA model (GRC–1942–C–01013).*

*Image 7: Lockheed F–94B Starfire on the tarmac at the center for a noise-reduction study in December 1959 (GRC–1959–C–52343).*

# Introduction

*"If you want to see the new world, come to Cleveland, come to Brook Park, come to this NASA facility."*

—Marcy Kaptur

*Image 8: Test engineers prepare an Atlas-Centaur model for separation tests in the Space Power Chambers (GRC–1963–C–66358).*

# Introduction

"You make the future. It's not predicting the future. That's what I have told people many times," former Center Director Abe Silverstein explained to historian Virginia Dawson in 1984. "People who say, 'how did you figure out what to do?' Well, you are making the future because the only thing that you have to go on when the future arrives is what you have stored up from the past."[1]

The National Aeronautics and Space Administration (NASA) Glenn Research Center in Cleveland, Ohio, has been making the future for 75 years. The center's work with aircraft engines, high-energy fuels, communications technology, electric propulsion, energy conversion and storage, and materials and structures has been, and continues to be, crucial to both the Agency and the region. Glenn has partnered with industry, universities, and other agencies to continually advance technologies that are propelling the nation's aerospace community into the future. Nonetheless these continued accomplishments would not be possible without the legacy of our first three decades of research, which led to over one hundred R&D 100 Awards, three Robert J. Collier Trophies, and an Emmy.

Glenn, which is located in Cleveland, Ohio, is 1 of 10 NASA field centers, and 1 of only 3 that stem from an earlier research organization—the National Advisory Committee for Aeronautics (NACA). Glenn began operation in 1942 as the NACA Aircraft Engine Research Laboratory (AERL). In 1947 the NACA renamed the lab the Flight Propulsion Laboratory to reflect the expansion of the research. In September 1948, following the death of the NACA's Director of Aeronautics, George Lewis, the NACA rededicated the lab as the Lewis Flight Propulsion Laboratory. On 1 October 1958, the lab was incorporated into the new NASA space agency and was renamed the NASA Lewis Research Center. Following John Glenn's return to space on the space shuttle, on 1 March 1999 the center name was changed once again, becoming the NASA John H. Glenn Research Center.[a]

---

[a]For simplicity in this book, during the early years Glenn is referred to as the "AERL," the "lab", or the "laboratory"; during the rest of the NACA years (before October 1958), Glenn is referred to as the "lab," the "laboratory," or "Lewis"; and during the early NASA years (between Oct. 1958 and Mar. 1999), Glenn is referred to as the "center" or "Lewis." After Mar. 1999, Glenn is referred to as the "center" or "Glenn." Whenever a name change is described, the actual names are used.

*Image 9: View of Glenn Research Center from the west in 1956 (GRC–1956–C–43664).*

## Overview

Although Congress established the NACA in 1915 as an advisory committee to coordinate U.S. aviation research, the NACA began to operate its own research laboratory at Langley Field, Virginia, in 1920. In the late 1930s, when the United States became aware that Germany was producing aircraft with speeds and altitudes superior to those of U.S. models, the NACA decided to establish two new research sites: Ames Aeronautic Laboratory in Sunnyvale, California, and the AERL in Cleveland, Ohio, alongside the Cleveland Municipal Airport. The AERL was unique in that it was dedicated entirely to issues concerning aircraft engines.

The NACA broke ground for the AERL on 23 January 1941 and initiated research activities 16 months later. The laboratory strove to improve the piston engines used to power Allied military aircraft during World War II, and researchers studied superchargers, compressors, turbines, fuels, lubrication, and entire engine systems in the AERL's new facilities. The lab also began to address the new issues associated with the introduction of the first jet engines.

The AERL reorganized after the war and added new, more powerful test facilities, concentrating nearly all of its resources on the emerging jet engine technology. Renamed the Lewis Flight Propulsion Laboratory in 1948, the lab began investigating rocket and ramjet engines for missile applications. Ramjets combust self-sustaining, high-velocity air intake to generate thrust. Researchers also began addressing flight safety issues such as crash survivability and ice formation, and analyzing high-energy propellants for rocket engines.

Lewis expanded its field of research even further in the 1950s. Advanced rocket propellant studies led to the determination that the lightweight, highly reactive liquid hydrogen could be safely used as a fuel.

*Image 10: A mechanic prepares a General Electric I–40 turbojet for testing (GRC–1946–C–15677).*

Researchers continued to investigate flight safety issues and improve turbojet and ramjet engines. In the mid-1950s Lewis researchers also began to study nuclear propulsion for both aircraft and rockets.

The Soviet Union's launch of the Sputnik satellite in October 1957 spurred the nation to pursue rockets and space missions more actively. The following year, Congress established the new NASA space agency with the three NACA laboratories serving as its foundation. For nearly 10 years, the center, renamed the NASA Lewis Research Center, concentrated all of its resources on the space program. Lewis's work on liquid-hydrogen systems and chemical rockets was critical to the success of the Saturn and Centaur upper-stage rockets. Lewis managed the Centaur Program and made significant advances in nuclear and electric propulsion, space power generation, and space communications during this period.

*Image 11: Associate Director Abe Silverstein converses with General Curtis LeMay in 1957 (GRC–1957–C–45199).*

While the Apollo Program completed a series of Moon landings in the late 1960s and early 1970s, Lewis refocused its efforts to address a new set of largely civilian aeronautical problems including noise abatement, emissions reduction, and supersonic transport. Lewis continued to manage Centaur (which was now launching both satellites and interplanetary spacecraft) and to work on nuclear propulsion and power.

The continual reduction of NASA's budget in the early 1970s and the development of the space shuttle, which Lewis did not play a primary role in, led to significant layoffs and program cancellations. In search of alternative research areas, Lewis introduced a series of new programs to remedy problems on Earth: pollution, renewable energy, energy-efficient engines, and communications. Lewis successfully applied its engine and space technology expertise to these new fields while maintaining its leadership in aeronautical research. Pulling itself up by the bootstraps in the early 1980s, Lewis acquired a diverse array of new programs involving Centaur, the new space station, communications satellites, and new efficient aircraft designs.

Lewis made significant contributions to the space station power system and the shuttle's microgravity experimental program in the 1990s, but federal budget deficits required serious cutbacks and restructuring across the Agency. Although the effects on the center were traumatic, Lewis remained resilient. In March 1999 the center was renamed the NASA John H. Glenn Research Center. Glenn was realigned in 2004 to contribute to the new Vision for Space Exploration and was active in the subsequent development of the Crew Exploration Vehicle for the Constellation Program. When Constellation was replaced in 2010, the center began contributing to the development of the new Crew Exploration Vehicle as the Agency prepared to take humans to Mars, asteroids, and beyond.

**History Lessons**

The three pillars of the center's success have been its robust physical assets, its astute leadership, and the accomplished staff. NASA is known for its test facilities, and Glenn is chocked full. Not only is nearly every square inch of Glenn's main Lewis Field campus near Cleveland, Ohio, occupied, but Plum Brook Station, its 6,000-acre remote testing facility in Sandusky, Ohio,

*Image 12: Lewis employees gather at the original picnic grounds in summer 1969 (GRC–2015–C-06537).*

contains several world-class testing facilities. The wind tunnels and engine test chambers on the Lewis Field campus were unparalleled in the 1940s and 1950s. In the 1960s, rocket stands, vacuum chambers, and microgravity facilities were introduced to address new requirements. Although the facilities added in recent decades have often been physically smaller, they are just as vital to emerging fields like computational studies and space communications. Glenn's achievements would not have been possible without its impressive array of facilities and accompanying infrastructure.

The center's leadership has frequently had the foresight to steer research into new areas of study just as aerospace technology and national needs were evolving. These changes have been abrupt and not always easy—completely refocusing on the turbojet in 1945, switching to space in 1958, returning to aeronautics in 1966, and emphasizing energy conversion in the 1970s. Since that time, the center's leadership has striven to provide a more diversified portfolio, balancing large space programs with continued aeronautics and communications work.

"Lewis Means Teamwork" was not just a slogan, it was a way of life. Research was a collaborative effort that ultimately started and ended with the researchers and scientists, but it required the assistance of many others. Engineers from the Test Installations Division integrated into the test facilities the experimental equipment that best suited the research goals. The mechanics and technicians installed the test hardware and made necessary adjustments. The analysts computed the test data. Many others made vital contributions, including the photographers who filmed the tests, the editors and graphic artists who prepared the reports, and the librarians who archived the research. The complementary efforts of a wide swath of Glenn employees has produced a legacy of 75 years of accomplishments.

Nearly all of Glenn's current core competencies can find their roots in the center's NACA period and the early 1960s. In some cases the lineage is easy to trace. For example, NASA's current ion engine is based on the engine that evolved from Harold Kaufman's early 1960s thruster design and eventually powered *Deep Space 1* and *Dawn*. Others, such as Brayton and

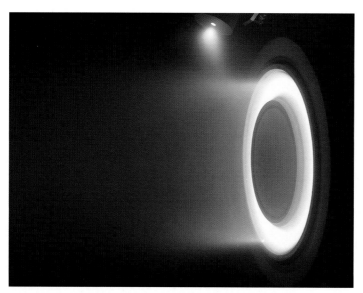

*Image 13: A Hall thruster is fired in the Electric Propulsion Laboratory (GRC–2000–C–01122).*

*Image 14: A technician prepares a test article in the Icing Research Tunnel (GRC–1985–C–09350).*

Stirling power-conversion devices, had their roots in the lab's aircraft compressor and turbine research during the 1940s and 1950s. The 1990s shuttle and space station microgravity experiments stemmed from early 1960s efforts to study liquid hydrogen in microgravity, which in turn, required the fluid dynamics experience from icing studies in the 1940s. Knowledge of these historical efforts can provide insights for current research and guide future progress.

Glenn's history not only reminds us of past accomplishments but of the people and efforts that led up to those feats. In particular, Glenn's history acknowledges those who came before us and affirms our membership in a community that extends well beyond the 3,000 colleagues currently at the center. Our history binds us to our predecessors. Many of us work in the offices where some of the center's greatest achievements were conceived, conduct research in the test cells where some of the early-generation jet engines were tested, walk by the hangar from which early pilots Howard Lilly and Joe Walker taxied their aircraft out onto the runway, or park next to a supersonic wind tunnel that analyzed early Saturn rocket designs. It is easy to get swept up with our daily responsibilities, but it is important to take time to recognize the exceptional accomplishments made by those who preceded us. Our history is all around us.

In 2004, former Center Director Julian Earls recalled the words of his father, " 'You can't come back from some place you've never been.' " He joked, "I'm not quite sure I understand what he meant by that even to this day, but what it does emphasize is the importance of taking a look backward, appreciating those who have paved the way for us to be here, for the accomplishments we make, and the accomplishments we will continue to make."[2]

### Endnotes for Introduction

1. Abe Silverstein interview, Cleveland, OH, by Virginia Dawson, 5 October 1984, NASA Glenn History Collection, Cleveland, OH, p. 14.

2. Julian Earls, "Introductory comments at the Realizing the Dream of Flight Symposium," 5 November 2003, Cleveland, OH.

Image 15: Construction of the Flight Research Building in summer 1941. The hangar was
the first structure at the new Aircraft Engine Research Laboratory (GRC–2001–C–00334).

# 1. Rising from the Mud

*"What only a few months ago was a mass of steel structures, concrete forms, and deep holes in the ground . . . rapidly is becoming the world's largest aircraft engine research laboratory."*

—Ray Sharp

Image 16: Visitors view the hangar construction in 1941 (GRC–2011–C–00331).

# Rising from the Mud

On Sunday, 3 September 1939, Britain and France declared war on Germany as Clevelanders gathered to watch the National Air Races at the future site of the National Aeronautics and Space Administration (NASA) Glenn Research Center. Over 70,000 people witnessed Art Chester's record-breaking win in the 200-mile Greve Trophy Competition. The following day a significantly larger crowd watched the intrepid Roscoe Turner capture his third and final Thompson Trophy.[3] The races, a Cleveland Labor Day tradition since 1929, provided an outlet for a public trying to grasp the ramifications of Germany's invasion of Poland that Friday. The United States would not enter the war for two more years but soon undertook preparations for that possibility. These measures led to the creation of what would become Glenn and its remote test site, Plum Brook Station.

Congress established the National Advisory Committee for Aeronautics (NACA) in 1915 to coordinate the nation's aeronautical research, which at the onset of World War I, seriously lagged behind its European counterparts.[4] The 12-member committee, composed of representatives from the military, industry, and other institutions, initially supported the military and aircraft industry in a purely advisory capacity. In 1917 the committee began constructing the Langley Memorial Aeronautical Laboratory in Virginia in order to conduct research of its own. Langley built a series of increasingly advanced wind tunnels during the ensuing years to support the research activities. George W. Lewis, the NACA's Director of Aeronautical Research, served as a liaison between the committee and the lab. By the mid-1930s NACA advances such as its eponymous engine cowling and collection of

Image 17: Crowds swarm to the edge of present-day NASA Glenn to view the 1932 National Air Races. The intensely popular event was the most evident sign of Cleveland's ties to aviation, but the city also possessed a strong aircraft manufacturing industry and the nation's largest and most innovative airport (The Cleveland Press Collection, Michael Schwartz Library, Cleveland State University).

wing shapes helped establish the preeminence of the U.S. airline industry.[5]

Spurred by the realization that German aviation technology was ahead of the United States, the NACA hurriedly expanded its research capabilities in the early 1940s. This resulted in the Ames Aeronautical Laboratory in Moffett Field, California, and the Aircraft Engine Research Laboratory (AERL) in Cleveland, Ohio, which is now known as Glenn. Wartime shortages and contractual concerns initially hindered the construction of the AERL. The AERL's first test facility began operation in May 1942, but construction of the lab continued for nearly two more years. The research staff began arriving from Langley in 1943, and the largest facility, the Altitude Wind Tunnel (AWT), was completed in January 1944. By that time the NACA's new engine laboratory was already contributing to the war effort.

## Groundwork for an Engine Laboratory

Famed aviator Charles Lindbergh was also a long-standing member of the NACA's Executive Committee. He and George Lewis made several trips to Europe in the mid-1930s to review foreign aeronautical research capabilities. They discovered that the German aircraft industry had not only restored itself after being decimated during World War I but had surpassed the United States in several areas.[6] Germany was developing aircraft that could fly higher and faster than U.S. aircraft and had a large, robust aeronautical research establishment. German engineers paid particular attention to engine research, an area that the NACA had largely ignored. Lewis and Lindbergh's findings, coupled with the increased belligerence of the Nazi regime, left many in the NACA and Congress anxious.[7]

The NACA formed a Future Research Facilities Special Committee in late 1938 to identify the types of facilities needed to expand its research. The Langley campus was too congested to accommodate these new facilities, so the NACA decided to build its new high-speed flight laboratory elsewhere. Lindbergh led a team that analyzed different locations vying for the site, including Cleveland.[8] The Cleveland Chamber of Commerce's bid emphasized the city's proximity to universities and natural resources, the eminence of the Cleveland Municipal Airport, and the

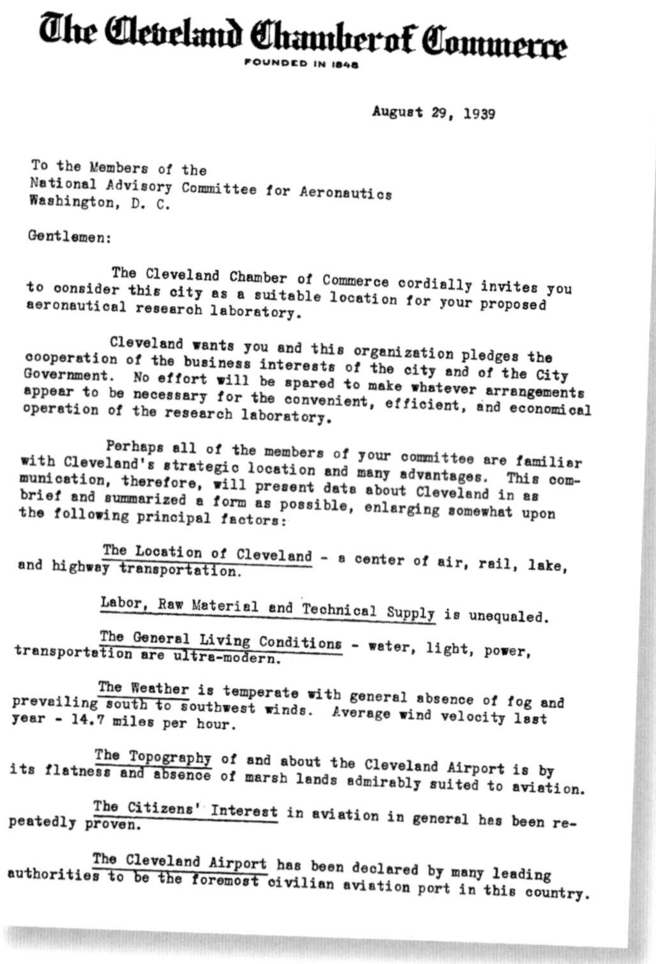

## The Cleveland Chamber of Commerce
FOUNDED IN 1848

August 29, 1939

To the Members of the
National Advisory Committee for Aeronautics
Washington, D. C.

Gentlemen:

      The Cleveland Chamber of Commerce cordially invites you to consider this city as a suitable location for your proposed aeronautical research laboratory.

      Cleveland wants you and this organization pledges the cooperation of the business interests of the city and of the City Government. No effort will be spared to make whatever arrangements appear to be necessary for the convenient, efficient, and economical operation of the research laboratory.

      Perhaps all of the members of your committee are familiar with Cleveland's strategic location and many advantages. This communication, therefore, will present data about Cleveland in as brief and summarized a form as possible, enlarging somewhat upon the following principal factors:

      The Location of Cleveland - a center of air, rail, lake, and highway transportation.

      Labor, Raw Material and Technical Supply is unequaled.

      The General Living Conditions - water, light, power, transportation are ultra-modern.

      The Weather is temperate with general absence of fog and prevailing south to southwest winds. Average wind velocity last year - 14.7 miles per hour.

      The Topography of and about the Cleveland Airport is by its flatness and absence of marsh lands admirably suited to aviation.

      The Citizens' Interest in aviation in general has been repeatedly proven.

      The Cleveland Airport has been declared by many leading authorities to be the foremost civilian aviation port in this country.

*Image 18: Cleveland Chamber of Commerce invitation to the NACA for what would become the Ames lab.*

city's continued support for aviation.[9] In the end, the NACA decided to build its new Ames Aeronautical Laboratory in Sunnyvale, California, but newspapers reported that Cleveland had been the second choice.[10]

The Lindbergh committee also reported on the nation's dearth of aircraft engine research. In 1938 only 12 or so of Langley's more than 160 researchers were working on engines.[11,12] In October 1939 the NACA began studying what types of facilities would be needed for an engine research laboratory. The report, issued on 23 January 1940, called for a new $10 million laboratory that would include an engine test stand, a fuels and lubrication facility, and—after some debate—a wind tunnel for engines.[13] Congress approved funds for the new laboratory in June 1940, just as the war in Europe escalated.[14]

After six months of comparatively modest levels of fighting, Germany had quickly conquered Denmark and Norway in April 1940, Belgium and the Netherlands in May, and France in June. Soon after, Italy declared war on France and Britain, and German U-boats began targeting ships ferrying supplies from the United States to Britain. The Battle of Britain, which commenced in July, turned into the aerial Blitz of London in early September.[15]

It was in this atmosphere that the NACA decided where to place its new engine laboratory. When 62 sites submitted bids in July 1940, the NACA quickly dismissed 16 for not meeting the prerequisite criteria. In August, a review team visited the top 20 potential sites, including 5 in Ohio. The group visited with local officials, inspected the locations, and discussed infrastructure and utilities.[16] They visited Cleveland on August 22.[17]

The NACA brought Rudolph Gagg in from Wright Aeronautical to supervise the design of the laboratory. During an 8 October 1940 meeting, Gagg strongly recommended the selection of Cleveland. At that point, Glenview, Illinois, had a slight lead in the NACA's intricate site-ranking scheme. The committee reconvened the following week in Cleveland, ostensibly to gather additional data, but in reality to secure commitments from city officials. The scientific ranking process had whittled the site list down to the point where politics could take over. Frederick Crawford and Clifford Gildersleeve, of the Cleveland Chamber of Commerce, led a negotiating group that consisted of officials from the Cleveland Municipal Airport, the Air Race Association, and the power company.[18]

There were several issues standing between Cleveland and the new lab, including the location, utility rates, and the grandstands for the air races. The airport agreed to relocate the proposed site from Brookpark Road to a more secluded plot between its northern fenceline and the Rocky River valley. This site served as a parking lot for the air races.[19] Crawford assured the NACA representatives that the air races, which were not run in 1940, were permanently over, and the stands along the edge of the property would be removed. Crawford also brokered a deal in which the electric company would provide discounted rates if the lab agreed to operate its large facilities overnight when demand for electricity was low.[20] On 24 October 1940, the

*Image 19: Aerial view of the National Air Races and parking lot in September 1938 (GRC–1991–C–01875).*

GROUNDBREAKING -- 1941

*Image 20: Groundbreaking ceremony for the AERL in January 1941. Left to right: William Hopkins, John Berry, Ray Sharp, Frederick Crawford, S. Paul Johnston, George Brett, Edward Warner, Sidney Kraus, Edward Blythin, and George Lewis (GRC–1982–C–06410).*

NACA formally selected Cleveland as the site for the new aircraft engine research lab. The press announced the selection four weeks later.[21]

## Building the Laboratory

On the blustery afternoon of 23 January 1941, several prominent Ohio politicians joined NACA officials and a handful of local reporters for the AERL groundbreaking ceremony. After a lunch downtown at the prestigious Hotel Cleveland, the group rode out to the airfield for a few brief remarks and a photo opportunity with the decorative pick and shovel. Despite the smiles, the war in Europe loomed in everyone's mind. When asked why the new lab was needed, NACA committee member Edward Warner replied, "The difference between winning a war and losing it may be the difference between [a] 1000 and 2000-horsepower motor, or the difference between [the] ability to fly at 20,000 feet or 30,000 feet." He darkly added,

*Image 21: Surveyors taking measurements for the NACA's engine lab in August 1941 (GRC–2011–C–00347).[22]*

*Image 22: Helen Ford (center) and Charles Herrmann (right) in front of the lab's first administrative building, the Farm House, in October 1941 (GRC–2006–C–01209).*

*Image 23: The radio shack office was located along Brookpark Road just east of the lab's entrance, in 1942 (GRC–2011–C–00346).*

"The consequences of the work done here may mean the continuance of our ability to exist."[23]

Charles Herrmann, a construction engineer from Langley, set up an office in a small radio shack on the edge of a frozen airfield. He was joined in the coming weeks by Helen Ford, an assistant administrator from NACA Headquarters, and inspector William Waite. The group considered themselves as pioneers establishing an outpost in the north. The once-teeming grandstands now appeared hazy in the wintery distance.[24]

Herrmann and Ford hired local construction crews to build the buildings and facilities, supervised the work of the inspectors, and interviewed potential employees. In addition to these tasks, Ford managed to keep the fire in the two-room radio house burning and supply coffee for the growing number of construction engineers. Herrmann and Ford regularly worked into the night to ensure that the projects continued on schedule.[25]

*Image 24: The AERL design team works in an office above Langley's Structural Research Laboratory during April 1941 (GRC 2007–C–02563).*

Meanwhile at Langley, the AERL design team was at work in an office above the Structural Research Laboratory designing the test facilities, laboratories, and offices that would soon populate the AERL site.[26] The six principal structures were the Engine Research Building, the hangar, the Fuels and Lubrication Building, the Administration Building, the Engine Propeller Research Building, and the AWT, with the Icing Research Tunnel (IRT) added in 1943.

James Braig arrived from Langley in the spring to supervise the storage of the incoming equipment and supplies underneath the only available structure, the grandstands. On 30 July 1941, the slowly growing staff of construction engineers and inspectors relocated to an empty residence, "the Farm House," along the main road entering the site.[27]

❖ ❖ ❖ ❖ ❖ ❖

Concurrently, another major construction effort was underway 60 miles to the west in Sandusky, Ohio. In late 1940 the U.S. War Department had begun making plans to build several dozen munitions manufacturing facilities. It would be another year before the

nation entered World War II, but the government did not wish to repeat the mistakes that it had made before World War I. The nation's failure to coordinate its ordnance manufacturing capabilities in advance of that conflict had hampered the military's effectiveness.

*Image 25: View northward toward the lab's entrance at Brookpark Road in August 1941. The Farm House is on the left (GRC–2011–C–00350).*

*Image 26: A barn along the perimeter of the Plum Brook Ordnance Works (PBOW) in 1941. The area had been home to some of the most fertile land in the state (GRC–2015–C–06562).*

The government had taken steps to address these problems, but it had been difficult to fund the projects because of the public's isolationist tendencies during the interwar period.[28]

*Image 27: Crews remove a farmhouse from the PBOW site in summer 1941 (GRC–2015–C–06561).*

After watching Germany's rapid advance across Europe in spring 1940, Congress approved a program in July 1940 to supply the army with munitions.[29] The government began to convince private industries to transition into war material suppliers and to establish munitions and ordnance facilities on wide tracts of private property. For defensive purposes, they selected sites between the Appalachian and Sierra Nevada mountains and at least 200 miles from any border. The government preferred undeveloped areas close to a city with rail access and a good water supply. The War Department built 77 munitions plants and ordnance works between 1940 and 1942. These sprawling complexes required the seizure of 44 million acres of private property.[30]

The War Department selected a 9,000-acre swath of farmland in northwest Ohio for one such ordnance manufacturing facility—the Plum Brook Ordnance Works (PBOW). The location was relatively close to Lake Erie ports, railway lines, and highways.[31]

*Image 28: PBOW staff raise the flag in front of the PBOW Administration Building in 1941 (GRC–2015–C–06812).*

*Image 29: Construction of one of the 99 concrete bunkers used to store explosives, c1941 (GRC–2015–C–06817).*

*Image 30: Workers erect an elevated guard tower around the perimeter of the PBOW in 1941 (GRC–2015–C–06816).*

Officials informed the community of the decision during a meeting at a local hall on 7 January 1941. The War Department repeatedly emphasized that time was of the essence. The 150 farmers who owned the land had until March to agree to the sale and to mid-April to move both their families and equipment. Over 100 homes, several small businesses, the local town hall, and a cemetery had to be relocated for the plant.[32-34]

The community and public officials pushed back, particularly regarding the low offers that were proffered by the land agents, but the War Department remained adamant. The government acquired all but 10 of the 150 properties by the March deadline. The holdouts vacated the land while the courts reviewed their compensation offers. In the end all the farmers vacated their properties and only one plaintiff received a significant increase in payment.[35,36] The arguments caused delays, but the construction proceeded. Over the course of six months, construction crews transformed the patchwork farmland into a small industrial city. The first trinitrotoluene (TNT) line began operating in December 1941.

❖ ❖ ❖ ❖ ❖ ❖

Construction at the new NACA laboratory in Cleveland was proceeding slowly. It was imperative to complete the hangar in order to store equipment, house the shops, and shelter the staff that would be arriving from Langley. The NACA selected the local R.P. Carbone Construction Company to build the

hangar and the Engine Propeller Research Building.[37] The company's poor performance and the wartime shortages of building materials delayed work at both facilities.[38] Matters were compounded by Langley's lack of experience designing complex engine test facilities such as the AWT. The NACA brought in experts from Wright Field, engine manufacturing companies, and Carrier Corporation to assist.

In August 1941 Edward Raymond Sharp arrived from Langley to oversee the construction. He would spend nearly 20 years managing the lab. Sharp was a former World War I seaman who joined the NACA as an airship rigger at Langley in 1922. After earning his law degree in 1925, Sharp spent the next 15 years as Langley's administrative officer.[39] In 1940 the NACA detailed Sharp to Ames for five months to oversee that construction work. Afterward he accompanied Ernest Whitney on the scouting mission of potential sites for the engine laboratory. Sharp then spent nearly a year as Chief of Langley's Construction Division before being assigned to Cleveland.[40]

Sharp would later emerge as a fatherly figure who preferred to advise rather than order, but he worked relentlessly in the early 1940s to increase the pace of construction. The 12- to 16-hour days and seven-day workweeks kept his mind from his wife Vera and their children left behind in Virginia. He wrote to her shortly after Christmas 1941, "Honey, I haven't had

*Image 31: Westward view of the Steam Plant and general AERL construction area in 1942 (GRC–2007–C–02309).*

*Image 32: Ray Sharp at his desk in 1942 (GRC–2015–C–06568).*

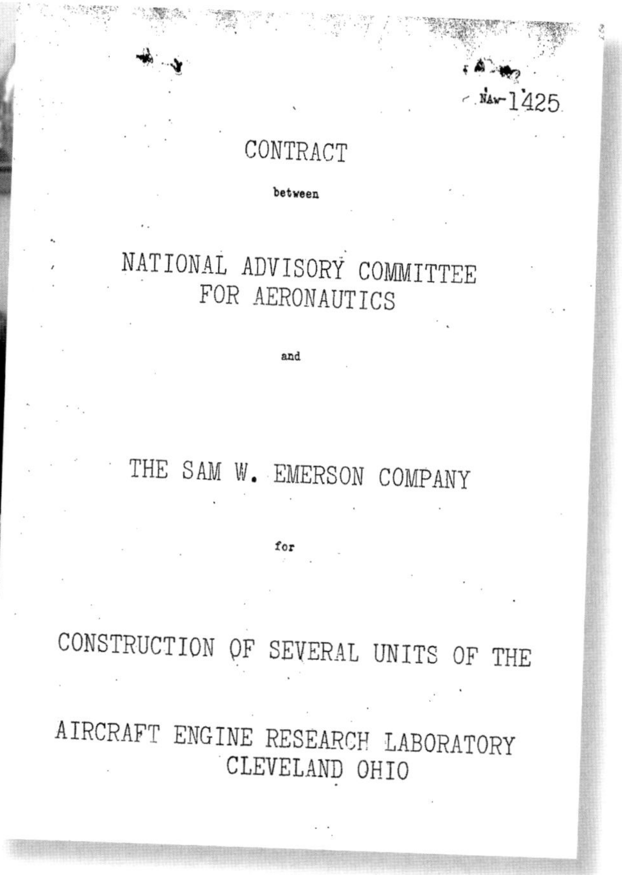

CONTRACT

between

NATIONAL ADVISORY COMMITTEE
FOR AERONAUTICS

and

THE SAM W. EMERSON COMPANY

for

CONSTRUCTION OF SEVERAL UNITS OF THE

AIRCRAFT ENGINE RESEARCH LABORATORY
CLEVELAND OHIO

*Image 33: NACA contract with the Emerson Company.*

much time to think, for it is a case of drive myself and drive others, but I missed my Sunday at home. Did Ray go to Sunday school? How is Brother's cold? Is Brother Edward getting lined up to leave? How long will it take you to leave? God, I get lonely up here without my wife and my family."[41]

In fall 1941 the AERL installed temporary offices inside the recently completed hangar to house the incoming personnel from Langley. The design team from Langley arrived in Cleveland on a snowy Monday, 15 December 1941.[42] The world had changed dramatically over the past week as they were making final arrangements for the transfer. Japan had attacked Pearl Harbor the previous Sunday and immediately invaded the Philippines. The United States and Great Britain declared war on Japan, which spurred Germany to declare war on the United States. After watching the violence from afar for over two years, the United States now had two wars to fight. It was clear that the NACA had to step up its efforts to complete the AERL construction and begin resolving problems facing military aircraft.

On New Year's Eve, three days after penning the letter to his wife, Sharp traveled to NACA Headquarters in Washington, DC, to hammer out the final details

of the contract with the prime construction firm for the remainder of the AERL structures, the Sam W. Emerson Company. In the early evening NACA Secretary John Victory and Sharp shuttled the contract to the White House where President Franklin Roosevelt approved it.[43] The pace of the construction accelerated almost immediately afterward.

By early February 1942 approximately 270 people occupied the offices on the hangar floor. In addition to the administrative staff, the hangar housed the mechanical and structural engineers, draftsmen, inspectors, mechanics, and technical service personnel.[44] The veterans from Langley were joined by journeyman laborers and untrained youth from the economically depressed Cleveland area. People were anxious for employment, even if they were not sure what work was to be done or even the location of the new laboratory. Ford recalled that some applicants showed up to apply at her and Herrmann's homes because they did not know where the lab was located.[45]

Image 34. In May 1942 the Prop House became the first operating facility at the AERL. It contained four test cells designed to study large reciprocating engines. Researchers tested the performance of fuels, turbochargers, water-injection, and cooling systems here during World War II. The facility was also used to investigate a captured German V–I buzz bomb during the war (GRC–1942–C–01134).

Image 35. The laboratory established its own fire department while the lab was being constructed in the early 1940s. The group, which was based at the Utilities Building, not only responded to emergencies but conducted safety inspections, checked fuel storage areas, and supervised evacuation drills and training. In addition, they frequently assisted local fire departments and responded to accidents at the adjacent Cleveland Municipal Airport (GRC–1943–C–04291).

*Images 36: Drafting staff members at work in the temporary hangar offices during 1942 (GRC–2015–C–06545).*

*Images 37: Temporary offices constructed inside the hangar to house the architectural and drafting personnel as well the machine shops (GRC–2015–C–06557).*

## Completing the Job

In early 1942 the Japanese captured large regions of the Pacific, including the Philippines, where 15,000 U.S. troops were taken prisoner in April. The pressure to expedite the construction of the lab increased. Military aircraft required improvements in fuels, lubrication, and engine cooling. In addition, the Air Corps was about to introduce its new superweapon—the Boeing B–29 Superfortress—and needed to test its engines in the AERL's unique facilities.

The Engine Propeller Research Building, better known as the Prop House, was the first test facility to be completed. The Prop House, which contained four atmospheric test stands to study full-scale piston engines, was set back near the woods on the far side of the lab to muffle the engine noise.[46]

On 8 May 1942, NACA management and local officials crowded in the control room to watch George Lewis activate a Wright Aeronautical R–2600 Cyclone engine and officially initiate research at the new laboratory.[47] AERL researchers used the R–2600 to develop a procedure for standardizing the evaluation of lubricating oils.

The ceremony, however, was largely symbolic because most of the lab was still under construction. The government immediately implemented a series of drastic measures to accelerate the work. General Henry "Hap" Arnold, Commander of the U.S. Army Air Forces, requested that the NACA's priority rating be elevated.[48] The military provided special supplies, the NACA signed new agreements with contractors and pressured them to meet deadlines, and Congress approved additional funds. George Lewis made weekly trips from Washington, DC, to examine the progress.[49,50]

Although a few researchers from Langley arrived in May 1942 for the Prop House opening, the engine research group did not begin migrating to Cleveland en masse until the fall. The Fuels and Lubrication staff arrived in mid-November, followed two weeks later by the Engine Analysis Section.[51] In summer 1942, the shops were moved from the hangar to the new Technical Services Building, and the Fuels and Lubrication Building and Engine Research Building opened. In late December the Administration Building was completed, and the personnel housed in the hangar transferred into new offices across the street.[52]

Now that most of the facilities were up and running, the NACA invited local officials and members of the nation's aeronautical upper crust to review the AERL on 20 May 1943. Attendees included Orville Wright, General Oliver Echols, Henry Reid, William Durand, Sam Emerson, and others. In addition, the NACA's

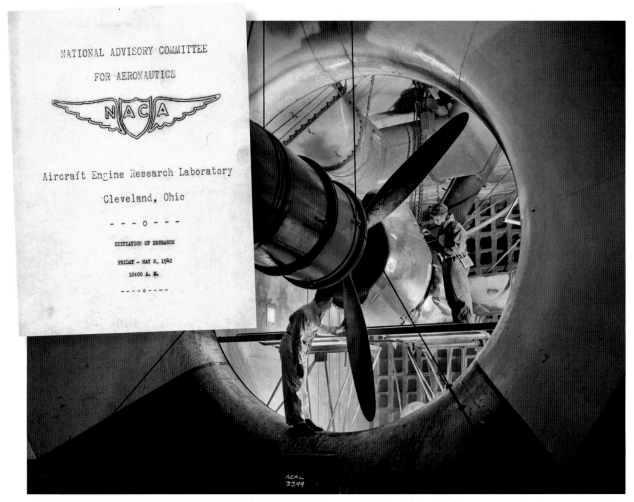

*Image 38 (above): Program for the initiation of research of the AERL. Image 39: AERL mechanics work on an engine installation on one of the Prop House's test stands in 1943 (GRC–1943–C–3349).*

Image 40: Receptionist Mary Louise Gosney enjoys the new
Administration Building in July 1943. She started at the lab in
November 1941 and spent an entire year in the hangar. She also
served as the lab's clearance officer and would later head the
Administrative Services Division (GRC–1943–C–01842).

The AERL staff was small and close-knit. They
celebrated the completion of each building with a
party. In October 1942, the AERL began issuing
a weekly newsletter, Wing Tips, whose "aim will
be to broadcast those events which are important
to the greatest number rather than strive merely
to entertain."[53] Wing Tips disseminated essential
information regarding administrative activities,
construction programs, and educational opportunities, but its tone and content reflected the lab's famil-
ial atmosphere. Popular features included "Lobby
Lines," which shared the comings and goings of
visitors and staff members, the classified ads, and
updates on intramural sports.

Image 41: Wing Tips kept the staff updated on the progress of the
construction.

executive committee made their first group visit to the AERL. The NACA officially dedicated the AERL with a flag-raising ceremony at noon. The visitors were served lunch in the cafeteria and toured the new test facilities in the afternoon. Afterward, Sharp and a few others drove the group downtown for dinner at the Hotel Cleveland.[54,55]

By this point the tide of the war had shifted to favor the Allied forces. Soviet troops beat back the German siege of Stalingrad in February 1943 and began the slow, determined push to Berlin; and U.S. forces were able to halt Japan's seizure of islands in the Pacific.[56] Nonetheless, there was much fighting to be done, and aircraft were an essential element. The lab's two wind tunnels were not yet in operation. The military was counting on the AWT to resolve serious engine cooling problems for the new Boeing B–29 bomber, and on the IRT to improve deicing techniques for military transport planes crossing the Himalaya mountains.

The AWT was the AERL's largest and most sophisticated test facility, and not surprisingly, the most difficult to design and build.[57] It was the first wind tunnel capable of operating full-scale engines under realistic altitude flight conditions, which were achieved by lowering the temperature and air pressure within the tunnel.[58] The tunnel shell consisted of two steel layers with a blanket of insulation between. The 1-inch-thick inner steel layer could withstand external atmospheric pressure when the tunnel interior was evacuated to high-altitude pressure levels. Massive exhauster equipment reduced the internal pressure levels, and a unique air scoop placed just beyond the test section prevented the engine exhaust from contaminating the airstream.[59]

The most difficult design aspect, however, was reducing the temperature of such a large volume of air to the specified –47 degrees Fahrenheit. After a great deal of struggle, the NACA contracted the renowned refrigeration experts at Carrier Corporation to design

*Image 42: Construction of the AWT in April 1943. The facility would be up and running in eight months (GRC–2008–C–00817).*

the refrigeration system. Carrier engineers developed a unique accordionlike shape for the cooling coils, which greatly increased their surface area. The result was the largest refrigeration system in the world and the pinnacle of renowned company owner Willis Carrier's personal career.[60] The system was robust enough to service both the AWT and the smaller IRT, which began operating in 1944.

After a sluggish start, construction of the AWT ramped up sharply in 1943. By the time that the massive project was completed in January 1944, researchers at other AERL facilities were well into their wartime investigations. The NACA had successfully accelerated the construction of the lab to meet the intended deadlines, but the cost had nearly doubled the original estimates.[61] The three-year struggle resulted in the most sophisticated engine research laboratory in the nation.

*Image 43 (above): Cartoon showing AWT's ability to simulate altitude conditions (Wing Tips). Image 44: Aerial view of the AERL in June 1945 (GRC–1945–C–10493).*

## Endnotes for Chapter 1

3. "Fliers Race for Coveted Trophy," *The Zanesville Signal* (4 September 1939).

4. Jerome Hunsaker, "Forty Years of Aeronautical Research" *Smithsonian Report for Aeronautics* (Washington, DC: Smithsonian Institution, 1956).

5. Richard P. Hallion, "America and the Air and Space Revolution: Past Perspectives and Present Challenges," National Aeronautical Systems and Technology Conference (13–15 May 2003), pp. 12–16 and 21.

6. George W. Lewis, "Report on Trip to Germany and Russia: September–October 1936," File 11059, 1936, NASA Headquarters Historical Collection, Washington, DC.

7. Alex Roland, *Model Research* (Washington, DC: NASA SP–4103, 1985), chap. 7.

8. John F. Victory, "Remarks at Cleveland Chamber of Commerce Luncheon Commemorating the Tenth Anniversary of Foundation of the Lewis Flight Propulsion Laboratory," 23 January 1951, NASA Glenn History Collection, Cleveland, OH.

9. Cleveland Chamber of Commerce, "An Invitation to the National Advisory Committee for Aeronautics," 29 August 1939, NASA Glenn History Collection, Directors' Collection, Cleveland, OH.

10. John D. Holmfeld, "The Site Selection for the NACA Engine Research Laboratory: A Meeting of Science and Politics" (Bachelor's thesis, Case Institute of Technology, 1967).

11. Benjamin Pinkel interview, Cleveland, OH, by Virginia Dawson, 4 August 1985, NASA Glenn History Collection, Oral History Collection, Cleveland, OH.

12. James Hansen, *A History of the Langley Aeronautical Laboratory, 1917–1958* (Washington, DC: NASA SP–4305, 1985).

13. "Report of the Special Committee on New Engine Research Facilities" File 11059, 24 January 1940, NASA Headquarters Historical Collection, Washington, DC.

14. Holmfeld, Bachelor's thesis.

15. Reg Grant, *World War II: Europe* (Milwaukee, WI: World Almanac Library, 2004).

16. Holmfeld, Bachelor's thesis.

17. "Cleveland Bids," *Cleveland Plain Dealer* (22 August 1940).

18. Holmfeld, Bachelor's thesis.

19. Holmfeld, Bachelor's thesis.

20. Holmfeld, Bachelor's thesis.

21. "City Gets $8,400,000 Plane Lab," *Cleveland Plain Dealer* (25 November 1940).

22. Lawrence Hawkins, "More Power to Us," *Cleveland Plain Dealer* (23 February 1941).

23. Hawkins, "More Power to Us."

24. Helen Ford, "From a Historical Viewpoint," c1944, NASA Glenn History Collection, Directors Collection, Cleveland, OH.

25. Ford, "From a Historical Viewpoint."

26. Ernest G. Whitney, "Lecture 22—Altitude Wind Tunnel at AERL," 23 June 1943, NASA Glenn History Collection, Cleveland, OH.

27. Helen Ford to Miss Scott, 23 December 1942, NASA Glenn History Collection, Directors' Collection, Cleveland, OH.

28. Steve Gaither and Kimberly Kane, "The World War II Ordnance Department's Government-Owned Contractor-Operated (GOCO) Industrial Facilities: Indiana Army Ammunition Plant Historic Investigation," December 1995, NASA Glenn History Collection, Plum Brook Ordnance Works Collection, Cleveland, OH.

29. Frank N. Shubert, "Mobilization: The U.S. Army in World War II, The 50th Anniversary," *www.army. mil/cmh-pg/documents/mobpam.htm* (accessed 11 June 2015).

30. Gaither, "The World War II Ordnance."

31. "Five Railroads Located in Plant Area," *Sandusky Star Journal* (6 January 1941).

32. "Purchasing Agent Tells Perkins Township Families of Acquiring of Properties," *Sandusky Star Journal* (8 January 1941).

33. "New Site for Perkins Cemetery," *Sandusky Star Journal* (17 January 1941).

34. "Production to Begin Soon at Plum Brook Powder Plant," *Cleveland Plain Dealer* (26 October 1941).

35. "Farmers Leave New TNT Area and County Roads Will Be Closed to Public October 12," *Sandusky Star Journal* (22 September 1941).

36. Mark Bowles, *Science in Flux: NASA's Nuclear Program at Plum Brook Station 1955–2005* (Washington, DC: NASA SP–2006–4317, 2006).

37. "Local Concerns Get Big Contracts," *Cleveland Plain Dealer* (19 January 1941).

38. Virginia Dawson, *Engines and Innovation: Lewis Laboratory and American Propulsion Technology* (Washington, DC: NASA SP–4306, 1991).

39. Michael Vaccaro to Allen Gamble, "Nomination of Dr. Edward R. Sharp for the First Annual Presentaton of the Career Service Award," 26 August 1955, NASA Glenn Collection, Sharp Collection, Cleveland, OH.

40. Henry Reid to Staff, "Reorganization of the Construction Division," 20 September 1940, NASA Glenn History Collection, Cleveland, OH.

41. Raymond Sharp to Vera Sharp, 28 December 1941, NASA Glenn Collection, Sharp Collection, Cleveland, OH.

42. Mary Louise Gosney, "Lobby Lines," *Wing Tips* (5 December 1956).

43. Edward R. Sharp, "Excerpt From Chronological Record of Events Relative to Cost-Plus-a-Fixed-Fee Contract Naw-1425," NASA Glenn History Collection, Cleveland, OH.

44. "Construction Program—Cleveland, Ohio," 10 February 1942, NASA Glenn History Collection, Dawson Collection, Cleveland, OH.

45. Ford, "From a Historical Viewpoint."

46. "Initiation of Research at the NACA Laboratory," press release, 8 May 1942, NASA Glenn Collection, Directors Collection, Cleveland, OH.

47. Dawson, *Engines and Innovation.*

48. Henry H. Arnold and J. H. Towers to Director, Bureau of the Budget, 11 May 1942, NASA Glenn History Collection, Cleveland, OH.

49. Jesse Hall interview, Cleveland, OH, by Bonnie Smith, 28 August 2002, NASA Glenn History Collection, Oral History Collection, Cleveland, OH.

50. Henry H. Arnold and J. H. Towers to Director.

51. "More LMAL Employees to Report Here," *Wing Tips* (24 November 1942).

52. "Offices Are Moved to New Administration Building," *Wing Tips* (15 December 1942).

53. "Folks, Meet Wing Tips," *Wing Tips* (27 October 1942).

54. Donald Schneider, "Orville Wright Hails Plane Lab," *Cleveland Plain Dealer* (21 May 1943).

55. "Flag Presented to Staff as Aviation Leaders Dedicate AERL to Research," *Wing Tips* (21 May 1943).

56. Grant, *World War II: Europe.*

57. Robert Arrighi, *Revolutionary Atmosphere: The Story of the Altitude Wind Tunnel and Space Power Chambers* (Washington, DC: NASA SP–2010–4319, 2010).

58. Whitney, "Lecture 22."

59. Whitney, "Lecture 22."

60. Margaret Ingels, *Willis Haviland Carrier: Father of Air Conditioning* (Garden City, NY: Country Life Press, 1952), pp. 97–101.

61. Roland, *Model Research,* chap. 8.

# 2. Keeping Them Flying

*"The service that you essential employees are rendering…*
*is of more value … in the winning and shortening of the war*
*than it would be were you in active military services."*

—John Victory

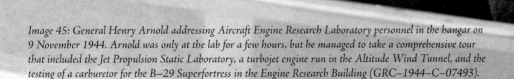

*Image 45: General Henry Arnold addressing Aircraft Engine Research Laboratory personnel in the hangar on 9 November 1944. Arnold was only at the lab for a few hours, but he managed to take a comprehensive tour that included the Jet Propulsion Static Laboratory, a turbojet engine run in the Altitude Wind Tunnel, and the testing of a carburetor for the B–29 Superfortress in the Engine Research Building (GRC–1944–C–07493).*

*Image 46: A Fuels and Lubrication Division researcher at work in August 1943 (GRC–1943–C–02124).*

The poor performance of U.S. aircraft during World War I had impressed upon the military the need to prioritize aerial bombing in World War II. The Army Air Corps relied on just 19 different aircraft models for all of its combat and transportation requirements during World War II, and there were only a handful of engine models available to power these aircraft. It was vital to improve the performance of these aircraft without massive redesigns. The NACA suspended its basic aeronautical research activities almost immediately after Pearl Harbor to address this issue. The Langley and Ames laboratories concentrated their efforts on reducing drag, dive control, ditching, and deicing. The Aircraft Engine Research Laboratory (AERL) was responsible for combatting problems with the piston engines that drove the aircraft, such as cooling, improved fuels, and combustion.[62]

The emergence of the turbojet engine midway through the war, however, significantly altered the AERL's focus. The jet engine had the potential to dramatically improve the performance of military aircraft but required improvements before it would be viable.[63] The AERL promptly made the leap to jet propulsion while continuing its wartime piston engine assignments. "You've

Image 47: AERL recruiting pamphlet.

Image 48: AERL staff members wish Ray Sharp happy holidays in December 1945 (GRC–1945–C–13948).

got a dual task," General Henry Arnold, Chief of the U.S. Army Air Corps, told AERL staff in December 1944. "You've got a job ahead of you to keep the army and the navy air forces equipped with the finest equipment that you can for this war. You also have the job of looking forward into the future and starting now those developments, those experiments, that are going to keep us in our present situation—ahead of the world in the air. And that is quite a large order, and I leave it right in your laps."[64]

The AERL, with its stable of new facilities and research staff, was ready to tackle specific engine problems and improve overall engine performance. The researchers could analyze superchargers, turbines, and compressors in the Engine Research Building test cells, study new fuel blends in the Fuels and Lubrication Building's laboratories, test complete engine systems in the Prop House or the Altitude Wind Tunnel (AWT), and study ice accumulation in the Icing Research Tunnel (IRT). After the dedication of the lab in May 1943, the AERL had roughly two and a half years to contribute to the war. The timing coincided with the completion of hundreds of new aircraft and engine manufacturing facilities across the nation and key Allied victories overseas.[65] The tide of the war was turning.

## Running the Lab

As the AERL facilities were beginning to come online in December 1942, the NACA Director of Aeronautical Research, George Lewis, asked construction manager Ray Sharp to stay on and administer the new lab.[66] Sharp was a natural leader who would serve as the public face of the lab for the next 20 years. Sharp was frequently out of his office visiting the test cells and offices. As a result he was familiar with nearly every employee and, despite his lack of a technical background, understood the work being done. He and his wife Vera were parental figures for the staff, skilled hosts of visitors and local officials, and fierce advocates for the researchers.[67] The staff expressed their affection for Sharp in December 1942 by presenting him with a linen scroll signed by every employee. It stated that they had "received word of [Sharp's] appointment as Manager with rejoicing and with reassurance."[68]

Headquarters also named Carlton Kemper the AERL's Executive Engineer. He was responsible for managing the lab's research. Kemper joined Langley in 1925 and was named Chief of the Powerplants (engines) Division four years later. The division performed fundamental research on engine efficiency, power, and fuel consumption. The Langley propulsion researchers had limited test facilities and generally conducted their investigations using just a single cylinder.[69,70] The Powerplants Division, which expanded dramatically in the late 1930s, provided the bulk of the initial research staff for the AERL. Addison Rothrock, who spent 16 years at Langley before transferring to Cleveland in November 1942, served as Acting Executive Engineer until Kemper's arrival in January 1943.[71]

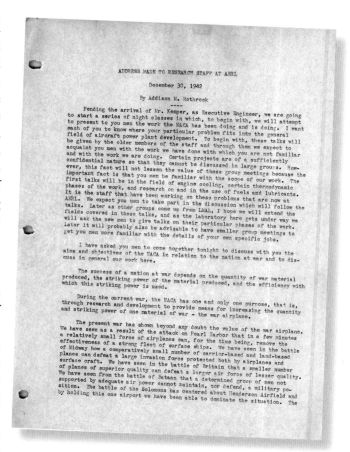

*Image 49: Addison Rothrock's speech to AERL staff in December 1942.*

In that time, Rothrock initiated a program of night classes for the employees to learn about other areas of research at the lab and how each group contributed to the broader goals of the AERL. A researcher or guest speaker would describe his work, followed by a general discussion afterward. At the initial meeting Rothrock described the history of the NACA, the importance of quality aircraft to the war effort, and the group's expectations of the staff members.[72] George Lewis, who was struck by the comprehensiveness of Rothrock's address, took steps to distribute copies to all new employees at each of the NACA laboratories.[73]

In early 1944, the AERL introduced a series of "smoker talks." Lab managers, headquarters staff, or visiting military officials discussed the broader role of NACA research. Guest speakers discussed topics related to the NACA or the war. The talks were followed by refreshments and informal group discussions on the topic. NACA Secretary John Victory spoke at the initial event in March 1944 and lab managers, military officials, and visiting researchers were featured in future events.[74]

The AERL originally included five hastily organized research divisions—the Fuels and Lubrication Division (Rothrock), Thermodynamics Division (Benjamin Pinkel), Engine Installation Division (Ernest Whitney), Engine Research Division (Charles Moore), Flight Research Division (Joseph Vensel), and Supercharger Division (Oscar Schey). The division chiefs, all Langley veterans, worked independently when organizing their personnel. As such, some divisions contained only a couple of heavily staffed sections, whereas other divisions had many sparsely populated sections. The lack of communication resulted in both inconsistency and duplication of effort.[75] The AERL would rectify the situation after the war, but it was crucial to get up and running as quickly as possible.

The military sponsored nearly all NACA research during the war. They submitted requests for different studies to NACA Headquarters. George Lewis then worked with officials at the NACA laboratories to develop methods of performing the work. In order to get large numbers of aircraft into the air as quickly as possible, the Army Air Corps insisted that the air war would be fought with proven technology—namely the piston engine and combat aircraft already in production. There was no time to develop new technologies as the nation rushed to mobilize and establish new production plants.[76]

This policy limited the military's options, particularly regarding engines. All U.S. military aircraft were powered by just seven different types of piston engines during World War II. The three most powerful were Pratt & Whitney's R–2800 and Wright Aeronautical's R–2600 and R–3350. Only the R–2600 had been in production prior to 1940.[77] The rush to transfer the new engines from the drawing board to the production line resulted in a multitude of difficulties that required remediation.

The military's policy of using existing aircraft models for the war made it necessary to improve aircraft and engine performance without any lengthy redesigns. The use of superchargers and turbosuperchargers, which pumped additional airflow into the combustion chamber, significantly improved engine operation particularly in the thinner air at higher altitudes. The devices, however, also included tradeoffs such as increased engine temperatures, which caused several ill effects, including engine knock.[78]

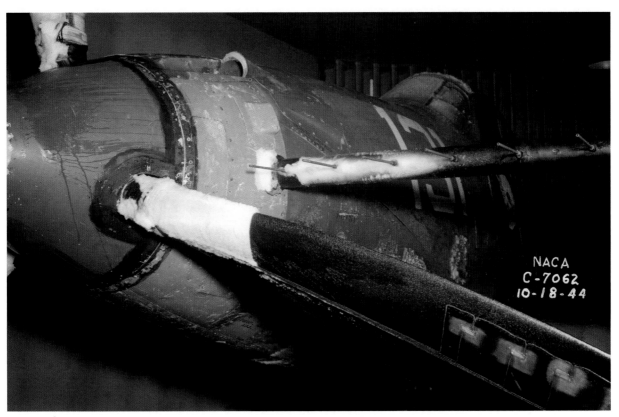

*Image 50: P–39 Mustang fuselage being tested in the IRT in October 1944 (GRC–1944–C–7062).*

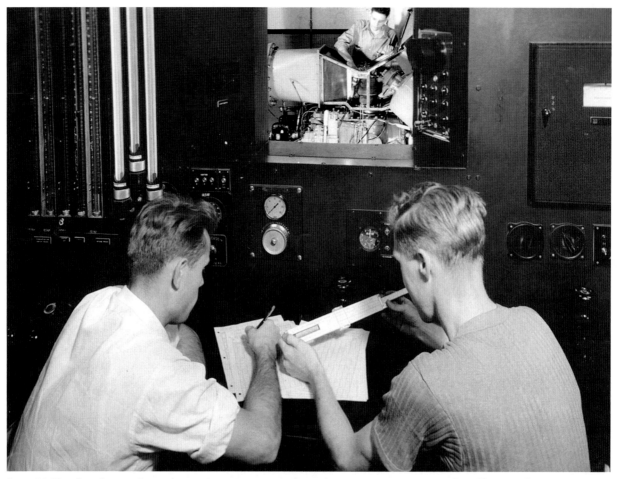

*Image 51: View from the control room during a June 1944 engine cooling investigation in an Engine Research Building test cell (GRC–1944–C–05498).*

The AERL was structured to address the primary concerns. The Supercharger Division sought to develop ways to drive more and more air into the engine's supercharger, the Fuels and Lubrication Division tried to reduce premature combustion of new fuel blends, and the Engine Research and Thermodynamics divisions worked to reduce the overheating caused by the resulting increased engine performance. These areas formed the basis for most of the AERL's research during the war.[79]

## Wartime in Ohio

In Sandusky, the Plum Brook Ordnance Works (PBOW) began operating on 16 December 1941, one week after Pearl Harbor.[80] The 9,000-acre facility was one of the largest ordnance plants in the nation. The Trojan Powder Company operated the two trinitrotoluene (TNT) and dinitrotoluene (DNT) plants and a pentolite production facility for the War

Department. PBOW contained scores of buildings, a maze of roads and railroad tracks, and extensive infrastructure. There were also 99 Quonset-hut-like concrete bunkers placed in several neat rows with sod and vegetation over the top. The site was protected by high fences and elevated guard houses.

PBOW operated around the clock for the duration of the war, churning out record amounts of powder to meet the military's continual requests for more and more explosives. The staff packed the powder into wooden crates that were stored in the concrete bunkers. The crates were then trucked approximately 80 miles to the Ravenna Arsenal, where others packed the powder into shells and bombs. The PBOW produced nearly one billion tons of explosives during its four years of operation—with the staff doubling and then tripling the anticipated output levels while maintaining a remarkable safety record.[81]

Image 52: PBOW staff gather for an October 1941 ceremony in front of the PBOW Administration Building (GRC–2015–C–06563).

Image 53: *One of three acid-producing facilities at the PBOW. The acid was used to manufacture TNT and DNT, c1941 (GRC–2015–C–06540).*

Image 54: *PBOW fire station located near Taylor and Columbus avenues, c1941 (GRC–2015–C–06564).*

*Image 55: PBOW News feature showing safety ceremony on a stage constructed with empty powder crates in May 1943.*

**WING TIPS, April 8, 1944**

# SERVICE STARS
### By Kay Hovanec
******

Something we had hoped not to see appeared in a Cleveland paper recently - a report of one of our AERL Service Stars being missing. He is LT. JOHN P. WISNER, formerly of the Machine Shop, missing in action over Germany since February 24. He was serving as a navigator in the Army Air Forces. Lt. Wisner entered service in June, 1942 and received his wings at Hondo, Texas in September, 1943.

His last letter home, dated February 22, told his parents that he was stationed in England.
******

What price profanity! It's $2.44 for the postage "kitty" from the Electrical people, who recently contributed a like amount as a result of penalizing the guilty.
******

PVT. ANTHONY FORTINI is now attending a Factory School at Consolidated Aircraft, San Diego, California, training to be a mechanic on the B-24.
******

Friends of GEORGE BENKO, Army Air Forces, who may be wondering about his present whereabouts should wonder no more. He is at San Antonio, Texas, Classification Center, after five months of training at Wittenberg College in Springfield, Ohio.

George says he has finished his exams and has been classified as a navigator.
******

*Image 56: "Service Stars" was a regular wartime column in the Wing Tips newsletter.*

With wartime labor shortages Trojan struggled to maintain the required 4,000-person workforce. Recruitment teams scoured Ohio and Kentucky towns with radio cars broadcasting the "highest paid jobs for men and women." PBOW employees ranged in age from 16 to nearly 80 and came from every U.S. state. Women made up greater percentages of the staff as the war progressed. Trojan tried to ease the harshness of life at the plant by providing dormitories, issuing an internal newspaper, and encouraging intramural sports teams, holiday functions, and other social activities.[82] The NACA was employing the same types of morale-boosting activities 50 miles to the east at the AERL.

✦ ✦ ✦ ✦ ✦ ✦

World War II permeated nearly every aspect of life at the AERL, not just the research. The greatest hardship was borne by the family, friends, and colleagues that were sacrificing a great deal more overseas. In 1943 the NACA mandated six-day workweeks without any overtime compensation and, in 1944, implemented a third shift.[83,84] There were also continual shortages of fuel, supplies, and equipment; elevated security levels; continual war bond drives; and a government-imposed 5-percent Victory Tax on wages. The AERL's 1,100-person staff was continually drained by the draft and enlistment.[85] To fill the openings, management urged its staff to recruit neighbors and relatives. NACA agents frequented campuses and industry sites, and they offered to train unskilled workers as mechanics, toolmakers, machinists, and others.[86]

As a result, women were given a rare opportunity to contribute in a variety of roles—from physicists to analysts to machinists. The AERL implemented a training program to quickly prepare inexperienced women for a variety of nonprofessional positions. There was debate at the lab regarding the effectiveness of women. Some protested that the great number of available jobs made it too easy for women to give up one job for another, thus wasting the organization's investment in their training.[87] Others were concerned that women would not give up their positions to returning veterans after the war.[88]

Nonetheless, the number of women employed at the AERL nearly doubled to 412 between 1943 and 1944.[89] The majority of these were clerks, but there were sizable numbers of laboratory aides and data processors. Professional positions such as engineers or

*Image 57: Women working alongside male colleagues in the AERL's Fabrication Shop (GRC–1944–C–05380).*

scientists were available, but few women were graduating with these degrees at that time. Nonetheless, one report describes the roles that women played in the research process. Women in the Drafting Section converted researchers' sketches into formal technical drawings and generated blueprints prior to the tests. Women in the Instrument Shop assisted mechanics with the installation of instruments for the tests. Afterward, women in the Publishing Section typed, edited, and printed research reports.[90]

Perhaps, the most significant female contributions were made in the Computing Section. The NACA introduced the concept of "computers" during World War II to relieve short-handed research engineers of some of the tedious data-taking work. The reliable gathering and processing of pressure, speed, temperature, and other data from test runs in the facilities was critical to the research method. The computers were young women who recorded test measurements and converted them into data that researchers could use to write their reports or modify the test program.[91]

*Image 58: Zella Morowitz worked in the AERL design office at Langley prior to her transfer to Cleveland in 1941, where she served as Ray Sharp's secretary for six years, c1943 (GRC–2015–C–06555).*

*Image 59: Memphis Belle crew Robert Hanson, Vincent Evans, and Charles Leighton; AERL Manager Raymond Sharp; Robert Morgan; William Holliday of the Cleveland Chamber of Commerce; Army Liaison Officer Colonel Edwin Page; Airport Commissioner John Berry; Cecil Scott; John Quinlan; and James Verinis. Kneeling are Harold Loch, Casimer Nastal, and Charles Winchell (GRC–1943–C–01870).*

*The NACA always maintained a close relationship with the military, and the AERL's contributions to the war effort were well-noted. As a result, some of the biggest names to emerge from the war visited the lab. These included Hap Arnold (November 1944), James Doolittle (October 1945 and 1946), Ike Eisenhower (April 1946), Curtis LeMay (August 1944), Chief of Air Materiel Command Edward Powers, Frank Whittle (July 1946), and new President Harry Truman (June 1944).[96] The crew of the Memphis Belle B–17 Flying Fortress were, perhaps, the most remembered guests of the era.*

*The Memphis Belle was the first U.S. bomber to complete 25 missions over Germany and France. Afterward the Air Corps assigned the bomber and crew to a three-month tour across the United States to sell war bonds. Captain Robert Morgan and the rest of the Memphis Belle aircrew arrived in Cleveland on a rainy 7 July 1943 for a three-day visit. The crew displayed the bomber for the public near the airport's fenceline and stored it in the NACA's hangar overnight.[97] A local company brought Morgan's family and his fiancé—the Memphis Belle's inspiration—to Cleveland to participate in the activities. The visit was a success with the public, but it strained Morgan's personal relationship with the "Belle." The couple ended their relationship days later.[98,99]*

The Computing Section contained roughly 100 women who were assigned to one of the five research divisions. The NACA insisted that the women have four years of high school math because it was necessary to understand the relationships of the information.[92]

The researchers relied on manometers to measure pressure levels inside a test facility. Manometers were mercury-filled glass tubes that looked and functioned like barometers. For each test, technicians installed dozens of pressure-sensing instruments that were connected to a unique manometer tube located inside the control room. Since readings were dynamic, researchers relied on cameras to capture the mercury levels at given points during the test. The computers examined the film and noted the pressure levels. The supervisors, women with math degrees, worked with the researchers to determine what output was desired and broke the mathematics down into steps that the computers could easily follow.[93] The computers, sitting in rows at long tables, then made their calculations with the adding machines. Each test had a multitude of parameters and readings to be noted and processed. The researchers often had to wait several weeks for the results, but the wait would have been exponentially longer without the Computing Section.[94]

In early 1944 the NACA was able to obtain deferments for its staff because of the criticality of their work. The military would induct the drafted employees as reserves, then immediately assign them to the NACA laboratory.[95] This new policy stabilized the staff for the duration of the war, but it did not diminish the contributions of the female employees.

## Wartime Investigations

As construction tapered off in 1943 the AERL found itself well positioned to tackle the military's engine difficulties. The areas of fuels and lubrication were of particular interest at the lab during the war. The quality of aviation fuel is critical to the operation of piston engines. Lower octane fuels result in engine knock, which can damage engine components and lower performance. Although this lowered performance is not readily apparent in automobiles, it poses significant problems in aircraft engines, which operate at full throttle most of the time. The fuel quality problem was not evident before the war when the nation had only a modest number of aircraft. The exponential growth in the number of military aircraft, however, taxed supplies of high-octane aviation fuel. The petroleum industry began creating synthetic fuels that could be mixed with traditional fuel, but these new fuels were difficult to produce and could only be made in limited quantities. The military had to find a middle ground between quality and quantity.[100]

The AERL had the most extensive fuel and lubrication research capabilities in the country. An entire division of researchers was dedicated to finding a way to produce larger quantities of high-octane fuel. In addition, researchers analyzed fuel composition in the lab's chemistry laboratories, studied the performance of different fuels in subscale and full-scale piston cylinders in the lab's test cells, and performed final analyses with actual flights in the lab's research aircraft.[101]

The Fuels and Lubrication Division conducted a variety of studies during the war to understand and eradicate engine knock, analyze new synthetic fuels, and resolve lubrication issues such as the foaming of oil. The group's most extensive effort was the evaluation of several new types of synthetic fuel. In the years leading up to World War II, the petroleum industry had developed methods of creating synthetic gasoline—including high-performance and antiknock fuels. To meet the intense wartime demand for these fuels, the military blended them with more available fuels.[102]

At the request of the Air Materiel Command, AERL researchers evaluated the characteristics of 16 different types of fuel blends during the war, including those incorporating the antiknock xylidine additive and

*Image 60: Researchers work with test setups in a Fuels and Lubrication Building lab room during March 1943 (GRC–1943–C–01370).*

the high-performance triptane fuel. It was necessary to determine if the synthetic fuels increased engine temperatures.[103] Young researcher Walter T. Olson led the efforts to rate the antiknock and performance characteristics of the 50 different compounds in the xylidine additive. Industry later used this information to create fuels for domestic U.S. aircraft.[104]

AERL researchers also tested triptane blends in small-scale engines. Once the optimal mixes were identified using single-cylinders, the researchers tested them on a full-scale engine and eventually ran them on the lab's Consolidated B–24D. The researchers found that the triptane mix performed 25 percent better than traditional fuel.[105] In 1944 a national fuel study group, the Coordinating Research Council, used the AERL studies to create a new antiknock rating scale that replaced octane with triptane. The council's award-winning research in the 1940s was heavily based on the efforts of the Fuels and Lubrication Division.[106]

❖ ❖ ❖ ❖ ❖

Tribology is an interdisciplinary field that encompasses the study of friction, lubrication, bearings, gears, and wear. This somewhat esoteric subject is essential to the reliability and assurance of long lifetimes of engines, pumps, and other turbomachinery. Edmund Bisson initiated the NACA tribology studies in 1939 at Langley. He and his colleagues transferred to the AERL in Cleveland in January 1943 as a section in the new Engine Components Research Division. The group performed most of its investigations on single-cylinder test rigs in the Engine Research Building, then passed its findings along to engine manufacturers who verified the data with full-scale engine tests. Bisson's group became the lab's most prolific report issuer during this period.[107]

During the war the section concentrated their efforts on the lubrication and wear of metal piston rings and cylinders. Piston rings fit around the piston to seal

Image 61: Bisson with physicist and mathematician Lucien C. Malavard (GRC–1949–C–24300).
Image 62: Reprint of "Fuels Talk" by Bisson.

*Image 63: A representative from the Allison Engine Company instructs AERL mechanics on the operation of a basic Allison powerplant. The staff was taught how to completely disassemble and reassemble the engine components and systems (GRC–1943–C–03045).*

the combustion chamber and transfer heat from the piston to the cylinder wall. The military was struggling to keep its engines clean when its aircraft used unpaved runways in North Africa. The sand and grit quickly wore down the rings and cylinders, necessitating lengthy engine overhauls. The military combatted the situation by making the cylinders larger in anticipation of degradation, but this was an inefficient response.[108]

Bisson and Bob Johnson investigated the possibility of restoring the damaged rings to the original specifications through chromium plating and refinement. After finding that the plating inhibited proper lubrication, they determined that a new process—interrupted-chromium plating—provided durability and reduced friction.[109] Bisson and Johnson also demonstrated that piston ring performance could be increased by infusing nitrogen into the metal to create a hardened surface over the softer steel.[110] After the war, the lab's fundamental research on friction, lubrication, and wear helped establish tribology as a scientific discipline.[111]

Engine cooling was, perhaps, the most pressing aircraft problem during World War II. Piston engines generate high levels of heat, of which only about a third is used for propulsion. The excess heat, if not dissipated, causes unnecessary fuel consumption, aerodynamic drag, and ultimately cylinder failures.[112] The widespread introduction of superchargers during the war only exacerbated the problem. The AERL conducted comprehensive cooling investigations on a number of engines during the war—most notably the Allison V–1710 and the Wright R–3350.

The Allison V–1710 was the only liquid-cooled engine used during World War II. Its liquid-cooled engines relied on a working fluid, not the airflow, to dissipate the engine heat. As such, they were more compact than large radial air-cooled engines.[113] In 1940, before Allison even completed development work on the V–1710, the military placed large orders for the engine. The Army Air Corps was relying on the V–1710, which outperformed similar air-cooled engines in the early 1940s, to power the Lockheed P–38 Lightning, Bell P–39 Airacobra, Curtiss P–40 Warhawk, and Bell P–63 King Cobra fighter aircraft.[114]

The military instructed Allison to incorporate a supercharger into the V–1710 to increase its performance. In fall 1942 General Oliver Echols asked the AERL to improve the performance of the V–1710's new two-stage supercharger at altitudes up to 20,000 feet. It was the AERL's first new research assignment.[115,116]

The V–1710 effort required the services of nearly every research group and test facility at the AERL. The military supplied the lab with a P–63A King Cobra and three of the Allison engines to facilitate the work, and Allison sent representatives to teach the NACA mechanics how to completely disassemble and reassemble the engine components and systems.[117]

The Supercharger Division enhanced the supercharger performance, the Fuels and Lubrication Division analyzed the effect of the increased heat on knock in the fuel, the Thermodynamics Division improved

the cooling system, and the Engine Installation Division analyzed the drag penalties associated with the modifications.[118] Much of the initial research was conducted on dynamotor test stands in the Engine Research Building. Once the researchers were satisfied with their improvements, the new supercharger and cooling components were test flown on the P–63A aircraft.[119] The AERL efforts improved the V–1710's performance, but they could not overcome the supercharger's design limitations. The military used the supercharger in the P–63s, but those aircraft had poor altitude performance and did not see combat.[120]

The AERL also analyzed engine cooling for air-cooled piston engines. The investigation of the Wright R–3350 radial engine was second only to the Allison study in terms of the number of groups and facilities participating. Four of the unproven R–3350s powered Boeing's state-of-the-art B–29 Superfortress. The B–29's ability to fly faster and higher than previous bombers despite being substantially larger made it the most significant Allied air weapon in the latter part of the war. The B–29 was intended to soar above antiaircraft fire and make pinpoint drops onto strategic

*Image 64: An AERL researcher demonstrates the improved fuel injection system for the R–3350 engine at a tour stop in the Engine Research Building in June 1945 (GRC–1945–C–10678).*

targets.[121] The 2,000-horsepower R–3350 was the only engine suitable for such an aircraft, but it frequently overheated.[122]

The B–29 was an extremely complicated aircraft that normally would have taken many years to develop. The military had ordered large quantities of the bomber before the first Superfortress took flight in September 1942. It was just two years after Boeing began design work and one year after the R–3350's initial flight test. The B–29 still had many problems that needed to be remedied, and Wright had not fully resolved a number of design issues that caused engines to operate at high temperatures.[123] As the bombers began rolling off the assembly line, the military added additional weight requirements that exceeded the aircraft's design. The issue was exacerbated by the military's rush to get the bombers into the air. The high altitudes and excess payloads strained the R–3350s and led to a rash of engine fires and crashes both in the United States and abroad, including one accident that resulted in 28 deaths in Seattle.[124,125] General Arnold, who was relying on the B–29s to carry out long-distance strikes on Japan from bases in China, was irate at the delays and the low number of available aircraft.[126]

The B–29 had many problems initially, but the overheating of the engines was the most significant. There were several factors contributing to the overheating. In an attempt to reduce aerodynamic drag, pilots

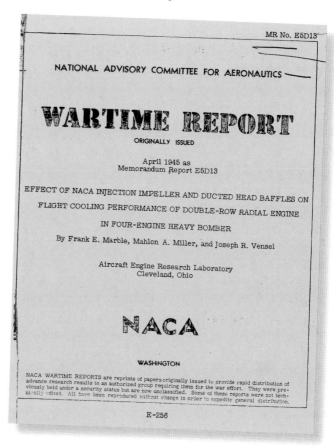

Image 66: NACA Wartime Reports were classified until after the war.

closed the engine flaps that, when opened, allowed an influx of external air to cool the engine. The R–3350's piston valves were also failing because they could not dissipate heat quickly enough. In addition, there was uneven fuel distribution to the various valves.[127] The military had been waiting for the AERL to begin operating so that it could address these issues.

As with the Allison studies the previous year, multiple divisions and test facilities contributed to the R–3350 effort. In fall 1943 researchers in the Engine Research Building, test cells modified the fuel-injection system so that it provided a uniform amount of fuel to all 18 of the engine's cylinders. This lowered the amount of fuel required. The researchers demonstrated the new fuel spray system on full-scale engine tests in the Prop House. The Engine Research Division staff also determined that a slight extension of the cylinder head provided sufficient additional surface area to expel heat and thus prevent cylinder failures.[128]

Image 65: Wright R–3350 installed in the AWT test section on 4 July 1944 (GRC–1944–C–05554).

General Arnold was particularly anxious for the AWT to come online.[129] The AWT, which was the nation's only wind tunnel capable of running engines in realistic simulated flight conditions, began operation in early February 1944. A few months earlier (fall 1943), the NACA transferred Abe Silverstein from Langley to manage the tunnel activities. Silverstein, who had begun his NACA career in 1929, had helped to design and then run Langley's massive Full Scale Tunnel. His specialty was improving airflow through engines—the very task that the military was now asking the AERL to address with the R–3350.[130]

In May 1944 AERL mechanics installed a full-scale R–3350 on a section of wing in the AWT test section. Silverstein and his colleagues quickly determined that the hottest portion of the engine, the exhaust area, was not receiving the cooling air. They were able to quickly add baffles inside the engine to duct the airflow to this area.[131] The AWT tests also led to reshaping the engine flaps, so that airflow increased without causing additional drag.[132]

The military brought a B–29 aircraft to the AERL in June 1944 to flight-test the modifications. Technicians installed the new fuel-injection system, cylinders, baffles, and cowling on the bomber's two left-wing engines. AERL researchers accompanied the military pilots as they conducted 11 test flights throughout July with various combinations of the modifications. The flight program verified the ground-based findings, and researchers concluded that the modifications could yield a 10,000-foot increase in altitude or comparable payload increases.[133,134] The testing in the AWT continued through mid-September.

The modifications arrived too late, though. In summer 1944, as the AERL staff was studying the R–3350 in the AWT, the Allies made two critical advances. Over 150,000 troops invaded the northern shores of France during D-Day, and U.S. Marines seized Guam and the Mariana Islands in the Pacific. The European invasion initiated a year-long push that would ultimately end in Berlin. The Mariana Islands provided the Allies with territory to build airfields from which to launch shorter B–29 sorties over the Japanese interior.

The B–29 pilots had difficulty hitting the desired targets during the early raids. That fall, the military decided to change the B–29's flight plans from high-altitude target bombing to low-altitude fire-bombing. This strategy reduced engine strain, permitted larger

*Image 67: A B–29 bomber on display in the hangar during June 1945 (GRC–1944–C–10587).*

*Image 68: P–38 Lightning fuselage in the IRT during March 1945 (GRC–1945–C–8832).*

payloads, and resulted in much wider destruction.[135] The new policy first manifested itself with the 9 March 1945 firebombing of Tokyo that left approximately 80,000 dead. Incendiary attacks were then launched against other Japanese cities.[136] The AERL improvements were implemented after the war, and the Superfortress went on to a long and successful postwar career.

❖ ❖ ❖ ❖ ❖ ❖

The buildup of ice during flight has been a serious concern for pilots for as long as aircraft have been able to attain weather-producing altitudes. Under certain weather conditions, the water droplets in clouds freeze upon contacting the solid surfaces of aircraft. The resulting ice accretion can add extra weight, disrupt aerodynamics, and block air intake into the engines. Consequently, Langley established a program in the 1930s to study ice buildup and prevention. In 1941 the NACA expanded its icing research and transferred the program to the new Ames laboratory. There Lewis Rodert perfected his award-winning ice-prevention system that used hot air from the engines to warm the wings and other vulnerable surfaces.[137]

The AERL initiated its own icing studies during the war using research aircraft and the new IRT, which took advantage of the AWT's massive refrigeration system to produce temperatures as cold as –47 degrees Fahrenheit. The IRT was an unpressurized tunnel with a spray bar system that released water droplets into the freezing airstream before it passed through its 6- by 9-foot test section. Researchers used the tunnel to study the buildup of ice and test different deicing systems. The lab's engineers struggled for many years to perfect the water droplet system.[138]

Icing tunnel testing began in August 1944 despite continuing problems with the spray system. Researchers initially studied ice buildup on propeller blades and antennas. The military also requested that the AERL investigate a new thermal pneumatic boot ice-prevention system and heated propeller blades. The researchers installed hollow gas-heated and solid electric-heated blades on a wingless Bell P–39 Airacobra in the IRT. The AERL tests demonstrated that both the heated boot and the heated propeller systems successfully prevented ice.[139,140]

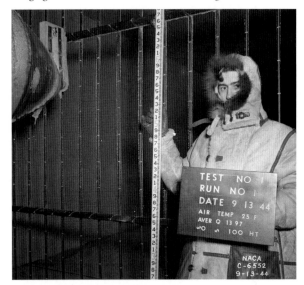

*Image 69: On 13 September 1944 an AERL technician prepares for the initial test run in the IRT (GRC–1944–C–06552).*

February 1945 marked the beginning of the AERL's most significant icing investigation of the war—the air scoop from the Curtiss C–46 Commando. The military was flying the C–46s regularly over the Himalaya mountains. At the high altitudes, the aircraft engines could not generate enough heat to prevent ice buildup on the engine intakes. This had resulted in stalled engines and hundreds of crashes.[141,142]

Throughout spring 1945, AERL researchers analyzed models of a standard C–46 air scoop and three variations in the IRT. They determined that the primary problem was the intake of freezing water—not icing. The NACA redesigned the scoop to limit the ingestion of freezing droplets without causing any performance losses. The modifications were later incorporated into the C–46 and Convair C–40.[143,144]

The AERL also acquired two Lockheed P–38J Lightnings and the North American XB–25E Mitchell to augment the work in the IRT. Researchers utilized one P–38 for ground tests on the hangar apron. They set up a blower in front of the aircraft to drive air and simulated rain into the engine. The studies revealed that ice accumulated on the carburetor and supercharger in a variety of meteorological conditions.[145] Flight tests with the other P–38 confirmed the researchers' theory that ice buildup could be reduced by closing the engine flap to block water ingestion and increase engine heat.[146] Researchers used the XB–25E Mitchell during the war to study ice buildup on components such as windshields, antennas, propellers, and engines. They measured the ice accretion then analyzed different anti-icing and deicing strategies. The lab would dramatically expand its icing research program after the war—both in the tunnel and in the skies.

*Image 70: A P–38J Lightning in front of the blower on the hangar apron that was utilized as a crude rain-simulating device. The blower was used extensively during the 1940s to supplement wind tunnel and flight research data (GRC–1945–C–09650).*

*Image 71: The Jet Static Propulsion Laboratory as it nears completion in August 1943. The secret facility was officially called the Supercharger Laboratory to disguise its true nature (GRC–2015–C–06544).*

## Emergence of the Turbojet

World War II spurred technological development in almost every field. Advances in propulsion, such as the turbojet, ramjet, and rocket, were the most significant for the AERL. Engineers in Germany and Britain developed the jet engine simultaneously in the 1930s, but the engines were not integrated into aircraft until the start of World War II. The British government invited Hap Arnold to witness the nation's first jet-powered flight in April 1941. Intrigued by the possibilities, Arnold made arrangements to secretly ship plans for the engine to the United States. Then the military tasked General Electric—which was not a major engine manufacturer at the time—with the job of creating a jet engine from these blueprints. The result was the I–A centrifugal-compressor engine. Bell Aircraft incorporated two of the I–As into the XP–59A aircraft, which secretly made its first test flight on 2 October 1942. The initial engines, however, including an upgraded I–16 version, did not perform as well as anticipated during that first year.[147]

In summer 1943, the military briefed a small group of NACA leaders on the U.S. jet engine work that had been carried out covertly during the past two years. They also unveiled General Electric drawings for a new facility—the Jet Propulsion Static Laboratory—to be constructed at the AERL. The NACA hurriedly built the new test stand along the airport fenceline to support General Electric's turbojet work. Disguised as the "Supercharger Facility," the Jet Propulsion Static Laboratory included two engine stands and two spin pits in which the General Electric turbojets could be operated.[148,149] In fall 1943, AERL researchers began clandestine tests of the I–A.[150]

Around this time military officials summoned Abe Silverstein to the General Electric plant in Massachusetts to make arrangements to install an I–16-powered Airacomet in the new AWT. General Electric was hoping to improve the engine's cooling airflow, an area of expertise for Silverstein.[151] In early 1944, mechanics installed a Bell XP–59A Airacomet in the AWT test section. Merritt Preston and his AERL colleagues exhaustively studied the aircraft and its I–16 engines from 4 February to 13 May 1944. They were able to modify the engine and its nacelle enclosure to improve the airflow distribution, which increased the I–16's performance by 25 percent.[152] The Air Corps used some Airacomets for pilot training.

In 1944, however, the German Messerschmitt 262 and British Gloster Meteor jet aircraft were beginning to participate in the war. Although these aircraft could fly significantly faster than piston engine aircraft, their impact on the war was muted. The first Messerschmitt test flight took place in August 1939, but the Germans did not mass-produce the fighter until mid-1944. By

*Image 72: The secret test of the Bell YP–59A Airacomet in the spring of 1944 was the first investigation in the new AWT. The Airacomet, which was powered by two General Electric I–A centrifugal turbojets, was the first U.S. jet aircraft (GRC–1944–C–04830).*

this time, aircraft fuel was in short supply and Allied bombing was decimating production facilities and airfields. The British only operated its Meteors over its own territory to prevent the capture of the new jet engine technology.[153]

Meanwhile, General Electric created the more powerful I–40 turbojet engine, which Lockheed integrated into the XP–80A Shooting Star. The Shooting Star was the first entirely U.S.-designed jet aircraft and the first U.S. aircraft to exceed 500 mph. It made its initial flights in January 1944.[154] The military asked the NACA to install a Shooting Star in the AWT and improve the engine performance. In spring 1945 AERL researchers tested the aircraft at simulated altitudes up to 50,000 feet. The AWT investigations allowed researchers to verify NACA formulas that predicted thrust level at altitudes based on sea-level

measurements.[155] The P–80 fighter and the I–40 engine became great engineering successes, just not in time to contribute to World War II.

*Image 73: Abe Silverstein, Head of the AWT, discusses the tunnel's research during the war, concentrating on the several General Electric and Westinghouse jet engines that were studied (GRC–1945–C–10661).*

*Image 74: Lockheed's YP–80A, powered by two General Electric I–40 turbojets, in the AWT test section in March 1945. The P–80 was the first U.S. aircraft to fly faster than 500 mph (GRC–1945–C–09576).*

General Electric's I–16 and I–40 had centrifugal compressors. These compressors consisted of a single large rotor that powerfully pushed the airflow outward past the rotor and into the engine. Engineers had also developed jet engines that employed axial-flow compressors. These consisted of a series of fans lined up on a shaft. Because this configuration allowed designers to increase the power by adding more stages without increasing the diameter of the engine, it quickly became the choice of engine designers.

Westinghouse employed an axial-flow compressor to create the first U.S.-designed turbojet. In March 1943 the 19A became the first operational turbojet designed in the United States. In January 1944 a Chance Vought FG–1 Corsair used the 19A to complement its piston

engines, making the 19A the only U.S. jet engine to be used in the war.[156] The Navy asked the NACA to test the engine's successor, the 19B, in the AWT. After the tests, run in September 1944, demonstrated good

TABLE 1
ALTITUDE WIND TUNNEL PROJECTS
(from initial date of operation to June 30, 1945)

| Model | Instrumentation time (Wind tunnel shop) | | Tunnel time | |
|---|---|---|---|---|
| | From | To | In | Out |
| Bell YP-59A airplane, General Electric I-16 engine | Jan. 4, 1944 | Feb. 3, 1944 | Feb. 4, 1944 | May 13, 1944 |
| Boeing B-29 nacelle, Curtiss-Wright R-3350 engine | March 15, 1944 (approx.) | May 14, 1944 | May 14, 1944 | Sept. 18, 1944 |
| Westinghouse 19-B jet engine, NACA stub wing | June 1, 1944 | Sept. 19, 1944 | Sept. 19, 1944 | Nov. 21, 1944 |
| Douglas XTB2D-1 airplane, Pratt Whitney R-4360 engine | Aug. 15, 1944 | Nov. 21, 1944 | Nov. 21, 1944 | Jan. 3, 1945 |
| General Electric TG-180 jet engine, NACA wing | Dec. 8, 1944 | Jan. 4, 1945 | Jan. 4, 1945 | Mar. 1, 1945 |
| Lockheed YP-80A airplane, General Electric I-40 jet engine | Dec. 30, 1944 | March 1, 1945 | March 1, 1945 | May 15, 1945 |
| NACA ram jet | April 15, 1945 | May 15, 1945 | May 15, 1945 | June 27, 1945 |
| Lockheed YP-80A airplane, General Electric I-40 jet engine | June 1, 1945 | June 27, 1945 | June 27, 1945 | Aug. 3, 1945 |

*Image 75: AWT wartime test schedule.*

*Image 76: On 8 May 1945 the staff awoke to news that Germany had surrendered. A mid-morning ceremony was held at the Administration Building, but work at the lab continued on (GRC–1945–C–09905).*

performance of the 10-stage engine at simulated altitudes, the Navy soon incorporated it into their first jet fighter, the McDonnell XFD–1 Phantom.[157]

❖ ❖ ❖ ❖ ❖ ❖

At 7 p.m. on 14 August 1945, President Harry Truman announced that Japan had accepted terms for surrender, and World War II came to a conclusion. The AERL staff had investigated nearly every existing engine model used in the war. The lab's work during this time included studies of new types of engines—the ramjet and turbojet, but the lab's greatest contribution to the war was improvement of turbosupercharged piston engines. The AERL utilized its new engine test facilities and flight research group to mitigate the overheating and engine knock frequently associated with the turbosupercharger. This groundbreaking work paved the way for the peacetime contributions that came next. The AERL staff would now turn their attention to improving the turbojet engine.

*Image 77: Final issue of the Plum Brook News from August 1945.*

*Image 78: An AERL metallurgist examines a supercharger in January 1944 (GRC–1944–C–03814).*

## Endnotes for Chapter 2

62. "Aviation Writers Association Inspection June 11, 1945 of the Aircraft Engine Research Laboratory," 1945, NASA Glenn History Collection, Cleveland, OH.

63. Edward Constant, *The Origins of the Turbojet Revolution* (Baltimore, MD: Johns Hopkins University Press, 1980).

64. Henry H. Arnold, "Transcript of General H. H. Arnold's Speech to AERL," File 4849, 9 November 1944, Historical Collection, NASA Headquarters, Washington, DC.

65. Tom Lilley, et al., *Problems of Accelerating Aircraft Production During World War II* (Cambridge, MA: Harvard University, 1970).

66. "Sharp Made Laboratory Head," *Wing Tips* (8 December 1942).

67. Billy Harrison interview, Cleveland, OH, 14 October 2005, NASA Glenn History Collection, Oral History Collection, Cleveland, OH.

68. "Sharp Made Laboratory Head."

69. Benjamin Pinkel interview, Cleveland, OH, by Virginia Dawson, 4 August 1985, NASA Glenn History Collection, Oral History Collection, Cleveland, OH.

70. Abe Silverstein interview, Cleveland, OH, by Walter T. Bonney, 21 October 1972 and 20 September 1973, NASA Glenn History Collection, Oral History Collection, Cleveland, OH.

71. Dana Lee, "Research Chief Is Welcomed," *Wing Tips* (8 January 1943).

72. Addison Rothrock, "Address Made to Research Staff at AERL," 30 December 1942, NASA Glenn History Collection, Directors Collection, Cleveland, OH.

73. "A. M. Rothrock Is Congratulated," *Wing Tips* (19 February 1943).

74. "Smoker," *Wing Tips* (19 February 1944).

75. Thomas Neill to George Lewis, "Personnel and Projects at AERL," 5 July 1943, NASA Glenn History Collection, Directors Collection, Cleveland, OH.

76. Frank N. Shubert, "Mobilization: The U.S. Army in World War II, The 50th Anniversary," *www.army. mil/cmh-pg/documents/mobpam.htm* (accessed 11 June 2015).

77. Lilley, *Problems of Accelerating Aircraft Production.*

78. R. K. Rajput, *Internal Combustion Engines* (New Delhi, India: Laxmi Publications, 2005).

79. Walter Olson interview, Cleveland, OH, by Virginia Dawson, 16 July 1984, NASA Glenn History Collection, Oral History Collection, Cleveland, OH.

80. Plum Brook Ordnance Works Restoration Advisory Board, "Quarterly Fact Sheet," January through March 2007, NASA Glenn History Collection, Plum Brook Ordnance Collection, Cleveland, OH.

81. "The Saga of Plum Brook Ordnance Works As a Two High-Explosives, Major Facility Comes to a Close," *The Plum Brook News* (18 August 1945).

82. "The Saga of Plum Brook Ordnance."

83. "It's Now Six-Day, 48-Hour Week," *Wing Tips* (22 January 1943).

84. "Women Offered Training at Vital Air-Lab," NASA press release, 17 May 1944, NASA Glenn History Collection, Directors Collection, Cleveland, OH.

85. Virginia Dawson, *Engines and Innovation: Lewis Laboratory and American Propulsion Technology* (Washington, DC: NASA SP–4306, 1991).

86. NACA, "NACA Looks to the Future" Pamphlet, c1943, NASA Glenn History Collection, Cleveland, OH.

87. Daniel Williams, "Women Inclined to Change Jobs," *Wing Tips* (12 February 1943).

88. Pearl Young, "AERL Family Forum," *Wing Tips* (12 February 1943).

89. "Utilization of Women at the Aircraft Engine Research Laboratory," June 1944, NASA Glenn History Collection, Directors Collection, Cleveland, OH.

90. "Utilization of Women."

91. Sylvia Stone memorandum for Staff, "Organization of the Computing Section, Administrative Services Division, Administrative Department," 16 November 1945, NASA Glenn History Collection, Directors Collection, Cleveland, OH.

92. Peggy Yohner interview, Cleveland, OH, by Virginia Dawson, 21 May 1985, NASA Glenn History Collection, Oral History Collection, Cleveland, OH.

93. Yohner interview by Dawson.

94. Yohner interview by Dawson.

95. M. G. White memorandum for the Adjutant General, "Voluntary Induction in the ACER of Essential Personnel of the National Advisory Committee for Aeronautics," 18 February 1944, NASA Glenn History Collection, Directors Collection, Cleveland, OH.

96. "Whittle, Jet Inventor Here," *Wing Tips* (19 July 1946).

97. "Memphis Belle, Vet Bomber, and Crew Pay Three-Day Visit to Lab," *Wing Tips* (9 July 1943).

98. Elaine de Man, "The Two Memphis Belles," *Air and Space* (1 May 1990).

99. "Memphis Belle's Pilot Has Surprise Reunion Here," *Cleveland Plain Dealer* (8 July 1943).

100. Addison Rothrock, "Address to AERL Smoker," 12 April 1944, NASA Glenn History Collection, Directors' Collection, Cleveland, OH.

101. Rothrock, "Address to AERL Smoker."

102. George Gray, *Frontiers of Flight: The Story of NACA Research* (New York, NY: Alfred Knopf, 1948).

103. H. Jack White, Calvin Blackman, and Milton Werner, *Flight and Test Stand Investigation of High-Performance Fuels in Double-Row Air-Cooled Engines, II—Flight Knock Data and Comparison of Fuel Knock Limits With Engine Cooling Limits in Flight* (Cleveland, OH: NACA MR–E4L30, 1944).

104. Gray, *Frontiers of Flight.*

105. H. Jack White, Philip C. Pragliola, and Calvin C. Blackman, *Flight and Test Stand Investigation of High-Performance Fuels in Modified Double-Row Radial Air Cooled Engines, II—Flight Knock Data and Comparison of Fuel Knock Limits With Engine Cooling Limits in Flight* (Washington, DC: NACA MR–E5H04, 1945).

106. "Fuels Work Done Here Commended," *Wing Tips* (16 August 1946).

107. Edmond Bisson interview, Cleveland, OH, by Virginia Dawson, 22 March 1985, NASA Glenn History Collection, Oral History Collection, Cleveland, OH.

108. Bisson interview by Dawson.

109. E. V. Zaretsky, editor, *Tribology for Aerospace Applications* (Park Ridge, IL: Society of Tribologists and Lubrication Engineers SP–37, 1997).

110. "Bisson Tells Research Staff of Urgent Need for Piston-Ring Research," *Wing Tips* (9 April 1943).

111. Bisson interview by Dawson.

112. Benjamin Pinkel, "Thermodynamics Division," 24 May 1944, NASA Glenn History Collection, Directors' Collection, Cleveland, OH.

113. Robert Schlaifer, *Development of Aircraft Engines: Two Studies of Relations Between Government and Business* (Cambridge, MA: Harvard University, 1950).

114. "Allison Engine," *Life Magazine* (13 January 1941).

115. Gray, *Frontiers of Flight.*

116. Richard Downing and Harold Finger, *Effect of Three Modifications on Performance of Auxiliary Stage Supercharger for V-1710-93 Engine* (Washington, DC: NACA RM–E6J18, 1946).

117. Bruce Lundin, John Povolny, and Louis Chelko, *Correlation of Cylinder-Head Temperatures and Coolant Heat Rejection of a Multi-cylinder Liquid-Cooled Engine of 1710-Cubic Inch Displacement* (Washington, DC: NACA Report 931, 1949).

118. Pinkel, "Thermodynamics Division."

119. Pinkel, "Thermodynamics Division."

120. Schlaifer, Development of Aircraft Engines.

121. Robert F. Dorr, *Superfortress Units of World War 2* (Osceola, WI: Osprey Publishing, 2002).

122. Walter Boyne, "The B-29's Battle of Kansas," *Air Force Magazine* (February 2012).

123. Boyne, "The B-29's Battle."

124. Gray, *Frontiers of Flight.*

125. Dorr, *Superfortress Units.*

126. Boyne, "The B-29's Battle."

127. Gray, *Frontiers of Flight.*

128. Gray, *Frontiers of Flight.*

129. Robert Arrighi, *Revolutionary Atmosphere: The Story of the Altitude Wind Tunnel and Space Power Chambers* (Washington, DC: NASA SP–2010–4319, 2010).

130. Silverstein interview by Bonney.

131. Abe Silverstein interview, Cleveland, OH, by Virginia Dawson, 5 October 1984, NASA Glenn History Collection, Cleveland, OH.

132. DeMarquis Wyatt and William Conrad, *An Investigation of Cowl-Flap and Cowl-Outlet Designs for the B–29 Power Plant Installation* (Washington, DC: NACA Wartime Report E–205, 1946).

133. Frank Marble, Mahlon Miller, and E. Barton Bell, *Analysis of Cooling Limitations and Effect of Engine-Cooling Improvements on Level Flight Cruising Performance of Four-Engine Heavy Bomber* (Washington, DC: NACA No. 860, 1948).

134. Frank Marble, Mahlon Miller, and Joseph Vensel, *Effect of NACA Injection Impeller and Ducted Head Baffles on Flight Cooling Performance of Double-Row Radial Engine of Four Engine Heavy Bomber* (Washington, DC: NACA MR–E5D13, 1945).

135. Dorr, *Superfortress Units.*

136. Clayton Chun, *Japan 1945: From Operation Downfall to Hiroshima and Nagasaki* (Osceola, WI: Osprey Publishing, 2008).

137. Michael Gorn, *Expanding the Envelope: Flight Research at NACA and NASA* (Lexington, KY: University Press of Kentucky, 2001).

138. *An International Historic Mechanical Engineering Landmark: Icing Research Tunnel* (Washington, DC: American Society of Mechanical Engineers, 20 May 1987).

139. Mark G. Potapczuk, "Aircraft Icing Research at NASA Glenn Research Center," *Journal of Aerospace Engineering* 26 (2013): 260–276.

140. William Leary, *We Freeze to Please: A History of NASA's Icing Research Tunnel and the Quest for Flight Safety* (Washington, DC: NASA SP–2002–4226, 2002).

141. Leary, *We Freeze to Please.*

142. Gray, *Frontiers of Flight.*

143. *An International Historic Mechanical Engineering Landmark.*

144. Uwe von Glahn and Clark Renner, *Development of a Protected Air Scoop for the Reduction of Induction-System Icing* (Washington, DC: NACA TN–1134, 1946).

145. Henry Essex, Edward Zlotowski, and Carl Ellisman, *Investigation of Ice Formation in the Induction System of an Aircraft Engine, I—Ground Tests* (Washington, DC: NACA MR–E6B28, 1946).

146. Henry Essex, Edward Zlotowski, and Carl Ellisman, *Investigation of Ice Formation in the Induction System of an Aircraft Engine, II—Flight Tests* (Washington, DC: NACA MR– E6E16, 1946).

147. General Electric Company, *Seven Decades of Progress: A Heritage of Aircraft Turbine Technology* (Fallbrook, CA: Aero Publishers, Inc., 1979), p. 48.

148. Virginia Dawson, "The Turbojet Revolution at Lewis Research Center," Annual Meeting of the Society for the History of Technology (4 November 1984).

149. Dawson, *Engines and Innovation*, chap. 3.

150. Pinkel interview, by Dawson, 4 August 1985.

151. Abe Silverstein interview, Cleveland, OH, by John Sloop, 29 May 1974, Glenn History Collection, Oral History Collection, Cleveland, OH.

152. Merritt Preston, Fred O. Black, and James M. Jagger, *Altitude-Wind-Tunnel Tests of Power-Plant Installation in Jet-Propelled Fighter* (Washington, DC: NACA MR–E5L17, 1946).

153. Laurence K. Loftkin, Jr., *Quest for Performance: The Evolution of Modern Aircraft* (Washington, DC: NASA SP–468, 1985), chap. 11, *http://www. hq.nasa.gov/pao/History/SP-468/ch11-2.htm* (accessed 30 July 2009).

154. Loftkin, *Quest for Performance.*

155. Abe Silverstein, "Altitude Wind Tunnel Investigations of Jet-Propulsion Engines," General Electric Gas Turbine Conference (31 May 1945), pp. 7–10.

156. Richard Leyes II and William A. Fleming, *The History of North American Small Gas Turbine Aircraft Engines* (Reston, VA: AIAA and Smithsonian Institution, 1999).

157. Leyes, *The History of North American Small Gas Turbine Aircraft Engines.*

Image 79: Mechanics lower an inlet duct for a Westinghouse J40 engine into the Altitude Wind Tunnel's 20-foot-diameter test section (GRC–1951–C–28463).

# 3. Setting Forth

*"When you're in the middle of a maelstrom,
I don't think you feel it the same way as when you're
on the outside looking in, because you're moving
with the stream, and that's about the way it was."*

—Abe Silverstein

Image 80: A mechanic inspects a General Electric I–40 turbojet engine. The lab had begun investigating jet engines during the war, but the "big switch" to jet propulsion began in October 1945 (GRC–1946–C–15674).

# Setting Forth

"Parking lots were unoccupied and offices were deserted. One stack of the Heating Plant was smoking, and the flag flew in a fresh breeze. In the Center Section of the Engine Research Building a large black and white cat strolled slowly down the deserted center aisle."[158] That is how *Wing Tips* described the scene at the Aircraft Engine Research Laboratory (AERL) during the two-day federal shutdown following Japan's surrender on 15 August 1945. The pacific atmosphere at the lab was countered by the euphoria in the streets and homes in Cleveland and across the nation. Loved ones and colleagues would be returning home, and everyday amenities like full tanks of gasoline, nylons, and cartons of cigarettes would be plentiful again.

The U.S. military emerged from World War II with new interest in jet aircraft, missiles, and nuclear propulsion. Great leaps in high-speed flight were within sight. The military would begin launching rockets at White Sands, New Mexico, in 1946; and the Bell XS–1 would break the sound barrier in 1947. The end of the war also signaled the NACA's return to its traditional research mission. The NACA stopped troubleshooting military aircraft and returned to research, focusing on the new technologies that had emerged during the war. The technologies came with a host of issues that had to be addressed, and they ushered in a time of growth and breakthroughs.

The arrival of the jet engine affected the technical direction of the AERL—the NACA's propulsion center—more dramatically than it did Ames and Langley, and the postwar period brought significant changes to the lab. The AERL's transition from wartime efforts to research began in 1944, but the transformation became more apparent with a major reorganization in October 1945. It was the first of several major shifts that the lab would go through over the years.

In 1946 the AERL increased the number of its employees to 2,600, where it would remain throughout the NACA period.[159] Three new large test facilities and several small supersonic wind tunnels would become operational within the next three years, and

*Image 81: Kathryn "Nicki" Crawford demonstrates that there are sufficient coins in the bucket to match her weight in 1946. The group of mechanics contributed the money to celebrate Crawford's upcoming marriage to their colleague Bill Harrison, who had recently returned from the Army Air Corps. The Harrisons spent the next 66 years together (GRC–2015–C–06814).*

the researchers quickly plunged into the new fields of supersonics, jet propulsion, and missiles. Peacetime also provided an opportunity to make aesthetic improvements at the AERL that had not been possible during the war, such as planting grass and almost 4,000 trees and shrubs.[160]

## Postwar Adjustments

The Plum Brook Ordnance Works ceased operations immediately after Japan's surrender in August 1945, and the Trojan Powder Company dismissed most of its 4,000 employees—retaining only a group to decontaminate and process the facilities. The 9,000-acre Plum Brook soon became a ghost town with only light security and maintenance contingents remaining. The War Assets Administration (WAA) assumed control in 1946 and used the Plum Brook Depot, as it was then referred to, to store ammunition.[161] The government continued to decontaminate the site in the postwar years. For the most part, however, Plum Brook was desolate.[162]

Only three years after the massive industrial site shut down, wildlife began reclaiming the enclosed area. Birds, deer, foxes, and other animals were soon present in large quantities. During winter 1947, groundhogs caused a potential hazard by burrowing into the sod

that covered the 99 bunkers that were still being used to store explosives. The countless burrows exposed the concrete roofs and increased the threat of explosion. The Ohio Conservation Division eventually began trapping and relocating the groundhogs.[163]

During the mid-1940s the WAA attempted to sell the Plum Brook site for either industrial or agricultural use.[164] It also sold roughly 3,000 acres of

*Image 82 (inset): 1949 advertisement seeking to sell the Plum Brook Ordnance Works. Image 83: Interior of 1 of the 99 Plum Brook bunkers that were used to store crates of trinitrotoluene (TNT) and dinitrotoluene (DNT) during the war, c1941 (GRC–2015–C–06565).*

**RESEARCH REORGANIZED**

A reorganization of the Research Department of the Laboratory into four new Divisions has been effected. The new Divisions are to be known as the Fuels and Thermodynamics Division, the Turbine and Compressor Division, the Engine Performance and Materials Division and the Wind Tunnel and Flight Division.

As Executive Engineer, Mr. Carlton Kemper heads the Research Department. Mr. Jesse H. Hall is Assistant to Mr. Kemper. As Chief of Research, Mr. Addison M. Rothrock will assist the Executive Engineer in following out and directing research projects. Dr. R. F. Selden will be assistant to Mr. Rothrock.

Mr. Benjamin Pinkel is Chief of the new Fuels and Thermodynamics Division. The three branches of this Division are: the Fuels Branch under Dr. L. C. Gibbons, the Combustion Branch under Dr. W. T. Olson and the Thermodynamics Branch under E. J. Manganiello.

In the Turbine and Compressor Division with Oscar W. Schey as Chief, there are two branches. Frank E. Marble will be in charge of Basic Turbine and Compressor Research and R. O. Bullock will be in charge of Applied Turbine and Compressor Research.

The Engine Performance and Materials Division under J. H. Collins, Jr., consists of three branches. These branches and their heads are: Engine Operation and Controls, A. E. Biermann; Jet and Turbo-Propeller Engines, J. C. Sanders; Materials, Dr. M. C. Shaw.

In the Wind Tunnel and Flight Division, Mr. Abe Silverstein is Chief. Dr. J. C. Evarrd, William Perl, I. I. Pinkel and N. D. Sanders will be assistants to the Division Chief in matters pertaining to the planning and carrying out of research. A. W. Young will direct
*Continued on last Page, Col. 1*

*Images 84 (left): The lab's new management team—Addison Rothrock (left) and Raymond Sharp (center)—with NACA Director of Research George Lewis (GRC–1945–C–12029). Image 85: Wing Tips article about AERL's reorganization.*

perimeter area to the public with a clause that gave the government the right to purchase the land back in 20 years. The General Services Administration took over Plum Brook in 1949 and began taking steps to prevent further deterioration of the remaining buildings and infrastructure.[165] In 1954 the Ravenna Arsenal assumed control as the NACA began expressing interest in using portions of the property.[166]

❖ ❖ ❖ ❖ ❖ ❖

The AERL's September 1945 reorganization virtually eliminated piston engine work and redirected researchers to address the technical challenges of turbojet, ramjet, and rocket engines. The division chiefs—Oscar Schey, Benjamin Pinkel, John Collins, Addison Rothrock, and Abe Silverstein—spearheaded the new plan without the input of manager Ray Sharp and Executive Engineer Carlton Kemper.[167] Sharp trusted his managers and usually did not interfere with technical decisions. Kemper had been on special assignment in Europe since April 1945 as part of the United States's effort to review and secure German technology following the fall of Berlin. The team interrogated researchers, inspected facilities, and captured documents.[168,169] During Kemper's absence, Addison Rothrock assumed control of the lab's research activities. When Kemper returned

to Cleveland in October, the reorganization was complete and his role as Executive Engineer became that of a consultant. As the new Chief of Research, Rothrock made most of the technical decisions.[170]

The new alignment included just four research divisions: Fuels and Thermodynamics, Compressor and Turbine, Engine Performance and Materials, and Wind Tunnels and Flight. The improvement of fuels was not as important for jet engines, so those researchers were placed in Ben Pinkel's Fuels and Thermodynamics Division to study heat transfer issues, combustion, and high-energy liquid propellants for rocket applications. This research was the germ of critical propellant advances in the ensuing decades. Jet engines did not employ superchargers, so the supercharger researchers were assigned to Oscar Schey's new Compressor and Turbine Division. John Collins's Engine Performance and Materials Division studied engine controls, performance, and new high-strength materials. Abe Silverstein's Wind Tunnels and Flight Division was responsible for engine testing in the

*Image 86: NACA Secretary John Victory (left) and Ray Sharp (right) lead General Dwight Eisenhower on a tour of the Cleveland lab on 11 April 1946. The former supreme commander of Allied Expeditionary Forces in Europe was visiting several U.S. cities at the time (GRC–1946–C–14688).*

Altitude Wind Tunnel (AWT), the expanded icing research program, flight testing, and the new supersonic research.

The military remained the NACA's primary sponsor after the war. Representatives worked closely with both NACA Headquarters and researchers at the lab to plan research projects. The military then made official research requests through headquarters. Nonetheless, AERL management, particularly Abe Silverstein, encouraged the researchers to steer the military into the most beneficial areas.[171] In addition, most of the basic research was internally driven. The resulting technical reports were made available to industry and military planners who did not have the time or facilities to undertake such fundamental efforts.[172] The AERL submitted a list of fundamental problems requiring research to headquarters in December 1945, including a variety of new engine types. There was pushback, however, from industry. The engine companies did not want to compete with the federally funded NACA in developing

new engines. In response, the NACA leaders decided to segregate the research by components, rather than by engine type. Therefore work was focused on subjects like turbines instead of turbojets.[173,174]

❖ ❖ ❖ ❖ ❖ ❖

During the final six months of the war, the once-secret laboratory began opening its doors to the press, aeronautic societies, servicemen, and industry leaders. The lab erected grandstands beside the AWT to provide a location to take photographs of the groups of visitors, and Ray Sharp made a point to have a photograph taken of individual visitors and their hosts outside of the Administration Building entrance.

In 1947 the NACA renamed the AERL the Flight Propulsion Research Laboratory to reflect the expansion of its research activities beyond engine research. That same year, the lab hosted its first Inspection. The NACA had held annual conferences, or Inspections, at Langley from 1926 to 1939. The NACA would invite hundreds of esteemed individuals from the manufacturing industry, military, and universities to visit the Langley laboratory on a specific

Visit of North Atlantic Treaty Organization Research and Development Conference to Lewis Laboratory, February 7, 1951. (Visitors listed on back).

*Image 87: Page from a photo album of visitors to the Administration Building in the postwar years (GRC–1951–C–27147).*

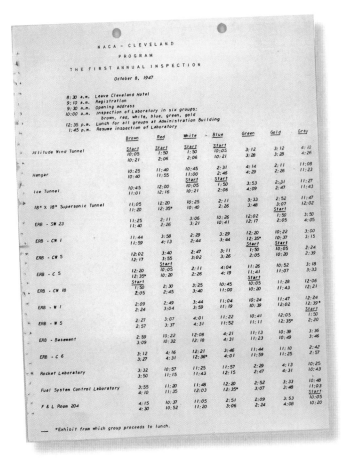

Images 88: Tour stop schedule for the 1947 Inspection.

date. Then the guests would be segregated into groups that would tour a series of stops, mostly at test facilities, where researchers would brief them on the latest efforts in their particular field. The events not only allowed the NACA to show off its activities, but also provided an opportunity for the NACA to receive feedback on what issues required research.[175,176]

The Inspections resumed in 1946 as multiday events at all three NACA laboratories. The Inspections were very important to the NACA, and great efforts were expended to ensure that they were carried out flawlessly. Everything—from the technical talks and tour schedule to the coffee service and transportation—was meticulously planned and rehearsed. The NACA Inspections always received rave reviews.

The Cleveland laboratory's 8–10 October 1947 Inspection featured the AWT and showcased the lab's compressor, turbine, fuels, high-altitude combustion, and materials research. The lab's P–61 Black Widow aircraft made a flyby with a ramjet engine running underneath its wing for the visitors at the Administration Building, and the 1,000 guests viewed the firing of small rocket engines in the new Rocket Lab test cells. In 1947 the lab also began the tradition of holding an open house for employees and their families on the Sunday following the event.[177]

Image 89: George W. Lewis (GRC–2015–C–06556). Image 90: Myrtle Lewis with her sons George, Jr., and Harvey during an October 1951 visit to the laboratory (GRC–1951–C–28570).

The 1947 Inspection in Cleveland included a conspicuous absence—George Lewis. He had joined the NACA Headquarters in 1919 and was named Director of Aeronautical Research five years later, serving as the liaison between the Executive Committee and the research laboratories. Lewis, who did not take a day of leave between the Pearl Harbor attack and the Armistice, began suffering health problems in 1945. These concerns forced him to miss the 1947 Inspection and to retire shortly thereafter. Lewis died in July 1948.[178] During the lab's second Inspection in September 1948, the NACA commemorated Lewis by changing the laboratory's name to the Lewis Flight Propulsion Laboratory.[179] Prior to his passing, Lewis was able to see his sons George, Jr., and Harvey begin 30-year careers as compressor and instrument researchers at the lab.

## Jet Propulsion

Lewis's primary emphasis in the postwar era was turbojet engines. The switch from piston to jet engines required the modification of test facilities and the training of personnel. The NACA researchers tackled all related issues—combustion at high altitudes, compressor and turbine design, controls, thrust augmentation, and new stronger materials. The component work was done in the Engine Research Building test cells, while the laboratory's unique altitude test facilities tested the entire engine system in realistic flight conditions.

Oscar Schey's Compressor and Turbine Division was the largest at the laboratory, and it issued the most technical reports during this period. The researchers were able to apply the experience they had gained previously in the Supercharger Division to the compressor and turbine work. Improvements in this area were a key aspect of the continual drive to advance engine performance while reducing weight and fuel consumption. In 1946 the laboratory added an entire new wing to the Engine Research Building for compressor and turbine testing.

Schey's group was primarily involved with axial-flow engines. Although axial-flow engines were more aerodynamic and lighter than centrifugal engines, they were much more complex. There were multiple stages and hundreds of specially shaped blades and stators to take into account, resulting in tens of thousands of measurements for each test run. Therefore the researchers

*Image 91: A mechanic examines compressor blades on a General Electric J47 engine (GRC–1949–C–22850).*

*Image 92: A failure of a Westinghouse J34 engine in the AWT test section (GRC–1950–C–26294).*

performed as much theoretical work as possible before studying single blades in small wind tunnels. Once that process was completed, they moved on to single-stage compressors, and then verified the findings with multistage compressors.[180]

The two main thrusts of the axial-flow research were the improvement of airflow through the compressor and turbine blades and the cooling of the turbine. The NACA also conducted basic research aimed toward the creation of new analytical design methods that would shorten the development schedule.[181]

Designers required a much better understanding of the interaction of the compressor blades, fan stages, and turbine. The complex stages also required new lightweight, high-strength materials that could endure the engine's stresses and high temperatures. These must be inexpensive, easily obtained materials.[182] At the time, Lewis was unique in its ability to study full-scale multistage compressors. The Compressor and Turbine Division made key advances in understanding how the compressor stages performed in relation to one another.[183]

The AWT was the laboratory's engine research workhorse in the 1940s and 1950s. Despite being designed for piston engines, the facility was robust enough that only slight modifications were required to test the more powerful turbojets. One of the most significant problems with the early turbojets was maintaining combustion at high altitudes, and the AWT was the nation's only facility in which the combustion and performance characteristics of a turbojet could be studied under altitude conditions.[184] Lewis researchers used the AWT to analyze almost every model of U.S. jet engine that emerged in the 1940s, including extensive studies of the Westinghouse J34 and General Electric J47, the nation's first commercially successful jet engines. The researchers investigated overall engine performance, operating range, acceleration, flameouts, fuel consumption, and amplification of thrust over a variety of altitudes and speeds.[185]

Another issue was the need for short bursts of power, particularly for takeoffs. Early jet engine nozzles were typically designed to operate at maximum speed, so their efficiency decreased at slower speeds. Consequently, early jet aircraft required longer

*Image 93: A mechanic with a fire extinguisher watches the firing of twin afterburners (GRC–1949–C–23744).*

runways and burned more fuel during takeoff than piston engines did. Lewis engineers developed several tools to address this, including the variable-area nozzle. The variable-area nozzle is an exhaust pipe that can expand or contract its diameter to quickly increase or decrease thrust levels. The nozzle improved control and fuel efficiency at lower speeds.[186] Researchers also found that the injection of water or water-alcohol mixture into the compressor reduced the engine temperature, allowing the compressor to pump additional air into the combustion chamber. Lewis studies indicated that this method increased thrust by 35 percent at takeoff speeds and by 50 percent in cruising conditions.[187]

A third method of thrust augmentation was the afterburner. Afterburners spray fuel into the nozzle where it combusts with the hot exhaust to produce additional thrust. They burn large quantities of fuel, however, so afterburners can only be operated for brief periods. Although the concept was developed in the 1930s, AERL researchers were the first to successfully operate

an afterburner.[188] The initial demonstration took place in 1945 when the researchers successfully ran a General Electric J47 engine with a low-velocity afterburner in the AWT.[189] They examined various different afterburner configurations during the 1940s, investigating each variable independently and over a variety of altitudes and speeds. As a result, J47 thrust increased by 37 percent at standstill and by 75 percent in normal flight.[190] By the mid 1950s, engine designers were including variable-area nozzle afterburners in nearly all jet engines. The use of afterburners evolved from merely assisting takeoff to being an integral part of supersonic flight.[191]

The AWT was such a success in the mid-1940s that there was a bottleneck of engines waiting to be tested. In 1946, the lab added a new Compressor and Turbine Wing to the Engine Research Building, including two 10-foot-diameter test chambers, known as the Four Burner Area, that could test full-scale engines in simulated altitudes up to 50,000 feet. In 1947 the Rolls Royce Nene became the first engine

*Image 94: Mechanics install a turbojet engine in a Four Burner Area test cell (GRC–1950–C–25120).*

tested in the two new test cells.[192] Over the next 10 years Lewis researchers conducted a variety of studies in the Four Burner Area with particular focus on the J47 and Pratt & Whitney's J65 turbojets.[193]

## Backbone of the Laboratory

The laboratory's attention to its technical force yielded significant dividends over the years. The technicians and mechanics in the Test Installations Division were a critical component of the research process. They were often as important to the achievement of the test's goals as the researchers or test engineers were. In general the researchers decided what they wanted tested—usually basing this on military or industry requests. They then obtained test equipment from engine companies or Lewis's Fabrication Shop. The researchers informed the Test Installation Division of the test program's objectives and parameters. Then the test engineers worked with the mechanics and technicians to install the test article, modify the facility's systems to achieve the desired test conditions, and install the extensive instrumentation and data-recording equipment. Although their names usually

do not appear on the research reports, the insights and modifications developed by the mechanics and technicians were critical to the ultimate success of a test. Engineers frequently incorporated the mechanics' ideas into the engineering drawings.[194,195]

*Image 95: Aircraft mechanics work on an early jet aircraft in the hangar (GRC–1946–C–14739).*

In 1942 the laboratory established a four-year Apprentice Program to train new employees to be highly skilled tradesmen and to facilitate the close interaction of the laboratory's engineers and scientists with the mechanics and technicians. The apprentice school covered a variety of trades including aircraft mechanic, electronics, instrumentation, machinist, and altitude systems mechanic.[196] The school quickly faltered when nearly all 70 of the original students entered the military. For several years the NACA hired many journeymen to fill these vacancies. In August 1947 the laboratory began taking steps to resurrect the Apprentice Program and inducted 47 apprentices into the program in January 1948.[197,198]

NASA-LeRC Apprentice Distribution Breakdown
(December 1949 - October 1, 1972)

| APPRENTICE | TRADE | GRADUATED | WG | WL | WS | GS | SEPARATED |
|---|---|---|---|---|---|---|---|
| Harrison, Billy R. | Flt. Prop. Mech. | 12/49 | | | | | |
| Herrlich, Walter | " | 12/49 | | | | 15 | |
| McComb, William H. | " | 12/49 | | | | 11 | 6/30/70 (Retired) |
| Schwab, William B. | " | 12/49 | 14 | | | | |
| Vladyka, George H. | Instrum. Mechanic | 12/49 | | | | | |
| Budak, Andrew | Machinist | | | | | | 4/30/72 |
| Gaydos, William J. | " | 12/49 | | | | | 6/06/52 |
| Guzik, Tadeusz H. | " | 12/49 | | | | | 9/28/51 |
| Mittermiller, Clarence G. | " | 12/49 | | | | 13 | |
| Nechvatal, Edward S. | " | 12/49 | | | | | 10/28/52 |
| Ruksenas, Joseph G. | " | 12/49 | | | | | 2/19/64 |
| Rys, Valentine J. | " | 12/49 | | | | | 1/05/51 |
| Sabovik, Stephen | " | 12/49 | | | | 12 | |
| *Skonieczny, Chester S. (Skony) | " | 12/49 | | | | | 1/02/52 |
| Szabo, William | " | 12/49 | | | | 13 | |
| | " | 12/49 | | | | | 1/02/51 |
| Mayher, Thomas J. | Aircraft Mechanic | 2/52 | 14 | | | | |
| Raymond, Wayne F. | " | 2/52 | | | | | 10/07/57 |
| Siggelkow, Richard A. | " | 2/52 | | 11 | | | |
| Tesar, Leonard J. | " | 2/52 | | 14 | | | |
| Vanta, Paul S. | " | 2/52 | | 14 | | | |
| Brichacek, James J. | Electrician | 2/52 | | 11 | | | |
| Wolfe, Wilbur F. | " | 2/52 | | | 11 | | |

\* Last name changed to name in parenthesis

- 1 -

*Image 96: Page from a compiled apprentice roster.*

The three-year program, which was certified by both the Department of Labor and the State of Ohio, included classroom lectures, the study of models, and hands-on work. The apprentices rotated through the various shops and facilities to obtain a well-rounded understanding of all of the work at the lab. The NACA held the apprentices to a higher standard than industry did. Participants had to pass written civil service exams before entering the program and possess some form of previous experience—either with mechanical model airplanes, radio transmission, or one year of trade school. The apprentices were promoted through a series of grades until they reached journeyman status. Those who excelled in the Apprentice Program would be considered for a separate five-year engineering draftsman program.

In December 1949 Lewis recognized 15 members of the World War II program with honorary degrees. The 1952 class contained the first official 46 graduates. This increased to 110 in 1957 and to over 600 by 1969.[199] The program remained strong for decades, and many of the laboratory's future managers began their careers as apprentices.[200]

*Image 97: Some apprentices take a break from their studies to pose for a photograph. Only 150 of the 2,000 hours of annual training were spent in the classroom (GRC–1956–C–43227).*

*Image 98: A Consolidated B–24D Liberator (left), Boeing B–29 Superfortress (background), and Lockheed RA–29 Hudson (foreground) parked inside the hangar. A P–47G Thunderbolt and P–63A King Cobra are visible in the background (GRC–1944–C–05416).*

## Chasing the Icing Clouds

Flight research has been a critical component of a vast array of the laboratory's research since the program's initiation in March 1943. During the war, the military provided the laboratory with the same types of aircraft that the research was intended to benefit. For example, new methods of cooling the Pratt & Whitney R–2800 engines were verified with test flights on the R–2800-powered Martin B–26. After the war, the research aircraft served as testbeds to investigate engines or systems that often had little to do with that particular aircraft. For instance, a B–29 was modified so that a ramjet could be lowered from its bomb bay and fired during flight. A near-endless procession of aircraft, mostly military, passed through Lewis's hangar during the 1940s.

In addition to the aircraft, Lewis had an elite corps of pilots: Howard Lilly, Bill Swann, Joe Walker, and William "Eb" Gough. Lilly, a young pilot with recent Navy experience, flew in the National Air Races when they were reinstituted in 1946. In July 1947 he transferred to the NACA's new flight research lab at Muroc Lake, where he became the first NACA pilot

*Image 99: Lewis pilot Howard Lilly poses with his P–63 King Cobra, which he flew in the 1946 National Air Races (GRC–2015–C–06813).*

*Image 100: AERL pilots during the final days of World War II: from left to right, Joseph Vensel, Howard Lilly, William Swann, and Joseph Walker. William "Eb" Gough joined the group months after this photograph. Vensel, a veteran pilot from Langley, was the Chief of Flight Operations and a voice of reason at the laboratory. In April 1947 Vensel was transferred to lead the new Muroc Flight Tests Unit in California until 1966 (GRC–1945–C–11397).*

to penetrate the sound barrier. On 3 May 1948, Lilly became the first NACA pilot to die in the line of duty. Swann was a young civilian pilot when he joined the NACA. He spent his entire career at the Cleveland laboratory and led the flight operations group from the early 1960s until 1979.

The AERL hired Walker, a former military pilot, as a physicist in early 1945, but he soon joined the pilot corps. Gough joined the AERL in December 1945 and was appointed head of the Flight Operations Section. In the Navy, he became the nation's fourth person to qualify on helicopters and the 35th on jets. Gough was stationed in the Pacific and tasked with the job of flying the reliable, but ungainly, Consolidated PBY Catalina flying boat. He survived being shot down and taken in by indigenous natives, and later rescued 10 nurses from Corregidor just before the island was captured by the Japanese.[201,202]

❖ ❖ ❖ ❖ ❖ ❖

The Icing Research Tunnel (IRT) functioned during World War II, but it was not the sophisticated research tool that its designers had intended.

*Image 101: A flight research member examines instrumentation in the B–24D during a 1945 icing flight (GRC–1945–C–10377).*

Hampered by the lack of data on the actual size of naturally occurring water droplets, engineers struggled for six years to develop a realistic water spray system. The spray bars had to generate small droplets and distribute them uniformly throughout the airstream while resisting freezing and blockage. NACA engineers meticulously developed and tested a variety of different designs before the system was perfected in 1950.[203]

Meanwhile, the Flight Research Branch performed most of the lab's icing research. The laboratory increased its activities and introduced two new areas of research after the war—turbojet icing and cloud physics studies. The lab performed a limited amount of icing research during World War II, but the effort was expanded significantly when the NACA transferred its official icing program from Ames to Lewis. Although the Lewis icing researchers utilized numerous aircraft during this period, the Consolidated B–24M Liberator and the North American XB–25E Mitchell were the primary tools. Both were heavily modified to permit flights through the worst icing conditions.

In January 1947, Wilson Hunter, head of the icing program, announced that research on piston engines, particularly propellers, was complete. Going forward, the branch would concentrate on studying the effects of ice buildup on jet engines. Although jet engines allowed aircraft to pass through inclement weather at high rates of speed, little was known about their susceptibility to ice. Lewis researchers made plans to study a General Electric I–16 centrifugal engine and a Westinghouse J34 axial-flow engine.[204]

The NACA researchers decided to mount the I–16 inside the rear area of their Consolidated B–24M bomber. Mechanics installed an air scoop on top of the aircraft to guide airflow to the engine inside. Then they placed spray nozzles in front of the scoop to simulate icing at the engine's inlet. Although ice blocked up to 70 percent of the inlet during the late 1946 research flights, the centrifugal engine exhibited no combustion problems.[205]

Axial-flow turbojets were another matter. The inlet and engine cover were particularly vulnerable, and any ice reaching the internal components could damage the compressor blades.[206] In April 1947 the icing researchers tested the J34 engine, which used a deicing

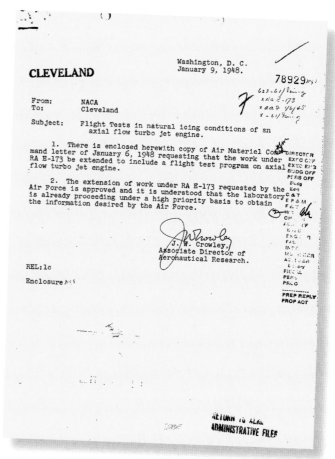

Image 102: NACA memo authorizing icing flight tests of jet engines.

Image 103: Abe Silverstein measures ice buildup on the Westinghouse J34 engine (GRC–1948–C–20836).

system to flow hot engine gases into the engine inlet. The test—the only icing research ever conducted in the AWT—demonstrated that, although the system could prevent ice buildup, it caused a decrease in thrust.[207] The researchers then examined several different ice-prevention techniques in the IRT. They concluded that the hot gas deicing system could in fact perform its task without degrading engine performance.[208]

With the numbers of jet aircraft growing, the military pressed for flight tests to determine the overall icing threat to axial-flow engines.[209] Lewis mechanics attached the J34 axial-flow engine to the wing of the B–24M aircraft and flew several hour-long missions through mild icing conditions in spring 1948. The engine did not stall, but the icing decreased thrust and increased nozzle temperatures.[210] When Lewis researchers redesigned the inlet area, they were able to successfully test three different heating systems that did not limit engine performance.[211]

❖ ❖ ❖ ❖ ❖ ❖

The NACA's icing research program seemed to be winding down in 1947 as the turbojet studies were carried out. Lewis's researchers, however, convinced management to undertake an even larger icing effort— the study of icing cloud physics. Rather than merely testing different anti-icing and deicing tools, the

icing group created an engineering field that thrives to this day, seeking to perfect weather-sensing equipment and systematically measure and categorize water droplets and clouds. At the time, it was hoped that aircraft designers would use the information to determine precisely what level of anti-icing protection would be best for a particular aircraft and to identify its performance and drag tradeoffs.[212]

Lewis pilots flew missions throughout the country to better understand the cloud conditions that cause ice formation. During the peak of these studies from 1946 to 1950, Lewis icing flight researchers documented and analyzed ice-producing clouds and developed numerous icing instruments, including Porter Perkins's icing rate meter. This lightweight device continuously recorded the frequency and intensity of icing conditions and warned the pilots when icing conditions were present.[213] The NACA was able to convince the military and several airlines to install icing rate meters on their normal flights and to send the data back to Cleveland for analysis.[214]

The NACA researchers and analysts studied natural icing cloud conditions for several years, making extensive measurements and conducting statistical analyses. The researchers were able to determine the general shape and extent of supercooled stratiform (thin,

*Image 104: The XB–25E Mitchell searches for icing clouds in January 1947. The aircraft, dubbed "Flamin Mamie," includes nose artwork depicting a fiery woman chasing off icing researchers (GRC–1947–C–17763).*

*Image 105: The wooded picnic grounds as it appeared in August 1945. The area was improved in the ensuing years (GRC–1945–C–12065).*

In the early 1940s the NACA constructed a recreational area in a wooded area along the edge of the lab. The area included handball courts, horseshoe pits, grills, tables, and a pavilion. The staff used the site for a variety of events, including the Lewis employee picnic held every July. The original recreational area was torn down in the late 1970s to make way for the Research Analysis Center. Since then, the picnic grounds have been located at different locations in the center's West Area.

horizontal) clouds and their relation to general weather conditions.[215] They found that stratiform clouds produced 95 percent of the icing encounters that pilots experienced during normal flights.[216]

The Federal Aviation Administration (FAA) used the NACA research to create mandatory ice protection criteria for aircraft manufacturers. These requirements were a balancing act between safety precautions and overburdening the manufacturers. It was assumed that some aircraft components could withstand brief periods of ice accretion. Those aircraft with less than maximum protection would have to rely on meteorological forecasting to navigate around severe icing conditions.[217] The Lewis icing program tapered off in the mid-1950s as these studies came to a conclusion, but it has reemerged as a robust element of today's aeronautical research program.

❖ ❖ ❖ ❖ ❖ ❖

Researchers needed to convert the data collected from the icing program, as well as other research, into meaningful information. The female computing staff performed that function during the war, but by the late 1940s Lewis had begun to acquire computational technology. One of the first notable examples was Lewis researcher Harold Mergler's differential analyzer. Mergler joined the Instrument Research Section in 1948, where he focused on the synthesis of analog computers with the machine tools used to create compressor and turbine blades for jet engines.

*Image 106: Harold Mergler with his differential analyzer (GRC–1951–C–27875).*

*Image 107: A computer at work in one of the three offices on the second story of the 8- by 6-Foot Supersonic Wind Tunnel (8×6) office building. The largest room housed approximately 35 women with advanced mathematical skills (GRC–1954–C–35057).*

Researchers had used differential analyzers since the 1930s to resolve computations up to the sixth order. Those devices, however, had to be rewired before each new computation.[218] In the late 1940's Mergler modified the differential analyzer, eliminating the need for rewiring. In four days Mergler's machine could calculate what previously had taken the computing staff weeks to accomplish, and some of the computer staff members transitioned to the new device. Icing researchers used the device extensively in the 1950s to calculate water droplet trajectories.[219]

Lewis's first electronic computer was the IBM 604 Electronic Calculating Punch. The 604, introduced in 1948, was IBM's first mass-produced electronic computer.[220,221] The programmable computer performed basic mathematical calculations at comparatively high speeds. This was supplanted by IBM's Card Programmed Electronic Calculator, which could be programmed by punch cards to allow programs of unlimited length.[222] The system recorded the manometer pressure readings from the test facilities directly into punch cards that were fed into the calculator.

Staffing requirements decreased as electronic data processing capabilities improved. The "computers" primarily female, quickly adapted and learned how to encode the punch cards. Computing remained one of the few technical professional areas at the NACA open to women at the time.[223] At the same time, many left to start families, while others earned mathematical degrees and moved into advanced positions.

After the war ended, the NACA removed or repurposed several of the AERL's temporary buildings. The Farm House, which had initially served as the main office, was dramatically modified in 1946 to increase its size, give it a more formal appearance, and add modern conveniences.[224] The new structure was built around the original. The building, renamed the Administrative Services Building, housed members of the Administrative Services Branch. These included the AERL's telephone operators, motor pool, travel agents, forms and records management staff, and mail service.

The AERL was then able to demolish the group's former location at the Tempo A Building. As the adjacent airport expanded in the early 1950s, the laboratory relocated the entire Administrative Services Building from its position near the main gate to the lot directly behind the Administration Building. The staff used the Administrative Services Building for 20 years before deterioration forced the center to demolish it in 1973.[225]

*Image 108 (top): The Farm House as it appeared shortly after the NACA took over in 1941 (GRC –2011–C–00345). Images 109 and 110 (center and bottom): The Administrative Services Building after the modifications (GRC–1946–C–15355) and after it was moved behind the Administration Building (GRC–1967–C–01234).*

## Supersonic Missile Research

Of all the new technologies to emerge from World War II, perhaps none offered as much military potential as guided missiles. The missiles' speed and ability to breach defense systems made them exceptionally useful for striking distant enemies. During World War II, the Germans demonstrated that missiles could be powered by airbreathing (V–1) or rocket-powered (V–2) engines. After the war, the U.S. Army famously brought the captured German rocket team to its research base at White Sands, New Mexico. In the early 1940s, Caltech researchers created a small rocket test site, known as the Jet Propulsion Laboratory (JPL), near Pasadena. A new U.S. rocket and missile industry took hold in Southern California.[226] The military was interested in developing rocket and ramjet engines and in using traditional propellants and experimental high-energy fuels.

The NACA founded a Special Committee on Self-Propelled Guided Missiles, which included Abe Silverstein, to track these developments. The NACA also established the Pilotless Aircraft Research Station at Wallops Island in 1945, and the High Speed Research Facility at Muroc Lake in 1946. The former provided Langley with a remote location from which to fire small missiles over the Atlantic, and the latter provided a large dry lake bed for testing experimental aircraft.[227,228]

The NACA also began constructing supersonic wind tunnels at its research laboratories. In 1942 Langley commenced work on a small supersonic tunnel with a 9-inch-diameter test section for studying swept-wing designs. The tunnel engineers, however, were not able to get the facility operating correctly until 1946.[229] The real supersonic wind tunnel push began in 1945 when the NACA approved the construction of a large supersonic wind tunnel at each of its three research laboratories. Almost immediately, Lewis began design work on the largest of these, the 8×6. These large facilities took several years to construct. In the meantime, the laboratories set out to rapidly build smaller supersonic facilities to study small-scale models.

As design work commenced on the 8×6, Lewis engineers quickly constructed several small supersonic tunnels that took advantage of the massive air-handling equipment at the AWT. The first

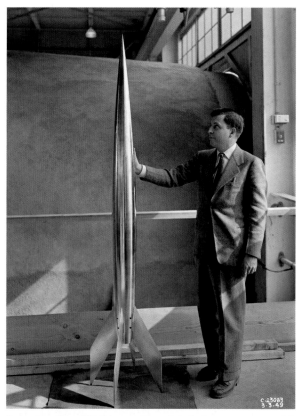

*Image 111: Harold Friedman with an 8-inch-diameter ramjet model (GRC–1949–C–23083).*

*Image 112: Construction of the lab's first supersonic tunnel. Eventually the building would house three small supersonic tunnels (GRC–1945–C–10764).*

of these, wedged between the AWT and the IRT, began operation in August 1945 after just 90 days of construction. Lewis added two other small tunnels to the facility in 1949 and 1951.[230] The three tunnels, with test sections ranging from 1.5 to 2 feet in diameter, were built vertically on top of each other and thus were referred to as the Stack Tunnels.

Researchers used the Stack Tunnels to study inlet configurations and boundary layer effects for jet engines.[231] The Duct Lab, a fourth AWT-based supersonic tunnel, was put into operation by November 1945, and the 1- by 1-Foot Supersonic Wind Tunnel in the Engine Research Building began running in 1951. Lewis used these small tunnels to study a variety of inlets, nozzles, and cones for missiles and scramjets.

❖ ❖ ❖ ❖ ❖ ❖

The ramjet's potential missile and aircraft applications spurred the Lewis Wind Tunnels and Flight Division to pursue an extensive research program in the 1940s and 1950s. Ramjets provide a very simple source of propulsion for missiles or aircraft. They are basically a tube in which high-velocity air is ingested and ignited. The heated air expands and is expelled at a significantly higher velocity for thrust. There are no moving components. Ramjets rely on the engine's forward motion to push the air through. This simplifies the design because no compressor or turbine is needed, but a booster stage of some sort is required to get the ramjet up to speed before it can be ignited. Ramjets perform better as the vehicle's speed increases. The concept was not new, but because manufacturers did not have facilities powerful enough to test the engines, development stalled.[232]

Lewis's missile work was initially segregated into three groups—inlet and nozzle studies in the small supersonic tunnels, full-scale engine tests in the AWT and launched from research aircraft, and theoretical aerodynamic research by John Evvard's Special Research Panel.[233,234] The Special Projects Panel initially designed a 20-inch-diameter ramjet

Image 113: A technician operates a Schlieren camera to view the airflow dynamics inside the 24- by 24-inch test section of one of the Stack Tunnels (GRC–1949–C–24977).

Image 114: John Evvard with a missile model in the 8×6 in February 1957 (GRC–1957–C–44223).

*Image 115: A B–29 bomber that was modified to serve as a ramjet testbed for Lewis researchers. The experimental ramjet was lowered from the bomb bay and fired (GRC–1948–C–21990).*

engine and analyzed the thrust levels and performance of each of the engine's components in the AWT during May 1945.[235,236] In 1946 and 1947 John Disher and his colleagues subjected the engine to a series of flight tests on the lab's Boeing B–29 aircraft. Flight mechanics mounted the ramjet on a retractable spar in the aircraft's bomb bay. Once the aircraft's proper altitude and speeds were achieved, they lowered the engine into the atmosphere and ignited it. The researchers were able to study changes in fuel flow, combustion efficiency, and fuel-air ratios.[237,238]

In March 1947 the army asked the NACA to increase its ramjet efforts, which now revolved around a 16-inch-diameter engine.[239,240] The researchers were able to augment their subsonic engine studies in the AWT with a series of supersonic missile drops from the B–29 and their F–82 Twin Mustang aircraft. Lewis pilots flew to Wallops Island where they released the ramjet missile from a high altitude. Initially, the missiles were boosted by a small rocket. The experimental ramjet took over as the missile began its downward trajectory and reached supersonic speeds. These missiles contained telemetry units

*Image 116: A North American XF–82 Twin Mustang prepares for flight with a ramjet missile under its right wing (GRC–1949–C–23330).*

*Image 117: The seven-stage axial compressor that powers the 8×6. The compressor was driven by three electric motors with a total output of 87,000 horsepower, resulting in airspeeds from Mach 0.36 to 2.0 (GRC–1949–C–23277).*

which recorded flight data that was relayed to tracking stations. The researchers analyzed modifications to different engine components and studied the effect of altitude and speed on performance, allowing them to determine the ramjet's performance and operational characteristics in the transonic range.[241]

The 8×6, which became operational in April 1949, was the most powerful of the NACA's three new large supersonic tunnels and was the only facility capable of running an engine at supersonic speeds.[242] Flexible sidewalls altered the tunnel's nozzle shape to vary the Mach number. A massive seven-stage axial compressor blew the airflow through the tubular facility at speeds from Mach 0.36 to 2.1 and loudly expelled it out the other end into the atmosphere.[243]

The tunnel's muffler successfully reduced the noise levels as static models were tested during the first few months of operation, but it proved inadequate when Lewis first ran an engine in the test section in January 1950. The sound roused residents from their sleep and spurred a rash of complaints from the community. The researchers, however, were thrilled with that

*Image 118: A researcher inspects a 16-inch-diameter ramjet engine in the 8×6 test section. Researchers studied the ramjet's performance at different speeds and varying angles of attack. The engine performed well, and the findings correlated with nonfueled studies in the smaller wind tunnels (GRC–1950–C–25776).*

initial run. It demonstrated, for the first time, that a jet engine could operate in an airflow faster than Mach 2.[244] The laboratory suspended operation of the tunnel and hired a local firm to design a new muffler system. The company conducted audio tests in the community and devised a large concrete resonator to enclose the end of the tunnel. The 8×6 was up and running again in less than a year.[245]

## Rocket Combustion

The fuels and lubrication researchers were incorporated into the new Fuels and Thermodynamics Division as part of the 1945 reorganization. Louis Gibbon's Fuels Branch compared the performance characteristics of different fuels in jet and ramjet engines. The studies during this period indicated that fuels with lower boiling points had superior combustion but did not operate as well at high altitudes.[246] Walter Olson's Combustion Branch sought to study the underlying physics of burning and to analyze the performance of different combustion chamber designs. The researchers investigated how combustors operate and identified key design criteria.[247]

Olson's branch included a small group of researchers led by John Dietrich that studied basic rocket engine combustion. This rocket group, ambiguously named the High Pressure Combustion Section, was located in the undeveloped area at the far west end of the laboratory. At the time NACA Headquarters was wary of sponsoring any research on weapons, which is what they considered missiles to be.[248] Lewis management, however, gave the group a wide latitude to pursue its studies on its own. The new rocket section initially studied issues such as combustion and cooling in solid propellant rockets. During a 1945 visit to the Jet Propulsion Laboratory in California, researchers

*Image 119: A researcher prepares a jet-assisted take off (JATO) rocket for a combustion study at the Rocket Lab (GRC–1945–C–10724).*

*Image 120: Firing of a nitric acid aniline JATO rocket at the Rocket Lab in March 1946. The Rocket Lab was expanded over the next 10 years and eventually included its own hydrogen liquefier (GRC–1946–C–14478).*

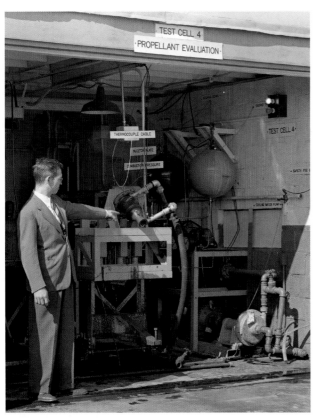

*Image 121: John Sloop demonstrates a small rocket setup in Cell 4 of the Rocket Lab (GRC–1947–C–19769).*

*Image 122: Proceedings from the NACA Conference on Fuels.*

briefed Olson on their extensive solid rocket efforts. To create their own niche, Olson and his new rocket group decided to concentrate on high-energy liquid propellants.[249]

❖ ❖ ❖ ❖ ❖ ❖

High-energy liquid propellants have been critical to the development of lightweight, high-thrust upper-stage rockets. Originally, rockets combusted gun powder or other solid propellant mixtures to create thrust. In 1926 Robert Goddard was the first to launch a liquid-fueled rocket. The next advance came with Wernher von Braun's V–2 missiles during World War II, which used turbopumps to pump liquid alcohol into the combustion chamber. After the war, Lewis researchers began exploring the use of liquids with lighter weight and more energy.

Paul Ordin led a Lewis team that explored the performance of virtually all available high-energy propellants, concentrating on the ones with the greatest combustion performance. Many of these were considered to be more exotic laboratory chemicals than actual propellants. As such, they were difficult to procure. There are stories of Lewis researchers personally transporting dangerous liquids on public trains or snarling local traffic with police-escorted caravans through the streets of Cleveland.[250]

The Lewis Fabrication Shop created almost all of the small rocket engines used in the program. The researchers operated the rockets in four cinder-block test cells in the High Pressure Combustion Laboratory better known as the Rocket Lab. The test cells were surrounded by earthen barriers to contain any blasts. Rockets were fired through the open doors of the cells into the atmosphere. Cell 22, which consisted of two test stands that could fire 5000-pound-thrust engines, was Lewis's premier rocket tool until late 1957. Lewis researchers studied combustion performance, injectors, nozzles, and cooling systems for different rocket configurations.

In May 1948 Lewis hosted a special conference to share its findings with military and industry representatives.[251,252] The conference led to the lab's first official rocket project. In 1949 the Navy asked Lewis

*Image 123: A ramjet installation in the 8×6 in May 1949 (GRC–1949–C–23522).*

to test a small rocket under consideration for the Lark missile. Although the test program did not prove to be consequential, the Navy project added legitimacy to the fledgling rocket group.[253]

The Lewis laboratory boomed, both literally and figuratively, during the postwar years. It had shrugged the wartime chains of quick fixes for the military and quickly hit its research stride. Backed by a battery of facilities to test the engines and their components, Lewis immersed itself in the study of turbojet and ramjet engines while getting a foothold in the incipient field of high-energy rocket propellants. The steady improvements of axial-flow turbojet engines from 1945 to 1949 remains one of the laboratory's most enduring aeronautical contributions. The 1940s engine developments would become apparent on the dramatically more powerful turbojets of the 1950s. Lewis researchers converted data from their altitude testing to create basic design principles that engine manufacturers used in developing successive generations of jet engines. The increased performance, safety, and efficiency led to applications on civilian aircraft and intercontinental flights.[254]

## Endnotes for Chapter 3

158. "War Is Over," *Wing Tips* (25 August 1945).
159. "A Brief Picture of the Flight Propulsion Research Laboratory," 24 September 1947, NASA Glenn History Collection, Cleveland, OH.
160. "Planting Improves Grounds," *Wing Tips* (29 November 1946).
161. Roelif Loveland, "Plum Brook Ordnance Works Is Now Silent and Desolate," *Cleveland Plain Dealer* (13 March 1946).
162. Tetra Tech, Inc., "Final Environmental Baseline Survey Report for the Plum Brook Reactor Facility Decommissioning Project," 28 February 2001, NASA Glenn History Collection, PBOW Collection, Cleveland, OH.
163. V. B. Gray, "Wildlife Rules Area Where Tons of Powder are Stored," *Cleveland Plain Dealer* (14 April 1948).
164. "War Plants for Sale," *Cleveland Plain Dealer* (19 December 1947).
165. David Vormelker, "Plum Brook Being Polished to Join Mothball Reserve," *Cleveland Plain Dealer* (11 June 1949).
166. Anonymous to John Weeks, "Plum Brook Historical Items," 10 March 1965, NASA Glenn History Office, Plum Brook Ordnance Works Collection, Cleveland, OH.
167. Virginia Dawson, "The Turbojet Revolution at Lewis Research Center," Annual Meeting of the Society for the History of Technology (4 November 1984).
168. "Kemper Reports on Foreign Labs," *Wing Tips* (2 November 1945).
169. *Operation LUSTY* (Wright-Patterson AFB, OH: National Museum of the United States Air Force), *http://www.nationalmuseum.af.mil/Visit/ MuseumExhibits/FactSheets/Display/tabid/509/ Article/196144/operation-lusty.aspx* (accessed 5 August 2015).
170. Virginia Dawson, *Engines and Innovation: Lewis Laboratory and American Propulsion Technology* (Washington, DC: NASA SP–4306, 1991).
171. Abe Silverstein interview, Cleveland, OH, by Virginia Dawson, 5 October 1984, NASA Glenn History Collection, Oral History Collection, Cleveland, OH.
172. Cindy Forman, "Term Project Case Study: Lewis Research Center," 16 October 1989, NASA Glenn History Collection, Directors Collection, Cleveland, OH.
173. Irving Pinkel interview, Cleveland, OH, by Virginia Dawson, 30 January 1985, NASA Glenn History Collection, Oral History Collection, Cleveland, OH.
174. Dhanireddy R. Reddy, "Seventy Years of Aeropropulsion Research at NASA Glenn Research Center," *Journal of Aerospace Engineering* (April 2013).
175. James Hansen, *A History of the Langley Aeronautical Laboratory, 1917–1958* (Washington, DC: NASA SP–4305, 1985).
176. Alex Roland, *Model Research* (Washington, DC: NASA SP–4103, 1985).
177. "NACA Inspection Book October 8–10, 1947," 1947, NASA Glenn History Office, Inspections Collection, Cleveland, OH.
178. William Durand, *Biographical Memoir of George William Lewis* (Washington, DC: National Academy of Sciences, 1949).
179. "George William Lewis Commemoration Ceremony," 28 September 1948, NASA Glenn History Collection, Inspections Collection, Cleveland, OH.
180. Robert English, "Compressor and Turbine Research: Scope and Methods," *Inspection at NACA Lewis Flight Propulsion Laboratory, 20-22 September 1949,* 1949, NASA Glenn History Collection, Inspections Collection, Cleveland, OH.
181. English, "Compressor and Turbine Research."
182. Robert Bullock, "Mixed Flow Compressor Research," *1947 Inspection,* 1947, NASA Glenn History Office, Inspections Collection, Cleveland, OH.

183. Walter Olson interview, Cleveland, OH, by Virginia Dawson, 16 July 1984, NASA Glenn History Collection, Oral History Collection, Cleveland, OH.

184. Carlton Kemper, "Dr. Kemper's Talk at Morning Session of First Annual Inspection," *Annual Inspection at NACA Lewis Flight Propulsion Laboratory, 8–10 October 1947*, 1947, NASA Glenn History Collection, Inspections Collection, Cleveland, OH.

185. W. Kent Hawkins and Carl L. Meyer, *Altitude Wind Tunnel Investigation of Operational Characteristics of Westinghouse X24C–4B Axial-Flow Turbojet Engine* (Washington, DC: NACA RM–E8J25, 1948).

186. George Gray, *Frontiers of Flight: The Story of NACA Research* (New York, NY: Alfred Knopf, 1948).

187. Gray, *Frontiers of Flight*.

188. "The Afterburner Story," *Ryan Reporter* 15, no. 2 (1 April 1954).

189. "The Afterburner Story."

190. Gray, *Frontiers of Flight*.

191. Bruce Lundin, David Gabriel, and William Fleming, *Summary of NACA Research on Afterburners for Turbojet Engines* (Washington, DC: NACA RM–E55L12, 1956).

192. Kemper, "Dr. Kemper's Talk."

193. Dawson, *Engines and Innovation*.

194. Howard Wine interview, Cleveland, OH, 4 September 2005, NASA Glenn History Collection, Oral History Collection, Cleveland, OH.

195. Larry Ross interview, Cleveland, OH, 1 March 2007, NASA Glenn History Collection, Oral History Collection, Cleveland, OH.

196. Mark Wright, "We Take Pleasure in Introducing the Apprentice School," *Wing Tips* (2 April 1943).

197. Billy Harrison and Nicki Harrison interview, Cleveland, OH, 14 October 2005, NASA Glenn History Collection, Oral History Collection, Cleveland, OH.

198. "47 Apprentices Enter Training," *Wing Tips* (6 February 1948).

199. "Apprentice Distribution Summary, December 1949–September 30, 1975," 1975, NASA Glenn History Collection, Apprentice Collection, Cleveland, OH.

200. Harrison interview.

201. Judith Bellafaire, "The Army Nurse Corps: A Commemoration of World War II Service," US Army Center of Military History, CMH Pub 72-14, c1993 *http://www.history.army.mil/catalog/pubs/72/72-14.html* (accessed 14 August 2015).

202. Helen Larson, "Gough Heads Pilots," *Wing Tips* (15 January 1946).

203. "Icing Research Tunnel History: 50 Years of Icing, 1944–1994," 9 September 1994, NASA Glenn History Office, Test Facilities Collection, Cleveland, OH.

204. William Leary, *We Freeze to Please: A History of NASA's Icing Research Tunnel and the Quest for Flight Safety* (Washington, DC: NASA SP–2002–4226, 2002).

205. Philip Pragiola and Milton Werner, *Flight Investigation of the Effects of Ice on an I-16 Jet Propulsion Engine* (Washington, DC: NACA RM–E7A20a, 1947).

206. Loren Acker and Kenneth Kleinknecht, *Effects of Inlet Icing on Performance of Axial-Flow Turbojet Engines in Natural Icing Conditions* (Washington, DC: NACA RM–E50C15, 1950).

207. William Fleming, "Hot-Gas Bleedback for Jet Engine Ice Protection," *NACA Conference on Aircraft Ice Protection* (1947), pp. 86–92.

208. Uwe von Glahn, Edmund Callaghan, and Vernon Gray, *NACA Investigations of Icing-Protection Systems for Turbojet Engine Installations* (Washington, DC: NACA RM–E51B12, 1951).

209. Colonel R. J. Minty to the NACA, "Flight Tests in Natural Icing Conditions of an Axial Flow Turbojet Engine," 6 January 1948, NASA Glenn History Collection, Flight Research Collection, Cleveland, OH.

210. Vernon Gray, "The Vulnerability of Turbojet Engines to Icing Conditions," *NACA Inspection Book September 28–30, 1948*, 1948, NASA Glenn History Office, Inspections Collection, Cleveland, OH.

211. Leary, *We Freeze to Please*.

212. Pinkel interview by Dawson, 30 January 1985.

213. Porter Perkins, Stuart McCullough, and Ralph Lewis, *A Simplified Instrument for Recording and Indicating Frequency and Intensity of Icing Conditions Encountered in Flight* (Washington, DC: NACA RM–E51E16, 1951).

214. William Lewis, Porter Perkins, and Rinaldo Brun, *Procedure for Measuring Liquid Water Content and Droplet Sizes in Supercooled Clouds by Rotating Multicylinder Method* (Washington, DC: NACA RM–E53D23, 1949).

215. Dwight Kline and Joseph Walker, *Meteorological Analysis of Icing Conditions Encountered in Low-Altitude Stratiform Clouds* (Washington, DC: NACA TN–2306, 1951).

216. Paul Hacker and Robert Dorsch, *Summary of Meteorological Conditions Associated With Aircraft Icing and a Proposed Method for Selecting Design Criterions for Ice-Protection Equipment* (Washington, DC: NACA TN–2569, 1951).

217. Hacker, *Summary of Meteorological Conditions*.

218. Boaz Tamir, "The Differential Analyzer," The Future of Things (2014), *http://thefutureofthings com/3710-the-differential-analyzer/* (accessed 17 March 2015).

219. Dawson, *Engines and Innovation*.

220. IBM, "Workhorse of Modern Industry: The IBM 650," IBM 650 (2015), *http://www-03.ibm.com/ibm/history/exhibits/650/650_intro.html* (accessed 17 March 2015).

221. Frank da Cruz, "The IBM 604 Electronic Calculating Punch" (2009), *http://www.columbia.edu/cu/computinghistory/604.html* (accessed 17 March 2015).

222. Frank da Cruz, "The IBM Card Programmed Calculator" (12 January 2009), *http://www.columbia.edu/cu/computinghistory/cpc.html* (accessed 17 March 2015).

223. Peggy Yohner interview, Cleveland, OH, by Virginia Dawson, 21 May 1985, NASA Glenn History Collection, Oral History Collection, Cleveland, OH.

224. "Research Reveals Farm House History," *Wing Tips* (16 August 1946).

225. "Early Farm House Served Lewis Well," *Lewis News* (1 March 1991).

226. Robert Kraemer, *Rocketdyne: Powering Humans Into Space* (Reston, VA: AIAA, 2006).

227. Harold D. Wallace, Jr., *Wallops Station and the Creation of an American Space Program* (Washington, DC: NASA SP–4311, 1997).

228. Richard P. Hallion, *On the Frontier: Flight Research at Dryden, 1946-1981* (Washington, DC: NASA SP–4303, 1984).

229. Hansen, *A History of the Langley Aeronautical Laboratory*, pp. 250 and 467.

230. "NACA Announces New Supersonic Wind Tunnel for Jet Propulsion Research," *Wing Tips* (11 August 1945).

231. *Major Research Facilities of the Lewis Flight Propulsion Laboratory. Wind Tunnels—Small Supersonic Wind Tunnels* (Cleveland, OH: NACA Lewis, 24 July 1956): 1.

232. "Ram-Jet Research," *NACA Inspection Book September 20–22, 1949*, 1949. NASA Glenn History Office, Inspections Collection, Cleveland, OH.

233. DeMarquis Wyatt and Irving Pinkel, "Talks on Supersonic Propulsion," *NACA Inspection Book September 28–30, 1948*, 1948, NASA Glenn History Office, Inspections Collection, Cleveland, OH.

234. Jesse Hall to Carlton Kemper, "Cleveland Laboratory Test Program for Investigating Ramjet Engines in Free Flight," 18 August 1947, NASA Glenn History Collection, Cleveland, OH.

235. NACA, *Thirty-second Annual Report of the National Advisory Committee for Aeronautics* (Washington, DC: Government Printing Office, 1946).

236. Abe Silverstein, "Ramjet Propulsion," c late 1940s, NASA Glenn History Collection, Directors Collection, Cleveland, OH.

237. H. Z. Bogert to Commanding General, Air Materiel Command, Wright Field, "Activities of AERL for the Week of 30 September to 4 October 1946,"

7 October 1946, NASA Glenn History Collection, Directors Collection, Cleveland, OH.

238. John Disher, *Flight Investigation of a 20-Inch Diameter Steady-Flow Ramjet* (Washington, DC: NACA RM E7I05a, 14 January 1948).

239. S. R. Brentall to George Lewis, 13 March 1947, NASA Glenn History Collection, Directors Collection, Cleveland, OH.

240. T. J. Nussdorfer, D. C. Sederstrom, and E. Perchonok, *Investigation of Combustion in 16-Inch Ram Jet Under Simulated Conditions of High Altitude and High Mach Number* (Washington, DC: NACA RM–E50D04, 1948), p. 1.

241. Warren North, *Summary of Free-Flight Performance of a Series of Ramjet Engines at Mach Numbers From 0.80 to 2.2* (Washington, DC: NACA RM–E53K17, 1954).

242. "First SWT Run is Successful," *Wing Tips* (15 April 1949).

243. Lewis Research Center, "8- by 6-Foot Supersonic Wind Tunnel," February 1959, NASA Glenn History Collection, Test Facilities Collection, Cleveland, OH.

244. James Hartshorne, "Tests Show Noise at Air Lab Hurts," *Cleveland Plain Dealer* (1949).

245. Dennis Huff, "NASA Glenn's Contributions to Aircraft Engine Noise Research," *Journal of Aerospace Engineering* (April 2013).

246. Louis Gibbons, "Outline of Proposed Talk on Fuels to be Presented at Manufacturers' Conference," *NACA Inspection Book, October 8–9, 1947*, 1947, NASA Glenn History Office, Inspections Collection, Cleveland, OH.

247. Walter Olson, "Combustion Research (For First Annual Inspection)," *NACA Inspection Book, October 8–9, 1947*, 1947, NASA Glenn History Office, Inspections Collection, Cleveland, OH.

248. John Sloop, "High-Energy Rocket Propellant Research at the NACA/NASA Lewis Research Center, 1945–60," Seventh International History of Astronautics Symposium Proceedings, 8 October 1973, NASA Glenn History Collection, Sloop Collection, Cleveland, OH.

249. Dawson, *Engines and Innovation.*

250. John Sloop, *Liquid Hydrogen as a Propulsion Fuel, 1945–1959* (Washington, DC: NASA SP–4404, 1977).

251. "Fuels Conference," *Wing Tips* (4 June 1948).

252. "NACA Conference on Fuels: A Compilation of the Papers Presented by NACA Staff Members," 26 May 1948, NASA Glenn History Collection, Sloop Collection, Cleveland, OH.

253. Sloop, "High-Energy Rocket Propellant Research."

254. Walter Olson, "Lewis at Forty: A Reminiscence," 11 September 1981.

Image 124: A 5,000-pound-thrust rocket engine is fired from the Rocket Lab's Cell 22 in January 1955. The series of tests proved to be Lewis's first successful liquid-hydrogen/liquid-oxygen runs (GRC–1955–C–37428).

# 4. Breakthrough

*"It was just a brilliant group of people, terrific problems,
a great deal of freedom, and terrific facilities and
computer capabilities."*

—Simon Ostrach

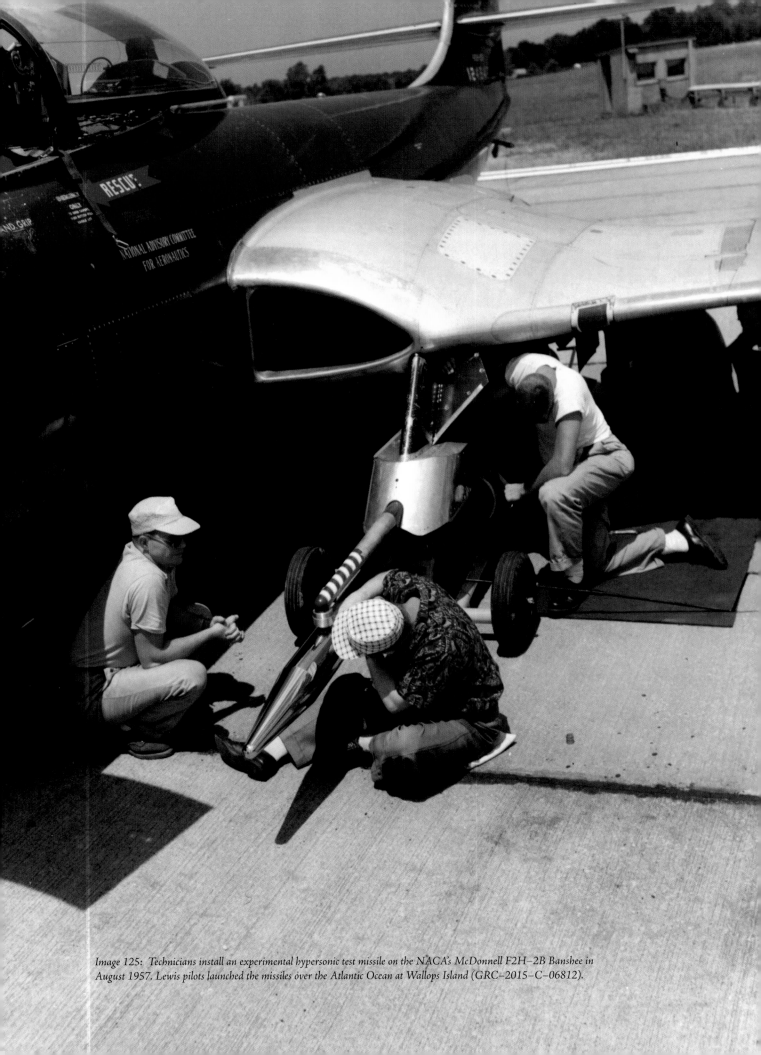

Image 125: *Technicians install an experimental hypersonic test missile on the NACA's McDonnell F2H–2B Banshee in August 1957. Lewis pilots launched the missiles over the Atlantic Ocean at Wallops Island (GRC–2015–C–06812).*

# Breakthrough

Two events took place on Monday, 29 August 1949, that significantly altered the future of the Lewis laboratory: Director Ray Sharp appointed Abe Silverstein as Chief of Research, and the Soviet Union detonated its first atomic weapon. Over the next eight years Silverstein would push the laboratory into new experimental research areas and become a prominent advocate of one of the cornerstones of the space program—liquid hydrogen.

Image 126: Sharp and Silverstein share a moment in 1958 (GRC–2015–C–06570).

The Soviet weapons test spurred President Harry Truman to initiate the development of a thermonuclear weapon—commonly referred to as the hydrogen bomb—in March 1950. Three months later, the Cold War reached new heights with the onset of the Korean War. As a result the military became keenly interested in new technologies that could rapidly improve their capabilities. The two that would affect the Lewis laboratory most were nuclear propulsion and liquid hydrogen. These fields, which would dominate Lewis's activities in the 1960s, had their genesis in the early 1950s. The military also called upon Lewis for a variety of missile work, including a long-term investigation of the ramjet engines for the Navaho missile, sophisticated tests of missile designs in the supersonic tunnels, and the continued launching of research missiles near Wallops Island.

In addition, Lewis's traditional aeronautical programs matured in the 1950s. The Icing Research Tunnel (IRT) yielded some of its best data to date, researchers designed the groundbreaking transonic compressor, and new generations of full-scale turbojet and ramjet engines were tested in Lewis's altitude facilities. The 8- by 6-Foot Supersonic Wind Tunnel (8×6) began operation, and the lab added two new unique facilities—the Propulsion Systems Laboratory (PSL) and the 10- by 10-Foot Supersonic Wind Tunnel (10×10). There were new endeavors, as well, such as the unique Crash Fire Test Program that sought to reduce fatalities in low-impact aircraft crashes.

## New Directions

In late 1947 when Lewis Chief of Research Addison Rothrock accepted a position at NACA Headquarters, Executive Engineer Carlton Kemper assumed responsibility for managing the laboratory's research efforts for nearly two years. In August 1949, to fill the vacant Chief of Research position, Director Ray Sharp appointed Abe Silverstein the head of the Wind Tunnels and Flight Division. Kemper, who had once been expected to serve as director of the entire laboratory, now served as a consultant within Silverstein's office.[255,256]

Just over two years later, Sharp expanded Silverstein's role by giving him the additional responsibility for managing and planning the test facilities. As such, Sharp replaced the Chief of Research position with the new title of Associate Director. Ray Sharp came from a managerial background and did not have any aeronautical training, so he handled the administrative aspects of running the laboratory, while Silverstein managed the research and test facilities. It was a winning partnership.[257]

Silverstein was supported by Oscar Schey (Chief of the Compressor and Turbine Division), and Benjamin Pinkel (Chief of Fuels and Thermodynamics Division). Sharp used his own skills to make sure that the researchers had the support and tools that they needed, such as a continuum of new test facilities.[258]

Silverstein initiated a series of organizational changes to address the expanding field of supersonic missiles and alternative methods of propulsion. The Compressor and Turbine and Engine Research divisions were slightly modified to take into account the new PSL and transonic compressor research, respectively.[259,260] In fall 1949 Silverstein created the new Supersonic Propulsion and Physics divisions out of his own former Wind Tunnels and Flight Division. The Supersonic Propulsion group managed research at the newly completed 8×6, and the Physics Division conducted icing, flight research, and flow physics activities.

Silverstein's most important move was dividing the Fuels and Thermodynamics Division into the Fuels and Combustion Division and the Materials and Thermodynamics Division. The latter studied heat transfer and materials for both traditional jet engines and the emerging nuclear aircraft concept. Silverstein increased the former's rocket fuels research. The division's once-hidden High Pressure Combustion Section was expanded and properly renamed the Rocket Branch.[261] The move coincided with the NACA's first real interest in rocket propulsion. Over the next few years, the group's work with hydrogen, oxygen, and fluorine provided the impetus for the eventual use of cryogenic propellants in the space program.

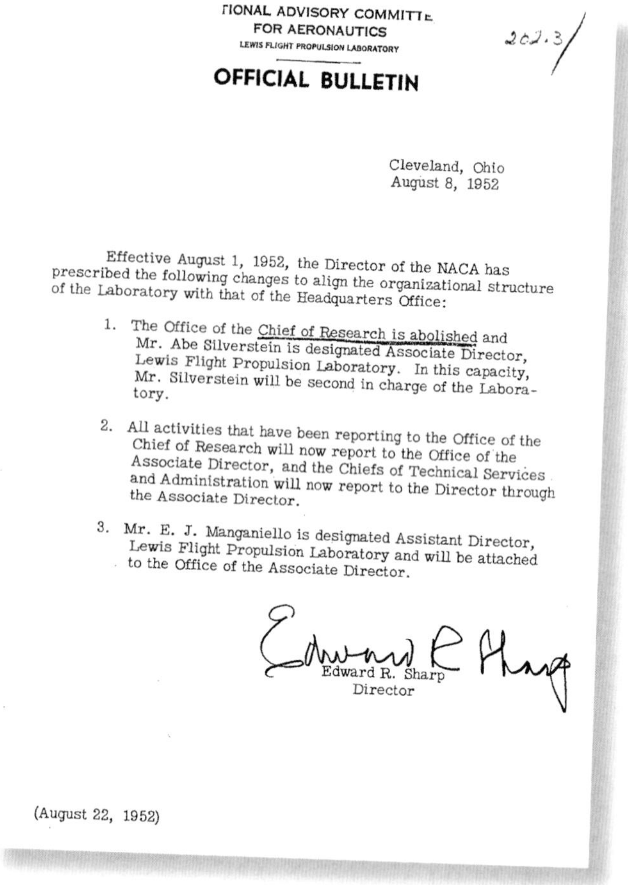

❖ ❖ ❖ ❖ ❖ ❖

In Room 213 of the 8×6 office building, Abe Silverstein assembled his basic research "brain trust," formally known as the Applied Mechanics Group. This independent group consisted of Simon Ostrach, Steve Maslen, Frank Moore, and Harold Mirels. Though the men had distinct personalities and approaches to research, they became fast friends and were informally known as the "Four Ms"—Mirels, Moore, Maslen, and [M]ostrach.[262] They each had recently returned to the lab after earning advanced degrees from prestigious universities and brought with them a theoretical approach to research that left them outside of the lab's predominant experimentally based research methodology. Silverstein called upon the men to work on special applied research problems but provided them the freedom to pursue their own individual basic research projects. Mirels, Moore, and Maslen performed research on boundary layers and flow issues related to high-speed flight. Ostrach concentrated on buoyancy-driven heat transfer.[263]

NATIONAL ADVISORY COMMITTEE
FOR AERONAUTICS
LEWIS FLIGHT PROPULSION LABORATORY

202.3/

**OFFICIAL BULLETIN**

Cleveland, Ohio
August 8, 1952

Effective August 1, 1952, the Director of the NACA has prescribed the following changes to align the organizational structure of the Laboratory with that of the Headquarters Office:

1. The Office of the <u>Chief of Research is abolished</u> and Mr. Abe Silverstein is designated Associate Director, Lewis Flight Propulsion Laboratory. In this capacity, Mr. Silverstein will be second in charge of the Laboratory.

2. All activities that have been reporting to the Office of the Chief of Research will now report to the Office of the Associate Director, and the Chiefs of Technical Services and Administration will now report to the Director through the Associate Director.

3. Mr. E. J. Manganiello is designated Assistant Director, Lewis Flight Propulsion Laboratory and will be attached to the Office of the Associate Director.

Edward R. Sharp
Director

(August 22, 1952)

*Image 127: Memo announcing Silverstein's promotion.*

Heat transfer is an aspect of the larger physics field of thermodynamics that deals with the transference of heat from one object to another. It has important applications to aeronautics, including engine cooling, lubricants, and aerodynamic heating. To address issues related to heat transfer, the laboratory established its Thermodynamics Division under Ben Pinkel during World War II.[264] Pinkel had begun his career in 1941 at Langley, where he headed the Engine Analysis Section. He transferred to Cleveland the following year to lead the Thermodynamics Division, where he and his colleagues studied the transfer of engine heat to cooling systems. Abe Silverstein merged the group with the materials researchers during his December 1949 reorganization. The new division continued to study heat-transfer issues, but it also began investigations into high-strength, high-temperature materials and propulsion systems for nuclear-powered aircraft.[265]

It was at this point that Silverstein brought heat-transfer expert Ernst Eckert to Lewis. Eckert was among the German scientists who came to the United States in the final days of World War II as part of Operation Paperclip. In 1946 Eckert agreed to a five-year contract with the Air Force and continued his research at Wright-Patterson Air Force Base in Dayton, Ohio. In 1949, Silverstein convinced the Air Force to allow Eckert to serve as an advisor for Lewis's heat-transfer personnel. Eckert traveled to Cleveland, where once a week he not only developed new theories for turbine cooling, but encouraged the staff to increase their fundamental research and strive to create solutions with long-term significance that were not tied to a specific application. Seeking more freedom, Eckert left the NACA in 1952 and created a new heat-transfer center at the University of Minnesota.[266]

*Image 128: Lewis researchers Harold Mirels, Franklin Moore, Stephen Maslen, and Simon Ostrach in September 1987 celebrating Maslen's induction into the National Academy of Engineering (GRC–2015–C–06552).*

Eckert influenced several key staff members, particularly Simon Ostrach and Robert Deissler. When Ostrach returned from Brown University in 1949 with new heat-transfer credentials, he fully expected to be assigned to Pinkel's Thermodynamics Division. Instead, Silverstein placed him in John Evvard's new Supersonic Propulsion Division, which supervised turbojet and supersonic aerodynamics research.[267] There Ostrach began studying buoyancy-driven convection on a vertical plate using a unique deductive process that he had developed at Brown.[268] For buoyancy-driven flows also known as natural convection, density differences (owed usually to temperature differences) drive fluid motion. Ostrach found that other mechanisms could also yield density differences and therefore drive fluid motion. He found that forces other than gravity could create buoyancy. Silverstein was initially dismissive of the subject, so Ostrach couched his research in terms of turbine-blade cooling for aircraft engines. Ostrach also was given an opportunity to demonstrate that natural convection was causing excessive heat transfer in the reactor control rods of the early nuclear submarines.[269]

Deissler started at the lab in 1947 in the Fuels and Thermodynamics Division, studying turbulent fluid flow and heat transfer in tubes. He was able to streamline calculations so that complex heat-transfer concepts could be verified through experiments and computational programs. His work helped to resolve

many of the early heat-transfer and fluid-flow issues related to nuclear propulsion.[270] Deissler's group was bolstered by the addition of Emphraim Sparrow and Robert Siegel in 1953 and 1955, respectively. In a 1957 reorganization, the group became the Heat Transfer Branch in the new Nuclear Reactor Division, with Deissler as branch chief. This core group along with Ostrach made Lewis the world's premier heat-transfer center for a number of years.

*Image 129: Robert Deissler receives an NACA Exceptional Service Award from NACA Director Hugh Dryden in October 1957. Deissler was cited for "achieving significant scientific results in the solution of fluid flow and heat-transfer problems associated with aircraft nuclear propulsion" (GRC–1957–C–46286).*

The Four Ms broke up in the early 1960s as Lewis's basic research efforts declined. Frank Moore left for Cornell, Harold Mirels for Aerospace Corporation, Steve Maslen for Martin Marietta, and Simon Ostrach for Case Western Reserve University, where he eventually became renowned in the field of microgravity. All four were later inducted into the National Academy of Engineering.[271] Sparrow left in 1960 to join Eckert at the University of Minnesota and has since published hundreds of research papers. Deissler and Siegel continued their careers at Lewis until the 1990s. Deissler's high-temperature research became a standard reference. In 1968 Siegel and colleague John Howell published the nation's first and most-enduring textbook on radiation heat transfer, which today is in its sixth edition.[272] Individually, these men received many awards over the years and have been internationally recognized for their research.

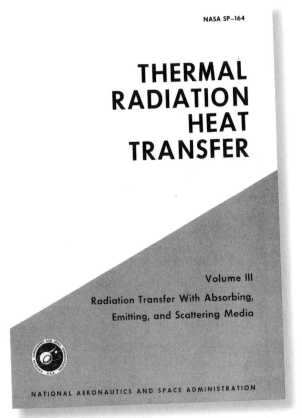

NASA SP-164

# THERMAL RADIATION HEAT TRANSFER

Volume III

Radiation Transfer With Absorbing, Emitting, and Scattering Media

NATIONAL AERONAUTICS AND SPACE ADMINISTRATION

*Image 130: Thermal Radiation Heat Transfer. Bob Siegel and John Howell started putting together notes for in-house classes to teach fellow employees about heat transfer in the early 1960s. The researchers fleshed out the information and published it in 1968 as NASA SP–164. Siegel and Howell updated the Special Publication (SP) three times, then published it as a textbook in the mid-1970s. The textbook has become the standard heat transfer textbook. It has been translated into numerous languages and was recently issued in its sixth edition.*

## Nuclear Aircraft

Engineers had considered utilizing the heat from nuclear decay for propulsion in the past, but it was not technically feasible until the Manhattan Project produced the first chain reactions in the early 1940s.[273] Following World War II, the military, the Atomic Energy Commission (AEC), and the NACA became interested in the use of atomic energy for propulsion and power. In theory, the performance and endurance of nuclear-powered aircraft would be limited only by the stamina of the crew. Long-duration missions were a military necessity before the advent of cruise missiles.

Lewis's involvement in nuclear propulsion began in the final months of World War II when researchers unsuccessfully lobbied NACA management for permission to investigate the use of nuclear fission to heat aircraft fuels.[274] NACA Headquarters granted Lewis consent shortly after the end of the war to study conceptual nuclear aircraft issues, but the staff did not have access to the top-secret information needed to design a nuclear engine system. Instead, a small group in Pinkel's division focused their efforts on high-energy heat transfer, shielding, and the effect of radiation on different materials.[275]

The Air Force sponsored a series of studies on the feasibility of using nuclear fission to power an aircraft. The Nuclear Energy for the Propulsion of Aircraft (NEPA) program at the Oak Ridge National Laboratory was the largest effort.[276] In 1948 the Air Force asked the NACA to participate in the struggling NEPA effort. Lewis assigned several engineers to a training program at the Oak Ridge National Laboratory to obtain expertise in high-temperature heat transfer and advanced materials technology.[277]

In 1949 Lewis contracted with General Electric to build a cyclotron behind the Materials and Structures Laboratory. Cyclotrons have two large electromagnets that cause the radioactive materials placed between them to emit charged particles. The charged particles, or ions, rotate around two semicircular cavities—one positively charged and one negatively charged.[278] Lewis researchers removed the high-energy particles and used them to irradiate sample materials. They then analyzed the effect of the radiation on the materials to determine which types would work best in a nuclear engine or airframe. In the mid-1950s Lewis modified the cyclotron to conduct nuclear physics studies.[279]

Image 131: Frank Rom was one of Lewis's chief nuclear propulsion researchers. He designed nuclear aircraft, pursued tungsten-based reactors for the nuclear rocket program, and helped design the Plum Brook Reactor Facility (GRC–1957–C–43739).

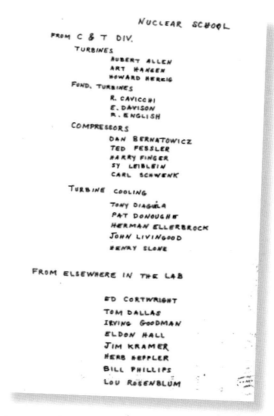

Image 132: A c1956 roster of participants in Lewis's in-house "nuclear school" and the branches from which they came.

The NEPA program concluded that a nuclear aircraft was feasible, but it did not create the technological breakthroughs necessary to overcome the expense and difficult design issues. The Air Force canceled the effort in 1950.[280] The AEC began its own Aircraft Nuclear Propulsion (ANP) program in 1949 with contributions from the Lewis laboratory. The Air Force began partnering on the program in December 1951.[281] In 1950 Abe Silverstein created a new 80-person Nuclear Reactor Division and established an in-house nuclear school where a handful of physicists trained the aeronautics researchers in the basics of atomic physics.[282,283]

Lewis researchers worked on aircraft and engine designs, studied heat transfer issues, and tested materials with the cyclotron. In 1952 Lewis sought to expand its nuclear

Image 133: The General Electric-designed cyclotron in the extended basement of the Materials and Stresses Building (GRC–1957–C–45988).

research capabilities by building a large test reactor. The reactor would help determine the best materials with which to build the nuclear aircraft engine and methods to reduce the heavy weight of the crew's protective shield. In October 1954 the AEC approved the NACA's request for a reactor license.[284] The new 60-megawatt facility was too large to build at the Lewis campus, so the NACA examined a number of remote sites. In March 1955 Lewis management decided to lease 500 acres of the unused Plum Brook Ordnance Works in Sandusky.[285] Construction of the Plum Brook Reactor Facility began in September 1956.[286] This was the beginning of Lewis's decades-long use of Plum Brook as a remote testing station.

*Image 134: Ray Sharp and Congressman Albert Baumhart break ground for the Plum Brook Reactor Facility in September 1956. The pick and shovel were the same as those used for the AERL groundbreaking in January 1941 (GRC–1956–C–43033).*

*Image 135: The Plum Brook Ordnance Works site as it appeared in April 1956 (GRC–1956–C–41679).*

*Image 136: Tornado damage included a collapsed roof on the Trunnion Building (GRC–1953–C–32966).*

*At approximately 9:45 p.m. Monday, 8 June 1953, a tornado swept through the southwestern section of the Lewis laboratory. In nearby neighborhoods, eight people were killed and over 200 injured before the tornado made its way to Downtown Cleveland and finally out over Lake Erie. The storm destroyed 50 homes and damaged nearly 2,000. Large areas of the city lost electrical power, and the roads were littered with debris for weeks.[287] The Cleveland tornado was one of dozens of deadly tornados spawned over a three-day period by the historic Flint-Worcester outbreak.[288,289]*

*At Lewis, there were no injuries among the approximately 100 people on duty at the time, but the storm caused $100,000 in damage. The wind ripped the roofs off of the Barrel Storage Building and Trunnion Building, and it blew out the windows and doors of the Fabrication Shop. The Research Equipment Building and 10×10 drive motor building were also damaged. As a precaution, facility engineers sealed the fuel lines leading to several of the large test facilities.[290,291]*

## Flight Safety

As the laboratory expanded its scope in the postwar years, it branched out into new fields not directly tied to engines, including flight safety. Irving Pinkel, Associate Chief of the Physics Division, led the division's efforts in the 1950s to study lightning strikes, hydroplaning on wet runways, reverse thrusters, the jettisoning of fuel, and low-altitude crash survivability. Pinkel began his NACA career in 1940 at Langley, where his older brother Benjamin was an engine researcher. The brothers transferred to Cleveland when the new lab opened in 1942. He initially investigated lubrication systems, then moved on to jet engine combustors and supersonic nozzles. Abe Silverstein assigned Pinkel to the icing research program in the mid-1940s before promoting him to Associate Chief of the Physics Division in 1949.[292]

The laboratory's initial flight safety effort, the icing research program, was in full force in the early 1950s. The Icing Branch continued to use the Consolidated B–24M and the North American Aviation XB–25E aircraft to study the properties of icing clouds for several years. Silverstein finally canceled the effort after

*Image 137: After years of experimentation, Lewis engineers finally perfected the IRT's spray bar system in 1949 (GRC–1949–C–24017).*

the researchers thoroughly identified the characteristics of these clouds and projected the chances of aircraft encountering those conditions.[293] In addition, Lewis engineers had finally resolved the spray bar dilemma in the IRT. The tunnel was used to study ice buildup on a variety of aircraft components, jet engine icing, and new deicing techniques. Researchers had demonstrated in the 1940s that the continual heating of leading edges was effective at removing ice from aircraft; but in the 1950s they made significant advances with a new system that employed cyclical heating. The intermittent heating reduced the amount of energy required to prevent ice accumulation. Lewis researchers also worked to combat ice buildup on radome antennae and jet engines in the 1950s.[294]

❖ ❖ ❖ ❖ ❖ ❖

One of the laboratory's most well-known flight safety endeavors is the Crash Fire Program. The nearly 10-year-long effort examined a variety of issues that contributed to the high casualty rates during runway or low-altitude aircraft crashes. In these incidents, an aircraft generally suffered damage to its fuel system and other components, but it was structurally

survivable. A rash of high-publicity passenger aircraft crashes during 1946 and 1947 threatened to discourage the expanding civil airline business. President Harry Truman called for an investigatory board that included the Civil Aeronautics Board, the military, and the NACA.[295]

Abe Silverstein created a panel at Lewis in October 1947 to assess the pertinent crash issues and outline a plan to address them. The group divided the fire research into three elements—segregation, extinguishment, and prevention.[296] In March 1948 NACA Headquarters approved Lewis's proposed program to investigate the characteristics of flammable materials, create fire detection and extinguishing systems, and identify fire prevention methods.[297] Silverstein assigned Irving Pinkel the responsibility for carrying out a crash program.[298]

Pinkel and his team developed a program that would use surplus military aircraft to conduct their research. At the nearby Ravenna Arsenal, Lewis engineers created a 1,700-foot test runway that included poles, embankments, and other obstacles to simulate

*Image 138: Irving Pinkel examines a crash from a camera tower along the track (GRC–2015–C–06548).*

was visible as it poured out of the damaged aircraft. Sometimes the fuel would run down the wings until it contacted the hot engine. Other times a cloud of fuel vapor released along the track would slowly continue moving forward until it caught up with the motionless aircraft. In these situations, ignition might not occur for several minutes after the vehicle had come to rest. In others, the aircraft burst into a ball of flame before it even reached the end of the track. The researchers determined that the best solution was the prevention of ignition altogether. Lewis provided a local company with the information needed to design an inerting system that automatically blocked fuel flow to the engine, sprayed cooling water on ignition sources, and disconnected the electrical system.[300]

The use of high-speed photography and motion picture film was essential to the Crash Fire Program. One of the reasons that the causes of crash fires were so difficult to determine was the inability of the human eye to view the ignition. A cadre of Lewis photographers painstakingly set up high-speed cameras and a variety of instrumentation in the aircraft and synchronized them with the seven camera stations along the track. Lewis photographer Bill Wynne developed a method for inserting images of timekeeping devices that were able to show time to a 1000th of a second on the film. The photographers installed clocks in front of the cameras and used mirrors to project the time onto the lens.[301]

situations in which an aircraft would fail to become airborne during takeoff. The researchers initially used Curtiss C–46 and Fairchild C–82 military transport aircraft, but later broadened the program's scope to include smaller aircraft and fighter jets. They mounted the aircraft to a rail that ran down the length of the runway. Then the test engineers started the engines and released the anchor pin. The highly instrumented aircraft would race down the runway at take-off speed before crashing into or through the various barriers. Photographers set up a series of high-speed cameras along the track to capture every aspect of the aircraft's demise.[299]

The program's preliminary phase, which began in 1949, sought to identify potential ignition sources and analyze the spread of fuel and oil. The team dyed the fuel red so that it

*Image 139: Lewis researchers crash a transport aircraft through barriers at the Ravenna Arsenal. Incandescent particles are visible from the explosion (GRC–1957–C–43929).*

As the program stretched into the 1950s, Lewis researchers began investigating fires associated with jet aircraft. They focused on aspects that were unique to jet engines, including the characteristics of jet fuel, the fuel tank location, and the larger quantities of fuel used.[302] Lewis performed studies using both a Republic F–84 Thunderjet and a Fairchild C–82 transport with jet engines strapped to its wings. The researchers discovered that the fuel flowed through the engine too quickly to ignite, but that fuel on the turbojet's hot metal surfaces did ignite. They used these data to modify their piston engine inerting system so that it could be used in jet aircraft. The Lewis team then verified the performance of the fire prevention system with a series of additional crashes.[303]

As the Crash Fire Program proceeded over the years, Lewis researchers expanded their investigations into new areas. The two most prominent were structural damage from impact and the design of safer passenger seats. Previous research indicated that humans could withstand greater stress levels than those associated with low-altitude crashes. Despite the studies, impact loads were causing deaths in many instances. Lewis conducted a series of crash tests with instrumented dummies in the pilot and passenger areas. Engineers would set up cameras and instrumentation inside the cabin to record the data as the aircraft crashed

Image 140: *A 1955 example of Bill Wynne's innovative visual timekeeping method for the Crash Fire Program (GRC–C–1955–38196).*

directly into a dirt mound.[304] The team was able to identify which components were most important to the aircraft's structural integrity and to provide that information to airframe designers.[305]

The Lewis team then turned their attention to the effect of seats and restraints. The procedure was similar to the one used for the structural-integrity tests.

Image 141: *Members of the Flight Research Section investigate the crash wreckage of an aircraft in the late 1940s (GRC–2015–C–06542).*

After determining the different loads and their effects on the passengers, the NACA researchers began designing new types of seats and restraints. They found that the passengers who were in the commonly used rigid seats received two-thirds higher g-forces than those in flexible NACA-designed seats.[306,307] One of the more controversial findings was that rearward-facing seats increased crash survivability. Some people in the aviation industry argued that the opposite was true, but both sides agreed that stronger, better secured seats were the most important concern.[308,309]

The Crash Fire Program concluded in 1957. The airline industry, however, did not adopt many of Lewis's findings. They contended that conspicuous modifications, such as the rearward-facing seats, might give passengers the impression that the company was not confident in its aircraft. The less obvious tools, such as the fire prevention system and stronger seats, required an investment to install and added extra weight that could be used for payload or additional passengers.[310-312] There were three runway crashes in early 1960 with large numbers of deaths that officials claimed could have been prevented by the NACA inerting system.[313]

*The Lewis Photo Lab, established in 1942, has been essential to much of the lab's research over the years. The photographers and specialized equipment accompany pilots on test flights. They use high-speed cameras to capture fleeting processes like combustion, and work with technology, such as the Schlieren camera, which captures supersonic flow. In addition, the group documents construction projects, performs publicity work, creates images for reports, and photographs data-recording equipment.*

*Arthur Laufman joined the Photo Lab staff in 1948 and began producing full-length technical films as a tool to educate those outside of the Agency on the research being conducted at Lewis. He worked with engineers to determine proper subjects for these films and develop a script. In addition to filming tests, Laufman shot footage of facilities, models, and staff members. He then edited the footage and added audio, visuals, and narration.[314]*

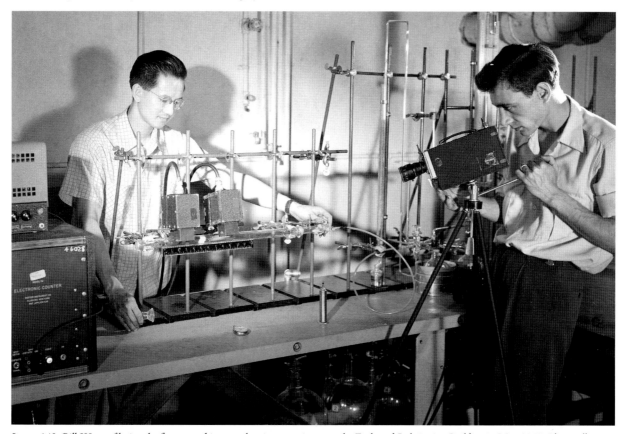

*Image 142: Bill Wynne filming the flame speed in a combustion experiment at the Fuels and Lubrication Building in May 1949. Photo cells above the tube measure the rate of the flame travel (GRC–1949–C–23407).*

*Image 143: A group of visitors views an Engine Research Building shop area crowded with jet engines that have been tested in the Four Burner Area (GRC–1957–C–45046).*

## Jet Propulsion

The size and performance of jet engines improved dramatically over the course of just a few years. Thrust had increased from 1,600 pounds in the early 1940s to over 10,000 pounds by the end of the decade.[315] The Lewis laboratory required new and improved test facilities to study these emerging engines. Almost immediately after the Four Burner Area began operation in 1947, Lewis management began designing a similar, but more powerful, facility known as the PSL. In addition to its own test capabilities, the PSL served as an essential component of a comprehensive plan to improve the lab's overall altitude testing capabilities by linking the exhaust, refrigeration, and combustion air systems from all the major test facilities.[316]

The PSL, which came online in October 1952, was the most powerful engine test facility in the nation and was designed to test more powerful engines than those in existence. The facility contained two 14-foot-diameter test chambers that provided simulated altitude conditions and high-speed airflow. Lewis researchers utilized one PSL test cell and the Altitude Wind Tunnel (AWT), which underwent a major upgrade in 1952, to test a new generation of engines such as General Electric's J79, the Rolls Royce's Nene, and Pratt & Whitney's J57.[317] By the end of the decade, these engines would power the first U.S. jet airliners.

Engine noise was a serious concern as the airline industry prepared to introduce jet aircraft to its fleet. Jet engines produce a loud high-pitched sound in comparison to the low rumble of piston engines. Preliminary tests showed that the source of the loudest noise was not the engine itself, but the interaction of the engine's high-velocity exhaust with the surrounding atmosphere. The pressures resulting from this turbulence produced sound waves. Lewis researchers undertook a variety of noise-reduction studies involving engine design, throttling procedures, and noise

*Image 144: The overhead air-handling line between the PSL and the 8×6 is installed in 1954 (GRC–2007–C–025662).*

suppressors.[318] Then researchers subjected a Pratt & Whitney J57 to an array of tests in the AWT to analyze the effect of different nozzle designs on the engine's noise levels. They found that the nozzles reduced the noise but also impeded engine performance. Further testing revealed that the addition of an NACA-developed ejector successfully mixed the exhaust and air, which reduced the noise levels without diminishing thrust.[319]

Lewis researchers also used the other PSL test cell to test a variety of ramjet and small rocket engines. In 1952 they began a multiyear study of the Wright Aeronautical XRJ–47–W–5 ramjet propulsion system for the North American Navaho missile. The Navaho was a reusable winged missile that was intended to transport a nuclear warhead up to 3,000 miles. The Navaho relied on a rocket-propelled booster vehicle to reach the high speeds required to ignite the two 48-inch-diameter ramjet engines. These very large engines were riddled with design problems.[320]

In 1951 the military asked Lewis to run a full examination of the XRJ–47–W–5. PSL was the only facility large enough to test the engine in simulated altitude conditions. Over the course of five years, Lewis researchers investigated the engine's ignition, the combustion chamber shell, fuel flow controls, different igniter configurations, and overall engine performance. They even tested the engine with experimental pentaborane fuel.[321] Concurrently, North American began a series of Navaho test flights, which yielded poor results. The military canceled the program in 1957

*Image 145: A Wright Aeronautical XRJ–47–W–5 ramjet installed in a PSL test chamber for the Navaho program (GRC–1952–C–30961).*

after another Navaho failure and the first successful launch of the Atlas missile—an alternative nuclear weapon delivery system.[322]

❖ ❖ ❖ ❖ ❖ ❖

Oscar Schey's Compressor and Turbine Division was the bastion of the laboratory's jet engine work in the 1940s and 1950s. The improvement of compressors and turbines was critical to the continual drive to design more efficient and powerful engines while reducing their weight. The researchers investigated topics such as the required number of compressor stages; the shape, angle, and thickness of the stator blades; and cooling of the turbine.

At the time, engines with compressors were operating successfully at subsonic and supersonic speeds. However, designers were unsure about the performance of compressors at transonic speeds—where the air at the base of the blades was high subsonic and the air at the blade tips was low supersonic. Designers of early supersonic jet engines believed that there were limits to compressor inlet speeds and the amount of pressure that the compressor could efficiently generate.[323]

In early 1950 Lewis researchers Seymour Lieblein, Irv Johnsen, and Robert Bullock began experimenting with a compressor designed specifically for the transonic realm.[324] They introduced new blade shapes, advanced flow schemes, and improved blade-loading principles that permitted higher pressure rises in each stage. This reduced the number of required stages and resulted in a smaller, lighter engine.[325,326] The researchers' design handled 30 percent more airflow than existing compressors, which could increase the air intake of existing engines without increasing engine size. The engine manufacturing industry was immediately interested in the device.[327]

*Image 146: A researcher measures the turbine blades on a 12-stage axial-flow compressor in February 1955 (GRC–1955–C–37659).*

In the mid-1950s Lieblein and Johnsen designed single-stage and multistage transonic compressors for advanced turbojet engines. They were able to demonstrate that there were no preexisting theoretical limitations to compressor design. Instead, performance could be continually increased with proper design methods. This breakthrough eventually led to the development

of supersonic compressors.[328] In 1955 the group published a secret compressor design guide that was referred to as the "Compressor Bible," officially titled the "Aerodynamic Design of Axial-Flow Compressors."[329]

During this period, Lieblein developed a more reliable method for calculating compressor blade loads, known as the Diffusion Factor, or "D Factor."[330] The use of the D Factor resulted in reductions in compressor weight and manufacturing costs. General Electric was the first to utilize the tool during the design of its successful General Electric J85 turbojet in the late 1950s.[331] Lieblein, Bullock, and Johnsen received the prestigious Goddard Award in 1967—after their work had been declassified.

❖ ❖ ❖ ❖ ❖

Image 147: *Irv Johnsen, Seymour Lieblein, and Robert Bullock receive the 1967 AIAA Goddard Award for their compressor research at Lewis in the 1940s and 1950s (GRC–2015–C–06551).*

Tribology is another field that is essential for jet engine development. In the late 1940s and early 1950s Lewis researchers were able to understand the previously unknown fretting wear process and the difference between beneficial and harmful oxidation.[332] Lewis researchers also made advances in the development of unproven solid lubricants such as graphite and molybdenum disulfide. Solid lubricants are often lighter, more heat resistant, and in need of fewer seals than their liquid counterparts.[333] Lewis was able to

Image 148: *The high-speed, high-temperature test rig in the Fuels and Lubrication Building (GRC–1953–C–32722).*

analyze the effect of atmosphere, load, and temperature on these materials and use them to create a dry film that could be reliably used as a lubricant in engines. As engine capabilities increased, Lewis researchers continued to develop other solid lubricants that could withstand the increased temperatures. As a result, solid lubricants are now used widely in the design of aircraft engines.[334]

Lewis tribologists also sought to improve the strength of rolling element bearings during this period. These devices consist of two oval bands—usually the larger one fixed to the engine housing and the smaller inner one attached to the drive shaft. Ball bearings placed in grooves between the two bands allow the smooth rotation of the drive shaft.[335] Rolling element bearings have been around for centuries, but it was not until the advent of modern metallurgy and the incorporation of bearings into bicycles in the late 1800s that efforts were made to improve the consistency of the bearings during the manufacturing process.[336] Rolling element bearings became an essential element of the development of the jet engine in the 1940s. General Electric, which was a key manufacturer of jet engines, pioneered the development of high-performance bearings.[337]

As the nation sought more and more powerful jet engines, designers required new materials to withstand the greater temperatures and new bearing designs to handle greater engine speeds.[338] Although Lewis researchers Robert Johnson and William Anderson were able to substantially increase the number of rolling element bearing rotations during the 1940s, the development of bearing materials progressed somewhat slowly until the mid-1950s. Over the course of just a few years, most modern bearing materials were introduced and new manufacturing processes were developed.[339,340]

Industry engineers during that period sought to improve bearing hardness and durability by increasing the percentage of alloy in the steel. While conducting basic analyses in the late 1950s that included tension, compression, and rotating-beam tests, Lewis researchers found that the industry logic was incorrect. Further research at Lewis revealed that bearing life actually decreased as the percentage of alloy was increased.[341]

Lewis researchers conducted extensive research and testing on rolling element bearings in the 1950s and 1960s. Anderson, Erwin Zaretsky, and Richard Parker developed a method for relating rolling-element lifespan to permanent stresses caused by compression. They also increased the life of rolling contact bearings by up to 500 percent and identified the optimal hardness relationship between the bearing and the tracks. Then the researchers continued their efforts to apply the hardness differential concept to other bearing applications.[342,343]

Image 149: *Erwin Zaretsky, Bill Anderson, and Richard Parker receive awards from the NASA Inventions and Contributions Board for developing a method to improve the life and reliability of contact bearings (GRC–1966–C–02713).*

## Large Supersonic Wind Tunnels

After reviewing the German high-speed wind tunnels during the final months of World War II, members of the NACA began advocating for a new supersonic research center. In November 1945 Lewis Director Ray Sharp commented that water power was the only economical way to operate such a complex, and he proposed building a site near the Boulder Dam. The NACA Headquarters accepted Sharp's suggestions and initiated planning. Meanwhile the Air Force was considering a similar center of its own. The nation could only afford one such facility. The NACA felt that it was capable of handling all supersonic research, while the military felt responsible for anything associated with national defense. Industry, which could not afford to construct these types of facilities, pressured the government to make them available for private use.[344]

## *The Lewis Unitary Plan Wind Tunnel* ............

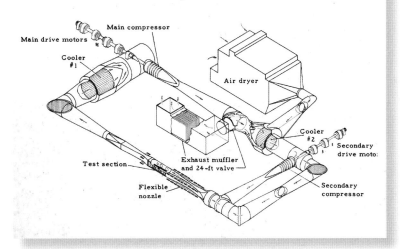

Image 150: Diagram of the 10×10.

In October 1949 Congress passed the Unitary Plan Wind Tunnel Act to satisfy all of these sectors of society without spending taxpayer money on redundant facilities. The act's most enduring contribution was the establishment of the military's Arnold Engineering Development Center (AEDC) in Tennessee. Instead of a new NACA supersonic laboratory, each of the three existing NACA laboratories would receive a new large wind tunnel and upgrades to existing facilities. Private industry and universities could use the facilities if they covered the operation costs.[345]

In 1950 Lewis engineers began work on the 10×10 by analyzing up to 50 different designs for a new supersonic propulsion wind tunnel that would surpass the brand new 8×6 (which Lewis used in the 1950s to test a variety of ramjet missiles and supersonic inlets). The resulting $33 million 10×10 was largest of the three NACA tunnels built under the Unitary Plan and the most powerful propulsion wind tunnel in the nation. Construction began in July 1952, Lewis invited over 150 guests from industry, other NACA laboratories, and the media to a special one-day Inspection and dedication of the tunnel on 22 May 1956.[346]

The 10×10 was designed to operate as a closed circuit for aerodynamic tests and as an open circuit for propulsion investigations. The 20-foot-diameter eight-stage axial-flow compressor initially generated airflows

up to Mach 2.5 through the test section, and when augmented by a secondary compressor, it could generate wind streams up to Mach 3.5. The 10-foot-square, 76-foot-long stainless steel nozzle section just upstream from the test section was designed to be adjusted to change the speed and composition of the airflow.[347]

In the late 1950s researchers used the 10×10 to obtain performance characteristics from the Convair B–58 Hustler engine pods and General Electric's J79 turbojet, as well as a variety of ramjet engines for missiles such as the Talos and Typhoon.[348] Lewis also completed a major upgrade of its 8×6 in 1956 and eventually incorporated a return leg. The configuration permitted operation as either an open system during propulsion system tests, with large doors venting directly to the atmosphere, or as a closed loop during aerodynamic

Image 151: A 16-inch-diameter ramjet being installed in the 10×10 test section during May 1956 (GRC–1956–C–42032).

Image 152: An engineer examines the main compressor for the 10×10 tunnel. The stainless steel compressor had 584 blades ranging in length from 1.8 to 3.25 feet (GRC–1955–C–39724).

tests. In 1956 Lewis engineers drilled 4,700 holes into a portion of the test section walls to accommodate transonic research. The perforations allow the airflow to exit the tunnel and prevent airflow blockages.[349]

❖ ❖ ❖ ❖ ❖ ❖

The Korean War spurred the development of large electronic computing systems in the United States, and the government sponsored an array of computational programs at universities and private industry.[350] This coincided with Lewis's increasing need for computing systems for its administrative tasks, analytical work, and processing of test data. As the laboratory grew and more facilities began operation, less and less computing time was available for business and scientific applications. In the mid-1950s Lewis dramatically increased its computing capabilities. The new 10×10 office building provided space for the laboratory's new Central Computer Facility.

In 1954 Lewis's initial electronic computing system, the IBM 604, was superseded by IBM's Card Programmed Electronic Calculator (CPC), which could be programmed by punch cards. This allowed programs of unlimited length.[351] The system recorded the manometer pressure readings from the test facilities directly into punch cards that were fed into the calculator. The CPC, however, was replaced the next year by two workhorses—the Engineering Research Associates UNIVAC 1103 and three IBM 650 Magnetic Drum Calculators.

The 10×10 computer center also included the Computer Automated Digital Encoder (CADDE). CADDE converted direct-current raw data from a test facility into digital format stored on magnetic tape. The machine was connected to several large facilities, but it could only record data from one at a time.[352,353] The data were sent to the 16- by 56-foot UNIVAC 1103, which processed all of the lab's experimental data. The 1103 was the first computer to use random access memory (RAM) storage and used magnetic tape storage. The scientific community embraced the 1103's high-speed processing and made it the first commercially successful computer.[354] The operators took the CADDE information and fed paper tape with the raw data into the 1103, which performed the needed calculations. Minutes later the results were sent back to the control room and were either printed or displayed on monitors. The 1103 also recorded the data on punch tape for later analysis.[355,356]

*Image 153: Lewis engineers operate the CADDE system in the 10×10 (GRC–1956–C–42021).*

Engineers at IBM responded to 1103's popularity by developing its first electronic computing machine, the 701 in 1952. The following year IBM came out with the business version of the machine called the 650 Magnetic Drum Calculator.[357] It was the most popular computer of the 1950s.[358] Lewis used the IBM 650s to perform analytical calculations and some data reduction. Demand for computation was such that the staff operated the 650s around the clock. In 1959 Lewis acquired an IBM 704 that quadrupled the power of the 650s; it was the center's first (FORTRAN) compiler.[359]

For jobs under 30 minutes, the Lewis researchers could insert the card decks on the 704 themselves and wait for results. Computer Services staff ran the longer projects overnight. The staff found that researchers often retrieved entire boxes of printouts just to look at the last page and discard the rest. In order to curtail the excesses and conservative time estimates, the staff began terminating operations as soon as the estimated time expired.[360]

## High-Energy Propellants

It is not surprising that one of Abe Silverstein's first actions as Chief of Research was to expand Lewis's small corps of rocket engineers and the Rocket Lab facility. As a member of the NACA's Special Committee on Self-Propelled Guided Missiles in the mid-1940s, Silverstein had observed the early U.S. missile programs on the West Coast.[361] During that period, the military had also sponsored small-scale research on the use of liquid hydrogen as a propellant for aircraft and rockets. Coincidentally, those efforts were canceled in the late 1940s just as Lewis increased its liquid propellant research.[362] The military remained wary of hydrogen's handling issues and could not identify an application that required its high power.[363]

Lewis's new Rocket Research Branch, led by John Sloop, began studying the properties of several high-energy fuels, but soon focused its efforts on the three with the most potential: hydrogen, boron, and beryllium. Hydrogen, with its high specific impulse and nontoxic exhaust, appeared to be the most promising, but Lewis researchers were also intrigued by the possibilities of other high-energy, but toxic, fuels. Lewis studied combustion, heat transfer, and mixtures for these fuels throughout the early 1950s.[364]

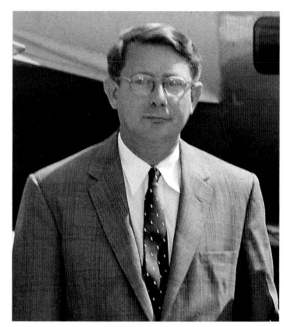

*Image 154: Abe Silverstein (GRC–1957–C–45195).*

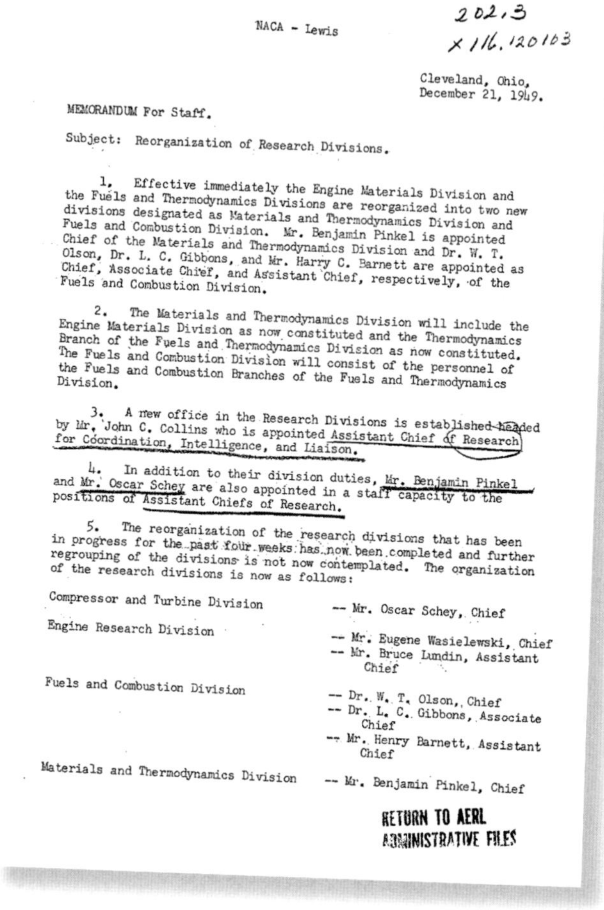

NACA - Lewis

202.3
X 116.120103

Cleveland, Ohio,
December 21, 1949.

MEMORANDUM For Staff.

Subject: Reorganization of Research Divisions.

1. Effective immediately the Engine Materials Division and the Fuels and Thermodynamics Divisions are reorganized into two new divisions designated as Materials and Thermodynamics Division and Fuels and Combustion Division. Mr. Benjamin Pinkel is appointed Chief of the Materials and Thermodynamics Division and Dr. W. T. Olson, Dr. L. C. Gibbons, and Mr. Harry C. Barnett are appointed as Chief, Associate Chief, and Assistant Chief, respectively, of the Fuels and Combustion Division.

2. The Materials and Thermodynamics Division will include the Engine Materials Division as now constituted and the Thermodynamics Branch of the Fuels and Thermodynamics Division as now constituted. The Fuels and Combustion Division will consist of the personnel of the Fuels and Combustion Branches of the Fuels and Thermodynamics Division.

3. A new office in the Research Divisions is established ~~headed~~ by Mr. John C. Collins who is appointed Assistant Chief of Research for Coordination, Intelligence, and Liaison.

4. In addition to their division duties, Mr. Benjamin Pinkel and Mr. Oscar Schey are also appointed in a staff capacity to the positions of Assistant Chiefs of Research.

5. The reorganization of the research divisions that has been in progress for the past four weeks has now been completed and further regrouping of the divisions is not now contemplated. The organization of the research divisions is now as follows:

| | |
|---|---|
| Compressor and Turbine Division | -- Mr. Oscar Schey, Chief |
| Engine Research Division | -- Mr. Eugene Wasielewski, Chief<br>-- Mr. Bruce Lundin, Assistant Chief |
| Fuels and Combustion Division | -- Dr. W. T. Olson, Chief<br>-- Dr. L. C. Gibbons, Associate Chief<br>-- Mr. Henry Barnett, Assistant Chief |
| Materials and Thermodynamics Division | -- Mr. Benjamin Pinkel, Chief |

RETURN TO AERL
ADMINISTRATIVE FILES

*Image 155: Silverstein memo announcing the reorganization of research divisions.*

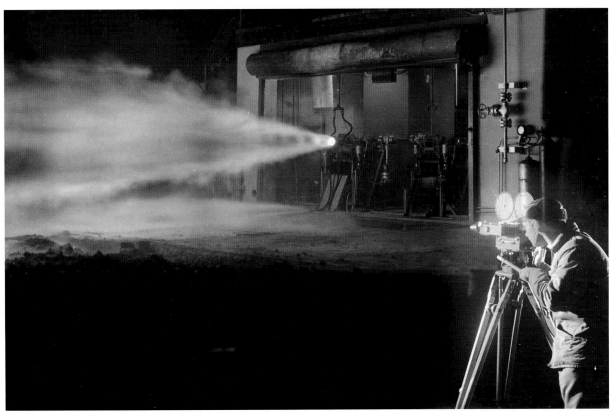

*Image 156: A photographer films the operation of a liquid-hydrogen/liquid-oxygen rocket engine in Cell 22 of the Rocket Lab. Tests were run in the evening when most of the lab was relatively vacant (GRC–1955–C–37427).*

In May 1951 the NACA created a Special Subcommittee on Rocket Engines. This was the NACA's first formal acknowledgment that rockets were a significant technology. Lewis took advantage of this new interest in rockets to successfully request funding for the RETF and upgrade the Rocket Lab. The Rocket Lab improvements allowed larger engines to be tested, and added a hydrogen liquefier and an exhaust scrubber that reduced emissions. The liquefier was critical because the lab had previously struggled to obtain enough hydrogen from commercial sources to conduct its research.[365] The RETF consisted of a single engine test stand that could fire 20,000-pound-thrust rocket engines using a wide array of propellants for up to 3 minutes. The facility also scrubbed pollutants from its exhaust and muffled the rocket's deafening noise. Lewis built the facility in a remote ravine to protect the staff from the explosive propellants being used. A facility operator controlled the tests from a control room 1,600 feet away. Construction of the RETF began in 1953.[366]

Meanwhile the Rocket Branch continued experimenting at the Rocket Lab with different propellants in its homemade rocket engines. Lewis staff studied virtually all propellant options but became focused on the use of hydrogen with either oxygen or fluorine as an oxidizer. Liquid hydrogen was not as powerful as some of the more exotic fuels, but it was not toxic and had a safe combustion rate. Hydrogen, however, had to be stored at −423 degrees Fahrenheit and required large tanks because of its low density.[367]

On 23 November 1954 Edward Rothenberg conducted Lewis's first successful liquid-hydrogen–liquid-oxygen run with a 5,000-pound-thrust rocket engine. The feat was repeated several more times over the next six weeks. The researchers then suspended their liquid-hydrogen testing for almost a year while they redesigned the injector and improved methods for starting and shutting down the engine.[368] Also in late 1954, Ed Jonash and his colleagues successfully experimented with the use of gaseous hydrogen in a turbojet engine.[369]

Hydrogen Runs     H₂-O₂   5000# Thrust

Lewis Research Center

R.O. _____ DATE _____

| Run No. | Patter Temp. (°F) | Inj. Temp. (°F) | Lox Flow (#/sec) | H₂ Flow (#/sec) | H₂ Density (#/ft³) | Po (psia) | O/F | I (2000) | c* (ft/sec) | F (#) | Injector | % Fuel | % Perform. I (Peak) | % Perform. I (Hd area) %/p | % Perform. c* | | |
|---|---|---|---|---|---|---|---|---|---|---|---|---|---|---|---|---|---|
| 1 | -165 | -305 | 0 | 6.48 | 4.60 | 0 | 0 | 0 | 0 | 0 | Hot L-on-L | 100 | 0 | 0 | 0 | Nov 3, 54 | |
| 2 | -167 | — | 10.60 | 4.20 | 4.55 | 607 | 2.58 | 843 | 7440 | 5025 | L-on-L | 28.4 | 89.5 | 89.8 | 90.7 | Nov 22, 54 | |
| 3 | -114 | — | 11.24 | 3.87 | 4.60 | 670 | 3.06 | ≈360 | 7720 | — | L-on-L | 21.6 | 94.0 | 94.0 | 94.2 | Dec 1, 54 | |
| 4 | -114 | — | 6.24 | 5.40 | 4.60 | ≈426 | 1.155 | 209 | ≈6630 | 3530 | L-on-L | 46.4 | 80.6 | 83.5 | 80.8 | Dec 2, 54 | |
| 5 | -116 | — | 12.68 | 5.53 | 4.65 | 671 | 2.89 | 862 | 5660 | 4770 | Spray ... | 80.25 | 69.4 | 69.8 | 69.0 | Jan 6, 55 | |
| 14 | | | | | | | | 345 | 7440 | | | 18.5 | | | | 2-29-56 | |
| 15 | | | | | | | | 331 | 7200 | | | 17.5 | | | | 1-24-56 | |
| 16 | | | | | | | | 347 | 7240 | | | 7.6 | | | | 2-24-56 | |
| I | | | | | | | | 356 | 7860 | | | 19.8 | | | | 3-9-5 | |

\* Estimated values.

*Image 157: Scan of the Cell 22 logbook with Lewis's first hydrogen-oxygen run highlighted.*

Almost immediately thereafter, Abe Silverstein became focused on the possibilities of hydrogen, initially for aircraft propulsion. In 1955 he and Eldon Hall coauthored a report which predicted that, for aircraft missions, liquid hydrogen would far exceed the performance of traditional hydrocarbon fuels.[370]

This coincided with the Air Force's renewed interest in the use of alternative fuels like liquid hydrogen. The creation of the hydrogen bomb had spurred the development of cryogenic storage equipment and large-scale liquefaction facilities; and proposed high-altitude long-distance military aircraft could benefit from the use of liquid hydrogen as a fuel.[371,372]

In fall 1955 the Air Force requested that Lewis examine the feasibility of converting a jet engine to run on liquid hydrogen. The military provided a Martin B–57B Canberra for the effort, referred to as "Project Bee." The aircraft was powered by two Wright J65 engines, one of which was modified so that it could be operated using either traditional jet fuel or liquid hydrogen.[373] Lewis personnel worked on pumping systems, insulation, and other related issues. They tested the system extensively in the AWT and the Four Burner Area test cells before installing it in the aircraft.[374,375]

The B–57B would take off using jet fuel, switch to liquid hydrogen while over Lake Erie, then after

*Image 158: As a converted B–57B prepares for a liquid-hydrogen flight over Lake Erie, black smoke emanates from the jet-fuel-powered engine. The hydrogen engine left a pronounced white contrail (GRC–1993–C–05546).*

Image 159: The new Rocket Engine Test Facility (RETF) is displayed at the 1957 Inspection (GRC–1957–C–45869).

*Image 160: Harrison Allen explains the benefits of high-energy aircraft fuels at the 1957 Inspection (GRC–1957–C–46151).*

❖ ❖ ❖ ❖ ❖ ❖

burning the hydrogen supply, switch back to jet fuel for the landing. Lewis pilots conducted several dry runs in fall 1956, but failed to make the switch to liquid hydrogen during the first two attempts in December 1956. The third attempt, in February 1957, was a success. The feat was repeated several times in spring 1957.[376]

Unbeknownst to the NACA, the military also asked Pratt & Whitney to design a new liquid-hydrogen aircraft engine for the secret Project Suntan program. Although the engine design proceeded well, there were disagreements about the aircraft's proposed range and the required hydrogen infrastructure. The military lost interest in the hydrogen aircraft engine in 1957, but Pratt & Whitney would convert the technology into its seminal RL–10 rocket engine. The Suntan program also resulted in the construction of several new industrial-size liquid-hydrogen production facilities that would be used by the space program.[377]

There were two key events planned for fall 1957 that would give the laboratory the opportunity to share its high-energy fuels work with a large number of industry and military professionals. The first was the NACA's Inspection in October.[378] The other was a classified Flight Propulsion Conference in November that concentrated specifically on Lewis's propulsion advances.[379] As the staff began the extensive preparations for these events, Silverstein made two decisions that would shape the future of the laboratory. In March he created an in-house advisory board, the Research Planning Council, to plan the future course of research and the necessary facilities for that research. It was clear that the emphasis would be on space-related research, so Abe Silverstein disbanded the Compressor and Turbine Division in July 1957.[380] The large group of researchers were shifted into the new Fluid Systems Combustion and Nuclear Reactor divisions. These two groups performed much of the research at Plum Brook Station in the 1960s.[381]

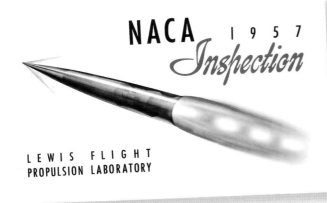

*Image 161: Brochure from the 1957 NACA Inspection.*

specific impulses (the amount of thrust for a given unit of propellant).[387] Lewis researchers had run hydrogen and fluorine together only once, in March 1955. That firing had lasted a mere 4 seconds and had not included nozzle cooling. Throughout 1957, Douglass, Glen Hennings, Edward Baehr, and Harold Price attempted to repeat the run using a regeneratively cooled engine that used the cold liquid hydrogen to prevent the nozzle from burning up from the high-temperature combustion.[388]

On 5 November, the group aborted its first attempt when a leak in the fluorine tank set the Rocket Lab's roof on fire. The test cell was quickly rebuilt, but time was running out. The team worked determinedly around the clock in the days leading up to the conference, but the difficulties persisted. Finally at 6 a.m. on Friday the 22d, they were able to get the hydrogen-fluorine rocket to fire. Price hurriedly crunched the data and delivered the figures in dramatic fashion to Douglass who was in the midst of his presentation to the conference.[389]

Although headquarters insisted that Lewis's rocket work was specifically for missiles, Lewis researchers were well aware that the new missiles would soon be capable of reaching space. In 1955 Walter Olson, head of the Fuels and Combustion Division, drafted a memo that advocated space exploration and encouraged the NACA to lead the effort.[382] This did not gain any traction at headquarters, which felt that the NACA should merely support the military in its missile development and would not suggest new uses like space travel. Congress had critically audited the NACA and Lewis during the Korean War, so management was wary of appearing to overstep its mission.[383]

As the Lewis team rehearsed their presentations for the upcoming Inspection, NACA officials ordered the removal of any mention of space missions.[384,385] On 4 October 1957, just days before the Inspection, the Soviet Union launched the world's first manmade satellite—Sputnik. Literally overnight, the NACA reversed its policy, and the references to space were reinserted into the talks. The 1,700 invited guests were very much interested in the application of Lewis's research to space as they toured the laboratory during the three-day Inspection. A highlight was the high-energy fuels talk at the recently completed RETF facility.[386]

In the meantime, the Soviets launched the even more impressive Sputnik II satellite, and the Rocket Branch raced to complete a hydrogen-fluorine rocket test in time for the 21–22 November 1957 Flight Propulsion Conference. Howard Douglass needed the data for a paper he would be presenting on fuels with high

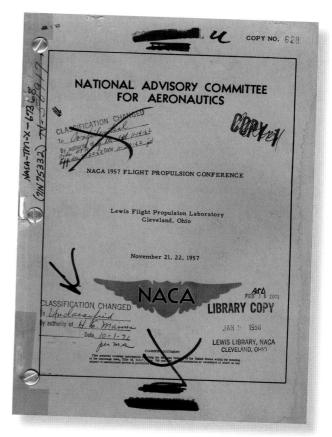

*Image 162: Proceedings from the NACA's 1957 Flight Propulsion Conference held at Lewis.*

Other papers discussed ramjet-powered missiles such as the Snark and Navaho and the B–58 supersonic bomber, but the critical portions of the conference dealt with high-energy propellants and the application of different propellants for a range of missions—including missiles, satellites, and a lunar landing. Lewis researchers discussed propellant options, turbopumps, and other propulsion issues.[390] Within months, the NACA would be transformed into a new agency dedicated to space, and Lewis would work to transform liquid hydrogen into a practical propellant in operational rocket engine systems.

## Endnotes for Chapter 4

255. Edward Sharp memorandum for Staff, "Designation of Laboratory Officials," 29 August 1949, NASA Glenn History Collection, Directors Collection, Cleveland, OH.

256. Virginia Dawson, *Engines and Innovation: Lewis Laboratory and American Propulsion Technology* (Washington, DC: NASA SP–4306, 1991).

257. Abe Silverstein interview, Cleveland, OH, by John Mauer, 10 March 1989, NASA Glenn History Collection, Cleveland, OH.

258. Silverstein interview by Mauer.

259. Eugene Wasielewski memorandum for Staff, "Organizational Changes Within the Engine Research Division," 26 July 1951, NASA Glenn History Collection, Directors Collection, Cleveland, OH.

260. Oscar Schey memorandum for Staff, "Organization of the Compressor & Turbine Research Division," 7 February 1952, NASA Glenn History Collection, Directors Collection, Cleveland, OH.

261. Abe Silverstein memorandum for Staff, "Reorganization of Research Divisions," 21 December 1949, NASA Glenn History Collection, Directors Collection, Cleveland, OH.

262. Simon Ostrach interview, Cleveland, OH, 2013, NASA Glenn History Collection, Oral History Collection, Cleveland, OH.

263. Ostrach interview.

264. "Research Divisions Organized," *Wing Tips* (5 February 1943).

265. Benjamin Pinkel interview, Cleveland, OH, by Virginia Dawson, 4 August 1985, NASA Glenn History Collection, Oral History Collection, Cleveland, OH.

266. Virginia Dawson, "From Braunschweig to Ohio: Eckert and Government Heat Transfer Research," *History of Heat Transfer: Essays in Honor of the 50th Anniversary of the ASME Heat Transfer Division*, Edwin Layton and John Leinhard, editors (New York, NY: ASME, 1988).

267. Public Information Office, "Biographical Sketch of John C. Evvard," 1972, NASA Glenn History Collection, Biographies Collection, Cleveland, OH.

268. Ostrach interview.

269. Simon Ostrach, "Heat in History Memoir on Buoyancy-Driven Convection" *Heat Transfer Engineering*, vol. 11, no. 3, 1990, pp. 19–26.

270. NACA Honors Two Lewis Scientists," *Wing Tips* (6 November 1957).

271. Ostrach interview.

272. John R. Howell, M. Pinar Menguc, and Robert Siegel, *Thermal Radiation Heat Transfer* (Boca Raton, FL: CRC Press, 2015).

273. Roger Launius, *To Reach the High Frontier: A History of U.S. Launch Vehicles* (Lexington, KY: University of Kentucky, 2003).

274. George Lewis to Cleveland, "Cleveland Memorandum for Director of Aeronautical Research, June 22, 1945," 1 August 1945, NASA Glenn History Collection, Nuclear Propulsion Collection, Cleveland, OH.

275. Pinkel interview, by Dawson, 4 August 1985.

276. John Kenton, "Who's Who in the Aircraft Nuclear Program," *Nucleonics* (August 1956).

277. "Simon—Dietrich at Oak Ridge," *Wing Tips* (18 October 1946).

278. Bonny McClain, "How Cyclotrons Work" (23 May 2007), *http://www.madehow.com/ Volume-7/Cyclotron.html* (accessed 3 April 2015).

279. "Cyclotron Data Tabulation," 1961, NASA Glenn History Collection, Directors Collection, Cleveland, OH.

280. Kenton, "Who's Who."

281. Aircraft Nuclear Propulsion Program Chronology, *Aircraft Nuclear Propulsion Program Hearing Before the Subcommittee on Research and Development of the Joint Committee on Atomic Energy, Congress of the United States, Eighty-sixth Congress, First Session*, 23 July 1959.

282. Frank Rom, *Review of Nuclear Rocket Research at NASA's Lewis Research Center From 1953 Thru 1973* (Reston, VA: AIAA–91–3500, 4 September 1991).

283. Jim Blue interview, Cleveland, OH, by Mark D. Bowles, 11 February 2002, NASA Glenn History Collection, Oral History Collection, Cleveland, OH.

284. Lewis Strauss to Jerome Hunsaker, 27 October 1954, NASA Glenn History Office, Nuclear Propulsion Collection, Cleveland, OH.

285. Nuclear Development Corporation of America, "Site Survey for NACA Research Reactor," 13 September 1955, NASA Glenn History Office, Plum Brook Ordnance Works Collection, Cleveland, OH.

286. Mark Bowles, *Science in Flux: NASA's Nuclear Program at Plum Brook Station 1955–2005* (Washington, DC: NASA SP–2006–4317, 2006).

287. "Tornado of 8 June 1953: The West Park Area," History of the West Park Neighborhood (14 June 2014), *http://westparkhistory.com/ tornado1953/tornado1953.htm* (accessed 12 April 2015).

288. Joshua Lietz, "Tornadoes on June 8, 1953," Tornado History Project, 2005–2015 (2015), *http://www. tornadohistoryproject.com/tornado/1953/6/8/map* (accessed 29 June 2015).

289. Lee Allyn Davis, *Natural Disasters* (New York, NY: Infobase Publishing, 2009).

290. "Twister Hits Lab Buildings," *Wing Tips* (12 June 1953).

291. "Tornado of 1953 Whips Through Lewis," *Lewis News* (12 April 1991).

292. Irving Pinkel, "Resume," 1970, NASA Glenn History Collection, Irving Pinkel Collection, Cleveland, OH.

293. Irving Pinkel interview, Cleveland, OH, by Virginia Dawson, 30 January 1985, NASA Glenn History Collection, Oral History Collection, Cleveland, OH.

294. William Leary, *We Freeze to Please: A History of NASA's Icing Research Tunnel and the Quest for Flight Safety* (Washington, DC: NASA SP–2002–4226, 2002).

295. Dwight Boyer and Max Gunther, "Taming the Killer Crash Fires." *True* (July 1960).

296. Gerard Pesman, "Minutes of Meeting of Panel on Reduction of Hazards Due to Aircraft Fires," 31 October 1947, NASA Glenn History Collection, Dawson Collection, Cleveland, OH.

297. William Littlewood to NACA Executive Committee, "Report of Committee on Operating Problems," 27 May 1948, NASA Glenn History Collection, Dawson Collection, Cleveland, OH.

298. Addison Rothrock to Cleveland, "Research on the Prevention of Fires in Aircraft Engine Nacelles," 12 August 1947, NASA Glenn History Collection, Directors Collection, Cleveland, OH.

299. Dugald Black, *Facilities and Methods Used in Full-Scale Airplane Crash Fire Investigation* (Washington, DC: NACA RM–E51L06, 1952).

300. Dugald Black and Jacob Moser, *Full-Scale Performance Study of a Prototype Crash Fire Protection System for Reciprocating Engine Powered Airplane* (Washington, DC: NACA RM–E55B11, 1955).

301. Bill Wynne interview, Cleveland, OH, 14 May 2013, NASA Glenn History Collection, Oral History Collection, Cleveland, OH.

302. Irving Pinkel, et al., Origin and Prevention of Crash Fires in Turbojet Aircraft (Washington, DC: NACA Report 1363, 1958).

303. Pinkel, *Origin and Prevention*.

304. A. Martin Eiband, Scott Simpkinson, and Dugald Black, *Accelerations and Passenger Harness Loads Measured in Full-Scale Light Airplane Crashes* (Washington, DC: NACA RN–2991, 1953).

305. G. Merritt Preston and Gerard Pesman, *Accelerations in Transport-Airplane Crashes* (Washington, DC: NACA TN–4158, 1958).

306. "Knowledge of Jet Fire, Impact Hazards Advanced by Crash Research." NASA press release, 27 June 1955, NASA Glenn History Collection, Cleveland, OH.

307. "Lewis Group Designs Forgiving Seat," *Wing Tips* (5 August 1955).

308. Irving Pinkel, *A Proposed Criterion for the Selection of Forward and Rearward-Facing Seats* (New York, NY: ASME 59–AV–28, 26 January 1959).

309. Wilson Hirschfeld, "Better Plane Seating Can Increase Safety," *Cleveland Plain Dealer* (6 February 1961).

310. Hirschfeld, "Better Plane Seating."

311. Charles Tracy, "NASA's Fire Douser Could've Saved 37," *Cleveland Plain Dealer* (22 January 1960).

312. Gene Methvin, "There's a Way to Prevent Fatal Plane Crash Fires," *Washington Daily News* (22 January 1960).

313. Hirschfeld, "Better Plane Seating."

314. "Lewis' Jack-of-All 4 Trades Filmmaker: Laufman Plays Key Role in 300-Tech Films," *Lewis News* (22 February 1985).

315. NACA, "Thirty-Eighth Annual Report of the National Advisory Committee for Aeronautics" (Washington, DC: Government Printing Office, 1954), p. 4.

316. Robert Arrighi, *Revolutionary Atmosphere: The Story of the Altitude Wind Tunnel and Space Power Chambers* (Washington, DC: NASA SP–2010–4319, 2010).

317. Robert Arrighi, *Pursuit of Power NASA's Propulsion Systems Laboratory No. 1 and 2* (Washington, DC: NASA SP–2012–4548, 2012).

318. "Jet Aircraft Noise Reduction," *Annual Inspection at NACA Lewis Flight Propulsion Laboratory, 7–10 October 1957,* 1957, NASA Glenn History Collection, Inspections Collection, Cleveland, OH.

319. J. S. Butz, Jr., "NACA Studies Ways to Soften Jet Noise," *Aviation Week* (4 November 1957).

320. "History and Organization of Wright Aeronautical Division Curtiss-Wright Corporation" (17 February 2009), *http://www.scribd.com/doc/12467677/ History-of-Curtis-Wright-Aeronautical-Company* (accessed 25 June 2015).

321. Warren Rayle, Dwight Reilly, and John Farley, *Performance and Operational Characteristics of Pentaborane Fuel in 48-Inch Diameter Ramjet*

*Engine* (Washington, DC: NACA RM–E55K28, 1957).

322. Ed Kyle, "Pilotless Bomber: US Space Technology Incubator," Space Launch Report (2005), *http://www.spacelaunchreport.com/navaho1.html* (accessed 10 April 2011).

323. "Compressors for Specific Tasks," *1951 Inspection, Lewis Flight Propulsion Laboratory*, 1951, NASA Glenn History Collection, Inspections Collection, Cleveland, OH.

324. "Compressors for Specific Tasks."

325. Jack Kerrebrock, "Flow in Transonic Compressors," *AIAA Journal* 19, no. 1 (1981): 4–19.

326. "Supersonic Turbojet Performance," *Annual Inspection at NACA Lewis Flight Propulsion Laboratory, 7–10 October 1957*, 1957, NASA Glenn History Collection, Inspections Collection, Cleveland, OH.

327. "A Summary of NACA Leadership in Transonic Compressor Research," 28 March 1952, NASA Glenn History Collection, Directors Collection, Cleveland, OH.

328. Seymour Lieblein, "Acceptance Remarks for the Goddard Award," 1967, NASA Glenn History Collection, Seymour Lieblein Collection, Cleveland, OH.

329. Compressor and Turbine Research Division, *Aerodynamic Design of Axial-Flow Compressors* (Washington, DC: NACA RM–E56BO3, 1958).

330. Seymour Lieblein, Francis C. Schwenk, and Robert L. Broderick, *Diffusion Factor for Estimating Losses and Limiting Blade Loadings in Axial-Flow Compressor Blade Elements* (Washington, DC: NACA RM–E53D01, 1953).

331. Leroy Smith, "Axial Compressor Aerodesign Evolution at General Electric," *Journal of Turbomachinery* 124 (July 2002).

332. "Lewis Contributions to Aeronautics in the Field of Tribology," NASA Glenn History Collection, Tribology Collection, Cleveland, OH.

333. Harold Sliney, *High Temperature Solid Lubricants— When and Where to Use Them* (Washington, DC: NASA TM X–68201, 1973).

334. "Lewis Contributions to Aeronautics."

335. Edmond Bisson, et al., *Lubricants, Bearings, and Seals* (Washington, DC: NASA N66–33673, 1966).

336. Bernard Hamrock and William Anderson, *Rolling Element Bearings* (Washington, DC: NASA RP–1105, 1983).

337. Erwin Zaretsky, *Rolling Bearing Steels—A Technical and Historical Perspective* (Washington, DC: NASA/TM—2012-217445, 2012).

338. Zaretsky, *Rolling Bearing Steels.*

339. Zaretsky, *Rolling Bearing Steels.*

340. Edmond Bisson interview, Cleveland, OH, by Virginia Dawson, 22 March 1985, NASA Glenn History Collection, Oral History Collection, Cleveland, OH.

341. Zaretsky, *Rolling Bearing Steels.*

342. E. V. Zaretsky, et al., *Effect of Component Differential Hardness on Residual Stress and Rolling Contact Fatigue* (Washington, DC: NASA TN D–2664, 1965).

343. "Bearings Researchers Report Major Advance," *Lewis News* (6 August 1965).

344. Roger Launius, Thomas Irvine, and E. A. Arrington, *NACA/NASA and the National Unitary Wind Tunnel Plan, 1945–1965* (Reston, VA: AIAA–2002–1142, 2002).

345. Donald Baals and William Corliss, *Wind Tunnels of NASA* (Washington, DC: NASA SP–440, 1981).

346. "Lewis Unitary Plan Wind Tunnel," 22 May 1956, NASA Glenn History Collection, Inspections Collection, Cleveland, OH.

347. "New Wind Tunnel Tests Full-Scale Engines at Supersonic Speeds," 22 May 1956, press release, NASA Glenn History Collection, Test Facilities Collection, Cleveland, OH.

348. Ronald Blaha, "Completed Schedules of NASA Lewis Wind Tunnels, Facilities and Aircraft 1944–1986," February 1987, NASA Glenn History Collection, Test Facilities Collection, Cleveland, OH.

349. "Changes at the 8×6 Supersonic Wind Tunnel," *Wing Tips* (15 August 1956).

350. IBM, "The IBM 700 Series Computing Comes to Business," *http://www-03.ibm.com/ibm/history ibm100/us/en/icons/ibm700series/* (accessed 17 March 2015).

351. Frank da Cruz, "The IBM Card Programmed Calculator" (2009), *http://www.columbia.edu/cu/computinghistory/cpc.html* (accessed 17 March 2015).

352. "Lewis Unitary Plan Tunnel," *Wing Tips* (23 May 1956).

353. Peggy Yohner interview, Cleveland, OH, by Virginia Dawson, 21 May 1985, Glenn History Collection, Oral History Collection, Cleveland, OH.

354. George Dalakov, "The Cray Computers of Seymour Cray" (2015), *http://history-computer.com/ModernComputer/Electronic/Cray.html* (accessed 17 March 2015).

355. "Lewis Unitary Plan Tunnel," *Wing Tips* (23 May 1956).

356. Yohner interview by Dawson.

357. IBM, "The IBM 700 Series," *http://www03.ibm.com/ibm/history/ibm100/us/en/icons/ibm700series/*

358. da Cruz, "The IBM 650," *http://www.columbia.edu/cu/computinghistory/650.html*

359. Mainframe Systems Branch, "History of Computing at Lewis Research Center," NASA B-0300, September 1990, Glenn History Collection, Computing Collection, Cleveland, OH.

360. Betty Jo Armstead, "My Years at NASA," *Denver Museum of Nature and Science* (27 January 2015).

361. Abe Silverstein interview, Cleveland, OH, by John Sloop, 29 May 1974, Glenn History Collection, Oral History Collection, Cleveland, OH.

362. John Sloop, "High-Energy Rocket Propellant Research at the NACA/NASA Lewis Research Center, 1945–60," Seventh International History of Astronautics Symposium Proceedings (8 October 1973), NASA Glenn History Collection, Sloop Collection, Cleveland, OH.

363. John Sloop, "Problems Faced by Early Space Transportation Planners," Fifteenth Space Congress, 26–28 April 1978, NASA Glenn History Collection, Sloop Collection, Cleveland, OH.

364. Walter Olson, ed, *Current Trends in Research and Development for Aircraft Powerplants* (Cleveland, OH: Lewis Flight Propulsion Laboratory, 4 September 1958).

365. John Sloop, "NACA High Energy Rocket Propellant Research in the Fifties" 28 October 1971, NASA Glenn History Collection, Dawson Collection, Cleveland, OH.

366. Virginia Dawson, *Ideas Into Hardware: A History of the Rocket Engine Test Facility at the NASA Glenn Research Center* (Washington, DC: NASA, 2004).

367. Thomas Jefferson National Accelerator Facility, Office of Science Education, "It's Elemental: The Element Hydrogen" (2014), *http://education.jlab.org/itselemental/ele001.html* (accessed 15 September 2014).

368. John Sloop, *Liquid Hydrogen as a Propulsion Fuel, 1945–1959* (Washington, DC: NASA SP-4404, 1977).

369. V. F. Hlavin, E. R. Jonash, and A. L. Smith, *Low Pressure Performance of a Cylindrical Combustor With Gaseous Hydrogen* (Washington, DC: NACA RM–E54L30A, 1955), pp. 9–10.

370. Abe Silverstein and Eldon Hall, *Liquid Hydrogen as a Jet Fuel for High-Altitude Aircraft* (Washington, DC: NACA RM–E55C28a, 1955).

371. Norman Appold, A statement by Colonel Norman C. Appold of the Sun Tan Project Office, December 1958, NASA Glenn History Collection, John Sloop Collection, Cleveland, OH.

372. Robert Miedel, statements by Robert Miedel of Directorate of Procurement GQ ARDC, and William Miller and Lt. Richard Doll of the Sun Tan Procurement Office, December 1958, NASA Glenn History Collection, John Sloop Collection, Cleveland, OH.

373. R. Mulholland, Joe Algranti, and Ed Gough, Jr., *Hydrogen for Turbojet and Ramjet Powered Flight* (Washington, DC: NACA RM–E57D23, 1957).

374. Harold R. Kaufman, High-Altitude Performance Investigation of J65–13–3 Turbojet Engine With Both JP–4 and Gaseous-Hydrogen Fuels (Washington, DC: NACA RM–E57A11, 1957), p. 2–3.

375. William A. Fleming, et al., *Turbojet Performance and Operation at High Altitudes With Hydrogen and JP–4 Fuels* (Washington, DC: NACA RM–E56E14, 1956), p. 18.

376. Paul Ordin, "Progress Report on Project Bee" 1957, NASA Glenn History Collection, Cleveland, OH.

377. Sloop, *Liquid Hydrogen as a Propulsion Fuel*.

378. "NACA Triennial Inspection 1957, Lewis Flight Propulsion Laboratory," 1957, NASA Glenn History Collection, Inspections Collection, Cleveland, OH.

379. *NACA 1957 Flight Propulsion Conference* (Washington, DC: NASA TM–X–67368, 1957).

380. "NASA Lewis Research Divisions and Branches," 1 September 1959, NASA Glenn History Collection, Directors, Collection, Cleveland, OH.

381. "NASA Lewis Research Divisions and Branches."

382. Walter T. Olson, "A Suggested Policy and Course of Action for NACA With Regard to Rocket Engine Propulsion," 6 May 1955, NASA Glenn History Collection, Directors Collection, Cleveland, OH.

383. Alex Roland, Model Research (Washington, DC: NASA SP-4103, 1985).

384. Robert Graham interview, Cleveland, OH, by Sandra Johnson, 30 September 2005, NASA Glenn History Collection, Oral History Collection, Cleveland, OH.

385. Dawson, *Engines and Innovation*, chap. 8.

386. "Triennial Inspection, Lewis Laboratory, High Energy Rocket Propellants," October 1957, NASA Glenn History Office, Inspections Collection, Cleveland, OH.

387. Tom Benson, "Specific Impulse" (5 April 2015), *http://www.grc.nasa.gov/WWW/k-12/airplane/specimp.html* (accessed 5 April 2015).

388. Howard Douglass interview, Cleveland, OH, by John Sloop, 28 May 1975, NASA Glenn History Collection, Oral History Collection, Cleveland, OH.

389. Sloop, *Liquid Hydrogen as a Propulsion Fuel*.

390. *NACA 1957 Flight Propulsion Conference*.

Image 163: Future Center Director (1982–1986) Andy Stofan views a small-scale tank built to study the sloshing characteristics of liquid hydrogen (GRC–1961–C–58299).

# 5. Challenges of Space

*"There was a growing feeling that the engine development had seen its day . . .*
*What came along and sort of saved the day was the space problem.*
*That provided a completely new set of problems."*

—Ben Pinkel

Image 164: Nuclear propulsion display at the Parade of Progress Event at the Cleveland Public Auditorium in August 1964. There is a model of a nuclear spacecraft in the foreground and a Plum Brook Reactor Facility display behind (GRC–1964–C–71686).

# Challenges of Space

On 2 October 1958 a team of painters climbed onto the hangar roof and painted an "S" over the "C" in the large white "N–A–C–A" lettering. The National Aeronautics and Space Administration (NASA) had officially come into being on October 1, and the NACA Lewis Flight Propulsion Laboratory had become the NASA Lewis Research Center. After nearly six months of negotiations, Congress decided to base the new civilian space agency on the existing NACA organization and resources. NASA raced to catch up with the Soviet Union's succession of new achievements in space, and the United States's initial human spaceflight effort, Project Mercury, was soon superseded by the Apollo Program's attempt to send a man to the Moon. Congress supplied NASA with unprecedented resources to accomplish these tasks.

The period of the late 1950s and early 1960s was one of the most dynamic periods in Lewis's history. Just as management had restructured the lab to concentrate all efforts on the jet engine in the mid-1940s, Lewis now shifted to address the burgeoning space program. The center constructed new test facilities, doubled its staff, and suspended almost all aeronautical research. For the first time, the center had responsibility for managing several developmental programs, and steps were taken to philosophically and physically segregate the research and development staff.

During this period, Lewis researched a myriad of space-related topics, including electric propulsion, space power systems, and perhaps most importantly, the use of liquid hydrogen as a propellant. NASA assigned Lewis the management of the Centaur and Agena upper-stage rockets, the hydrogen system for the Nuclear Engine for Rocket Vehicle Application (NERVA) program, and the massive M–1 engine.

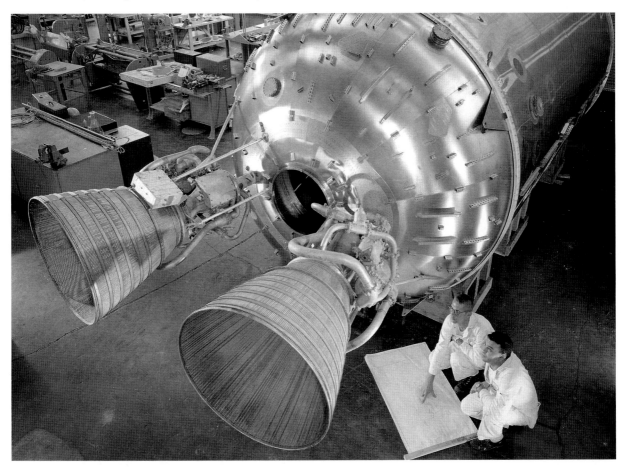

*Image 165: Lewis technicians examine a Centaur rocket in the Space Power Chambers shop (GRC–1964–C–71100).*

## Kickstarting the Space Program

The Soviet Union's launch of Sputnik in October 1957 did not provoke President Dwight Eisenhower as much as it did Congress.[391] The president addressed the uproar a month later by creating the new Special Assistant for Science and Technology position in his executive office and publically broaching the possibility of a new space agency in mid-November. Senator Lyndon Johnson led several weeks of heated Senate hearings on the state of U.S. missile and satellite development. It was in this environment that Lewis held a Flight Propulsion Conference discussing new propulsion alternatives for an array of missions. As President Eisenhower considered the nature of a new U.S. space agency, the NACA leadership struggled to decide what its role should be in these efforts.[392]

In the week following the Flight Propulsion Conference, Walter Olson updated his 1955 paper that suggested an NACA policy regarding spaceflight. The new 20-page document, issued on 2 December, emphasized the NACA's qualifications for participating in the national space effort and suggested that by fiscal year 1959 two-thirds of the NACA staff should be performing space work. He outlined 15 broad space topics that the NACA could contribute to and proposed a new laboratory dedicated to launching an orbiting space station.[393] Three days later Lewis's Research Planning Council modified

several of Olson's specific suggestions but approved the overall theme and scope.[394] The next day, Friday 6 December, Vanguard, the first United States attempt to launch a satellite, failed miserably.

Bruce Lundin, Chief of the Propulsion Systems Division, sat at his kitchen table on that Sunday to put down on paper the thoughts that had been forming in his mind.[395,396] There were many who agreed with Olson's suggestion that the NACA should remain a research group that would support whatever space agency finally emerged. Others proposed that the NACA focus its space efforts on a single large project, such as a space station. Lundin rejected both of these concepts.

In his resulting memo, "Some Remarks on a Future Policy and Course of Action for the NACA," Lundin not only recommended that the NACA lead the nation's space research, but called for the assimilation of other space research groups into the NACA. He stated, "[The NACA's] approach must obviously be bold, imaginative, aggressive, and visionary. The

Image 166: Bruce Lundin (left) and Walter Olson in May 1956 (GRC–1956–C–42155).

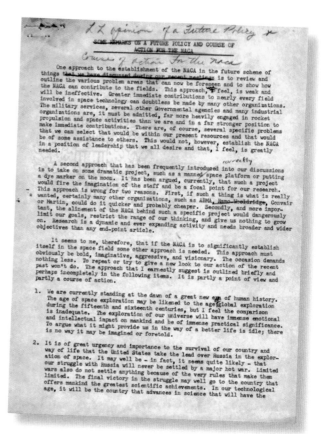

Image 167: Lundin's paper on the NACA's role in space, as marked up by Silverstein.

Image 168: *During the last few months of the NACA's existence, its leadership made a final tour of its three research laboratories. The group arrived in Cleveland on 24 June 1958. At one of the stops Lewis mechanic Leonard Tesar demonstrated the machining of a 20,000-pound-thrust rocket engine for the group in the Fabrication Shop. From left to right, Associate Director Eugene Manganiello, researcher Edward Baehr, NACA Chairman James Doolittle, NACA Executive Secretary John Victory, NACA Committee member Frederick Crawford, Tesar, Lewis Director Ray Sharp, and mechanic Curtis Strawn (GRC–1958–C–48117).*[397]

occasion demands nothing less. To repeat or try to give a new look to our action of the recent past won't do." The document advocated a broad range of space research to be coordinated by the NACA, the continued pursuit of aeronautical research, and the establishment of a new laboratory dedicated to space.[398] Lewis set up a Space Flight Laboratory Committee to establish the requirements for the NACA's proposed new space lab.[399]

Abe Silverstein updated Lundin's document—renamed the "Lewis Laboratory Opinion of a Future Policy and Course of Action for the NACA"—and presented the comments during a meeting of NACA leaders at headquarters on 18 December 1957. Silverstein was able to persuade the reluctant leaders from the other laboratories that the NACA should lead the space efforts. That evening at an informal dinner, younger NACA personnel further urged Hugh Dryden to be less careful regarding space.[400] Dryden took heed and adopted Lundin's recommendation that the NACA push to lead the nation's space effort. The minutes of Lewis's first Space Flight Laboratory

Committee meeting several days later indicate "we were asked to adopt the philosophy set forth in the 'Lewis Laboratory Opinion' document which was accepted with little modification as the official NACA opinion at a recent meeting at Headquarters."[401]

Lundin's argument for NACA leadership in space ultimately served as the basic template for the NACA's proposal in January 1958 to expand its space research and handle launching of all science missions. Dryden stated, "The NACA is capable, by rapid extension and expansion of its effort, of providing leadership in space technology."[402]

The President and Congress considered several alternatives for the space agency—most prominently the Air Force, the Atomic Energy Commission (AEC), and the NACA. Not wanting to expand the military industrial complex, President Eisenhower initiated legislation on 5 March 1958 to create a new civilian space agency firmly based on the NACA. That same day, the United States launched its first successful satellite, Explorer I. Congress modified the bill, replacing the NACA's committee structure with a single administrator. The new agency would also include nonmilitary rocket groups, National Science Foundation researchers, and the Jet Propulsion Laboratory.[403] Congress approved the bill known as the Space Act on 29 July 1958,[404] and NASA officially came into being on 1 October 1958.

❖ ❖ ❖ ❖ ❖ ❖

The new space agency required an official seal. In September 1958 NACA Headquarters solicited each of the laboratories for ideas. Jim Modarelli, head of Lewis's Technical Reports Division, and his illustrators began developing some concepts. Modarelli, who had been among a contingent of Lewis personnel who attended the Ames Inspection that summer, was intrigued by the display of an experimental high-speed aircraft design that included uniquely shaped wings. He later contacted the Ames researchers and their colleagues at Langley for more information and began sketching the twisted wing shapes. These red wings were the first element of the new seal. Modarelli then collaborated with Harry DeVoto at Ames on the seal's central blue oval, featuring the Earth and an orbiting spacecraft, surrounded by lettering around the border.[405]

New NASA Administrator T. Keith Glennan reviewed six proposals and quickly selected the Modarelli design.[406] The design passed through a required military review with only minor modifications, which included correcting the upside-down wings. President Eisenhower approved the new seal in December 1958, and the insignia began appearing in fall 1959.[407]

*Image 170: Jim Modarelli (GRC–1956–C–43683).*

The seal was intended strictly for formal NASA uses, so a second call went out for an informal version that could be used for everyday applications. Modarelli and Lewis artists Richard Schulke, Louise Fergus, and John Hopkins extracted the various components of the official seal and reapplied them in a simplified new logo, later informally dubbed "the meatball."[408] Headquarters not only approved the iconic design, but brought Modarelli onto their staff as Director of Exhibits in November 1959. After establishing the new agency's exhibits program, Modarelli returned to his former post at Lewis in April 1961.[409]

*Image 169: The official NASA seal.*

*Image 171: The NASA logo, often referred to as the "meatball."*

Image 172: *Silverstein represents the new space agency on CBS's "Face the Nation" television program on 8 March 1959 (GRC–2015–C–06538).*

❖ ❖ ❖ ❖ ❖ ❖

Abe Silverstein, who had been commuting between Cleveland and Washington, DC, for several months to assist with the agency planning, officially transferred to headquarters in May 1958. Silverstein served as a member of the small team that laid out the framework for NASA's initial missions and devised NASA's 1960 budget. Over the next year, he brought a number of Lewis managers with him to headquarters. These included DeMarquis Wyatt, Edgar Cortright, Harold Finger, George Low, and others.[410] Others such as Scott Simpkinson, Warren North, Glynn Lunney, and Merritt Preston joined the Space Task Group (STG) at Langley. The STG was responsible for planning Project Mercury and the ensuing human space programs.

As Chief of Space Flight Programs, Silverstein was third in command when NASA officially began operation on 1 October 1958. He was responsible for all spaceflight work done at the field centers and managed the personnel and budget decisions for the STG. Silverstein is credited with naming both the Mercury and Apollo programs. He also managed the nation's early weather and communications satellites and space probes such as *Ranger, Mariner, Surveyor,* and *Voyager.* NASA's first priority was Project Mercury."[411,412]

Image 173: *Orbit announcement of Lewis transfers to NASA Headquarters. During the transition to NASA, Wing Tips was redesigned and renamed Orbit. The name was changed to Lewis News in February 1964.*

## Project Mercury

The STG presented its plans for Project Mercury on 7 October 1958. The program sought to use an Atlas missile to launch a series of capsules with a single astronaut into space. After an extensive evaluation of hundreds of military pilots, in early April NASA selected what came to be known as the Mercury 7. Engineers also began the testing of dozens of capsule designs in NASA's wind tunnels. The STG defined the specifications based on those tests and hired McDonnell Aircraft Corporation to create the capsule.[413]

The STG sought a facility that could simulate high-altitude conditions for a series of *Mercury* thruster tests. In May 1959 Lewis agreed to modify its massive Altitude Wind Tunnel (AWT) to accommodate the program. The operators would not test the capsule in the tunnel's test section but just upstream inside the large leg. Engineers removed the turning vanes and refrigeration equipment in that area, leaving a 51-foot-diameter, 120-foot-long chamber that could simulate altitudes up to 80,000 feet. The STG also assigned

Image 175: *Technicians in the Fabrication Shop align the Mercury capsule afterbody with its pressure chamber in May 1959 (GRC–1959–C–50759).*

Image 174: *Gale Butler examines the Mercury capsule's retrograde rockets prior to a test run inside the AWT (GRC–1960–C–53146).*

Lewis the responsibility for testing the capsule's separation system and escape tower rockets and verifying the control system for the Big Joe capsule.[414]

Big Joe was a mock-up *Mercury* capsule designed to be launched to the edge of the atmosphere in order to simulate reentry without actually placing it in orbit. General Electric created the heatshield, and Langley designed the main portion of the capsule. Lewis was responsible for the capsule's lower section, which contained the electronics and retrorocket tanks, the automatic stabilization system, and assembly of the complete capsule.[415]

The stabilization system was critical to *Mercury's* success because the capsule would burn up during reentry if the heatshield was not positioned correctly. Lewis engineers installed an experimental rig with the lower capsule section inside the new AWT test chamber to verify the performance of the automatic controls and thrusters for Big Joe. The test operators set the rig spinning in three directions simultaneously, then activated the control system that automatically fired thrusters to stabilize the capsule. The series of AWT tests in spring 1959 successfully verified the stabilization system's performance.[416]

Lewis technicians in the Fabrication Shop then assembled all of the vehicle components into the final Big Joe capsule, which was flown to Cape Canaveral aboard a C–130 transport aircraft in June 1959. Forty-five Lewis personnel spent the summer in Florida preparing the capsule for its mission.[417] The 9 September 1959 Big Joe launch was successful despite a glitch in the separation from the Atlas. The capsule and the control system performed so well that NASA canceled plans for a second Big Joe launch.[418]

Lewis researchers also used the AWT to test the *Mercury* retrorocket package located in the lower section of the capsule. Three of these rockets separated the capsule from the booster, and three slowed the capsule for reentry into the atmosphere. The performance of these thrusters was critical because there was no backup system. In early 1960 the Lewis staff tested the *Mercury* capsule with simulated Atlas and Redstone boosters to ensure that the retrorockets did not damage the booster during separation.[419] A second series of tests verified the reliability of the igniter system for the braking retrorockets. These runs provided the researchers with an opportunity to calibrate the retrorockets so that they would not alter the position of the capsule when fired.[420]

The STG also sought to qualify the escape tower rocket motors to determine if their exhaust plume would present a danger to the spacecraft. The escape

*Image 176: Lewis technicians and engineers prepare the Big Joe capsule for launch from Cape Canaveral in 1959 (GRC–2009–C–02180).*

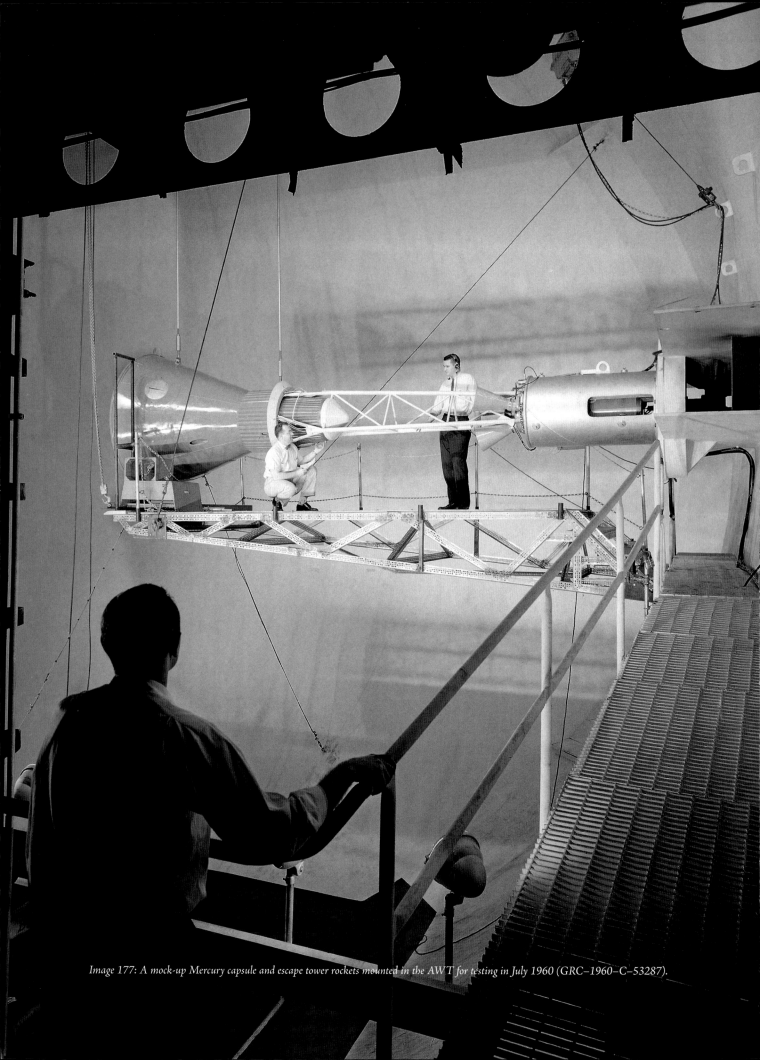

*Image 177: A mock-up Mercury capsule and escape tower rockets mounted in the AWT for testing in July 1960 (GRC–1960–C–53287).*

tower was a 10-foot steel rig attached to the nose of the *Mercury* capsule. The tower had its own propulsion system, which could jettison the astronaut to safety in the event of emergencies on the launch pad or during liftoff.[421] During summer 1960 mechanics mounted the escape tower and a mock-up *Mercury* capsule to the tunnel wall. After successfully firing three different motors at simulated altitudes up to 100,000 feet, the Lewis team determined that the plume was not a safety issue.[422-425]

❖ ❖ ❖ ❖ ❖ ❖

The most well-known of Lewis's work for Project Mercury was the Multi-Axis Space Test Inertia Facility (MASTIF) or gimbal rig. The exhaustive training regimen for the seven Mercury astronauts included instruction on how to bring a tumbling spacecraft under control. In late 1959 Lewis researcher James Useller decided to adapt the gimbal rig that had been developed to test the Big Joe control systems for this astronaut training. Mechanics installed a pilot's chair, hand controller, and instrument display at the center of the three-axis rig, as well as nitrogen thrusters on the outer cages as control devices.[426] Lewis pilot Joe Algranti spent months helping the researchers perfect and calibrate the rig.

In February and March 1960, the seven Project Mercury astronauts traveled to Cleveland to train on the MASTIF. The center sponsored a press conference and built a makeshift dressing room, the "Astro-Penthouse," in the AWT test section for the celebrated visitors. One by one the astronauts entered the AWT chamber and climbed into the rig. The test engineer began rotating the MASTIF on each of the three axes individually, then all three simultaneously. The number of rotations increased from just 2 per minute to 50. In turn, each astronaut used hand controls to activate the thrusters on the outer cages to slow the rotations and eventually bring it to a stop.[427,428]

The high-speed rotations not only disorientated the astronauts but blurred their vision. Researchers installed cameras and sensing equipment to study the phenomena. The MASTIF training was among the most demanding activities of the entire Mercury training program, and the astronauts often reached for the kill switch that shut the rig down.[429] Although several struggled with the rig initially, all seven managed to complete the training. They found that staring at a single object reduced their blurred vision.[430]

On 5 May 1961, just over a year after the conclusion of the AWT tests, Alan Shepard became the first American to enter space. Gus Grissom repeated the suborbital flight in July. The Mercury test program was Lewis's most direct involvement with the human spaceflight program of the 1960s. In a way it was similar to the troubleshooting work performed for the military during World War II. The effort accelerated the *Mercury* development and reduced expenses. The modifications made to the AWT for the tests, however, meant that the facility would never be used as a wind tunnel again.

Image 178: Lewis pilot Joe Algranti explains the MASTIF operation to Alan Shepard in February 1960. Shepard was the first astronaut to operate the MASTIF (GRC–1960–C–52706).

*Image 179: The MASTIF was erected in the wide end of the AWT, where the nitrogen thrusters generated a series of loud hisses as they were fired (GRC–1959–C–51723).*

Image 180: Displays at the November 1962 Space Science Fair at the Cleveland Public Auditorium (GRC–1962–C–62704).

## Reaching Forth

The initial *Mercury* flights whetted the public's appetite for information about the astronauts and the space program. Outreach was one of the core tenets of the new civilian space agency. This was in stark contrast to the restricted nature of the NACA, which held lavish Inspections for colleagues in the industry, but strictly limited access to the public. NASA expected that the interaction would encourage students to pursue a science- or technology-based education and, more importantly, garner public support and congressional backing for the Agency.[431]

For the first time in its history, Lewis made efforts in the early 1960s to engage the general public through a series of open houses and space fairs. In April 1961 the center contributed to a display of model satellites at the Case Institute of Technology that attracted more than 12,000 people over two weeks. After witnessing the positive reaction, Lewis decided to host its first public open house to mark the center's 20th anniversary. Approximately 17,000 people attended the two-day event in August 1962.[432] These events, however, were overshadowed by the massive Space Science Fair at the Cleveland Public Auditorium in November 1962. The event, cosponsored by NASA and the Cleveland Plain Dealer, included exhibits, films, and speakers from NASA centers, local universities, and private aerospace companies. Over 375,000 people attended the Space Science Fair over its 10-day run.[433] A similar event, the Cleveland Press Parade of Progress Exposition, was held in September 1964 at Public Hall. Lewis had the largest display of over 300 participants at the event.

After the Parade of Progress Exposition, however, the center stepped back from large events and focused

*Image 181: Lewis staff with one of the Spacemobile vehicles in October 1964 (GRC–C–1964–72829).*

its outreach efforts on connecting to smaller groups, particularly students. In 1961 NASA initiated the Spacemobile Program. Staff members used these vehicles to update educators and students on NASA's space exploration and aeronautic achievements. Lewis's four spacemobiles—each with an assigned lecturer, exhibits, and models—traveled throughout the Midwest. In 1966 the program conducted over 1,600 remote presentations.[434]

In 1963 Walter Olson was named Director of Public Affairs. Olson joined the lab during World War II and spent most of his career managing the center's fuels and propulsion chemistry work. His new responsibilities included the management of the center's technology transfer efforts and forging relations with universities.[435] Cal Weiss of the Technical Services Division approached Olson in 1963 to discuss the creation of an Educational Services Program to supply schools and the public with NASA materials and respond to requests from the community. The proposal was accepted, and Weiss was placed in charge of the effort. Weiss and his colleague Terry Horvath also initiated a formal speakers' club to discuss various NASA topics at local schools and social clubs.[436] The center also established a junior apprentice program to help high-school-age students prepare for mechanical engineering careers.[437]

## Expansion

Perhaps Abe Silverstein's greatest achievement during his time at headquarters was his ability to convince his colleagues to use liquid hydrogen as the primary propellant for upper-stage vehicles. Silverstein followed Lewis's progress with liquid hydrogen and was confident that it was the optimal propellant. The aerospace industry was also following Lewis's hydrogen work in the late 1950s. In 1957 General Dynamics began designing a second-stage rocket for the military based on the unique balloon tank design of its Atlas missile. Concurrently, but unrelated, Pratt & Whitney converted the technology from Project Suntan's hydrogen aircraft engine into the RL–10 rocket engine. The RL–10 would be the first commercially produced liquid-oxygen—liquid-hydrogen engine.[438]

In 1958 Silverstein led a committee that investigated the performance requirements for upper-stage rockets. The group concluded that the large missions which NASA was planning would require the use of high-energy propellants, namely liquid hydrogen. Almost immediately afterward, the military decided to merge General Dynamics's upper stage with Pratt & Whitney's RL–10 engines.[439] The result was the Centaur rocket. Meanwhile, renowned rocket engineer and director of the Development Operations Division of the Army Ballistic Missile Agency, Wernher von Braun, and his colleagues were designing the massive multistage launch vehicle that would become Saturn.[440]

In late 1959 Silverstein chaired the Saturn Vehicle Team, informally termed the "Silverstein Committee," that was responsible for selecting upper-stage designs for the Saturn rocket. He felt that it was foolish to develop propulsion systems for future missions using traditional rocket fuel combinations when the emerging high-energy propellants could offer significantly improved performance.[441] After much discussion, the Lewis contingent was able to convince a reluctant von Braun that liquid-hydrogen stages were the only practical alternative for Saturn. The initial Saturn configuration included an RL–10-powered second stage and the Centaur for the third stage.[442] Lewis's continuing efforts to develop liquid hydrogen would apply to both the Centaur and Saturn stages. In 1960 NASA contracted with Rocketdyne to develop a larger hydrogen-oxygen engine, the J–2, for the Saturn upper stages.[443]

In addition to its liquid-hydrogen research, Lewis supported the Saturn development in other ways. The Saturn program included the two-stage Saturn IB (S–IB), which was eventually used on some Skylab missions. The effort began in the late 1950s with a multiyear base-heating investigation of the S–IB's eight-engine booster in the 8- by 6-Foot and 10- by 10-Foot supersonic wind tunnels (8×6 and 10×10). The rocket engine's exhaust heat tended to recirculate in the nozzle area. The resulting overheating of the lower end, or base, of the booster could cause the engines to fail or could introduce aerodynamic concerns.

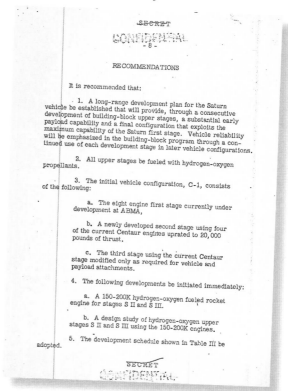

Image 183: *Recommendations of the Silverstein Committee regarding the use of liquid hydrogen in the Saturn upper stages.*

The Lewis study determined that turbine exhaust dramatically increased the base heating. Researchers found that the use of cooling air scoops and external flow deflectors decreased base heating significantly.[444]

Other efforts were focused on the Saturn V vehicle. These included new base-heating and engine-gimballing tests of the Saturn V second stage (S–IC), testing of the launch escape system, and microgravity studies of the S–IVB third-stage propellant tanks. A scale model of the S–IC was tested in both the 8×6 and 10×10 to determine the force required to gimbal the stage's five engines and the resulting flow patterns.[445] Researchers also determined the stability of the launch escape vehicle under simulated flight conditions in the 8×6.

❖ ❖ ❖ ❖ ❖ ❖

In early 1961 the United States seemed to be rapidly losing ground in the Cold War's new technology front. During President John Kennedy's first six months in office, the RL–10 engine and Centaur rocket were behind schedule, the Atlas boosters suffered repeated failures, and the failed nuclear aircraft program was canceled. During the same period, the

Image 182: *Gene Manganiello (right) welcomes Wernher von Braun to Lewis in December 1959 (GRC–1959–C–52148).*

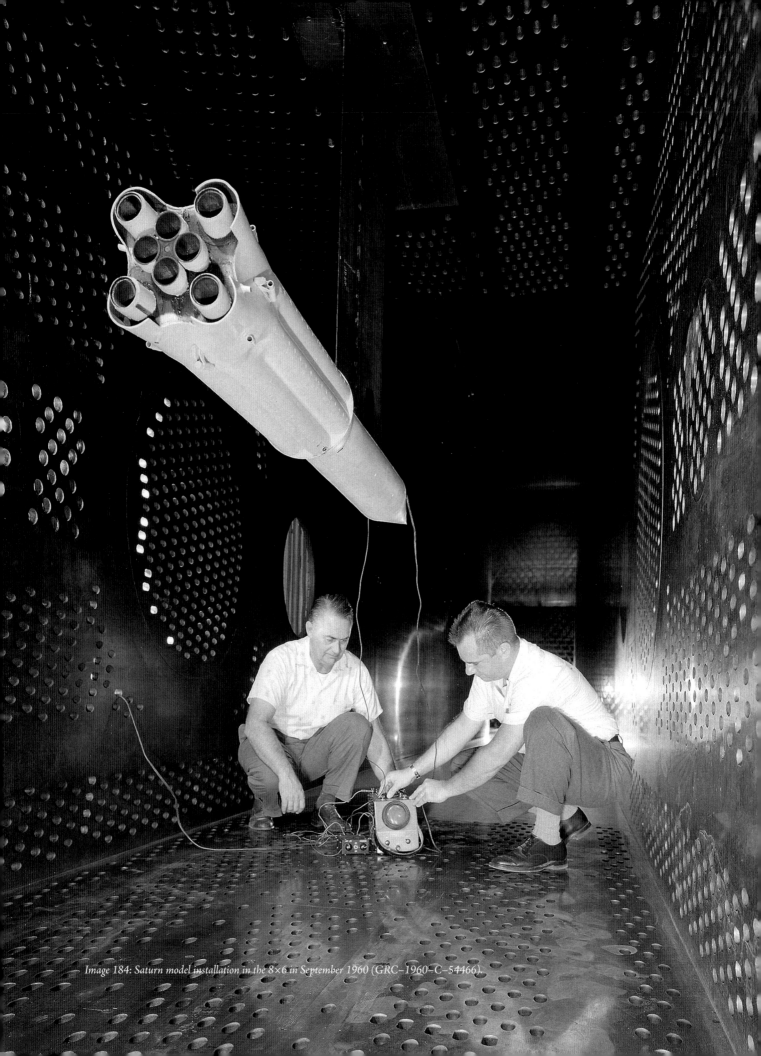

*Image 184: Saturn model installation in the 8×6 in September 1960 (GRC–1960–C–54466).*

Soviets launched animals into space and safely returned them to Earth. Most notably, on 12 April, cosmonaut Yuri Gagarin became the first human in space and the first to orbit Earth.

On 25 May 1961 President Kennedy informed the nation that he was accelerating the space program on two fronts—a human mission to the Moon and the development of nuclear rockets for the human exploration of Mars and beyond.[446] Lewis would make major contributions during the 1960s to both efforts. Congress increased NASA's budget dramatically to meet these goals. This led to a spike in staffing and a physical expansion that included the addition of the Johnson and Kennedy space centers. When NASA Headquarters reorganized its staff to accommodate the new Apollo Program, Abe Silverstein was displeased with the new chain of command. In October 1961 he returned to Cleveland to assume the director's position at Lewis. Ray Sharp had retired in January and passed away several months later. Eugene Manganiello, who had served as acting director in the interim, became Deputy Director.

Silverstein's first action as director was to hold a press conference announcing that Lewis was looking to hire young engineers, scientists, and technicians. People from aerospace companies, the military, and universities were hired during this national recruiting campaign. The rush for additional staffing resulted in the center's first hiring of journeymen since the reinstitution of the Apprentice School in 1949.[447] By 1964 Lewis had nearly 4,900 employees—more than any other NASA center except Marshall.[448]

Unlike the NACA, which developed technologies for the military to turn into actual engines or aircraft, NASA was now creating technology for its own use. The Agency's space centers—the Jet Propulsion Laboratory, Goddard, Johnson, and Marshall—were responsible for handling the bulk of that development. During the formation of NASA, the Agency sought to preserve the research nature of the NACA research laboratories—Lewis, Langley, and Ames—by assigning them to the Office of Advanced Research and Technology (OART). The OART centers were responsible for creating new knowledge and served

*Image 185: Silverstein holds a 3 November 1961 press conference announcing additional recruiting efforts. Over the previous months, the center had hired 135 new staff members, interviewed over 700 prospects, and had over 300 applications on file (GRC–1961–C–58359).*

*Image 186: Interior of the 20-foot-diameter vacuum tank in the Electric Propulsion Laboratory. The circular covers on the floor sealed the displacement pumps located beneath the chamber (GRC–1961–C–57748).*

as a measure of the nation's technical competency. Ray Bisplinghoff, head of OART at the time, said that the office was "the hard foundation upon which the nation's aeronautics and space programs rest."[449]

Abe Silverstein and his management corps debated whether or not to pursue any of the developmental programs that were emerging or to just continue supporting research in the NACA tradition. In the end, Bruce Lundin convinced Silverstein to do both. Silverstein reorganized the center so that the research and development divisions were completely segregated. He also initiated the construction of the Developmental Engineering Building (DEB) outside of the center's main gate to house the new developmental staff. The building, which opened in May 1964, even included its own cafeteria so that its staff would not have to enter the research portion of the center.

❖ ❖ ❖ ❖ ❖ ❖

The center also underwent substantial physical expansion during this period. Despite the Agency's increase

in funding, the Bureau of Budget still expected Lewis to convert as many of its aeronautics facilities to space research as possible before building new ones. Gene Manganiello, then acting director, noted that many of the facilities were already being converted and that the only remaining aeronautics area on campus was inside the hangar—and that was soon repurposed.[450]

In 1958 Lewis acquired the 115-acre West Area on the other side of Abram Creek from the main campus. When the funds for the land were requested in 1959, crews constructed a road through the ravine to link the property to the center. The land, purchased from several private owners and the Metropolitan Park Board, came with two homes which the center referred to as the Mitchell House and the Guerin House.[451] Lewis used the former as a training office and the latter for social activities. The primary purpose for the West Area, however, was the construction of new world-class facilities to support the center's incipient electric propulsion research program.

*Image 187: The Guerin House (GRC–1964–C–72264).*

Lewis's expansion was even more pronounced at Plum Brook. The center had already leased 500 acres at the site in 1956 to construct its nuclear test reactor. As Lewis expanded its research with high-energy propellants in the late 1950s, it was clear that it needed additional test facilities. Several incidents at the Rocket Lab and Rocket Engine Test Facility during this period made it apparent that increased testing on propellants was too risky at Lewis's congested Cleveland campus.

In 1957 Lewis made arrangements to lease an additional 3,100 acres at Plum Brook for these new facilities, collectively known as the Rocket Systems Area.[453] Plum Brook's large tracts of unpopulated space were perfect for the potentially dangerous research. The Rocket Systems Area consisted of a multitude of small test sites designed to tackle specific engine components such as turbopumps, turbines, and storage tanks. It also included the High Energy Rocket Engine Research Facility (B–1) and Nuclear Rocket Dynamics and Control Facility (B–3). These vertical test stands could test full-scale liquid-hydrogen fuel systems under simulated altitude conditions.

By 1961 the reactor and most of the Rocket Systems Area sites were beginning to operate, and several hundred Lewis employees were now permanently located at Plum Brook. In most cases, Lewis's research divisions developed the test programs for the Plum Brook facilities. The Plum Brook staff prepared the facilities, ran the tests, and collected the data. By this time Lewis had made arrangements to add the Cryogenic Tank Storage Site (K Site) and the Hydrogen Heat Transfer

*Image 188: The J Site crew on the "portable" rig on 13 August 1960 before the first test at Plum Brook Station (GRC–2015–C–06550).*

*Image 189 (inset): Initially, NASA let the PBOW structures stand. As more and more acres were acquired, however, workers began to destroy a large number of the buildings. In 1961 a local company was hired to raze all unusable structures and to dismantle three acid plants. However, a number of the nonmanufacturing structures were retained (GRC–2015–C–06565). Image 190: B Complex with the B–1 and B–3 test stands (GRC–1965–C–03012).*

Facility to the area, and by 1963 even larger test sites were in the works, including the Space Power Facility (SPF) and the Spacecraft Propulsion Research Facility (B–2).[454] In March 1963 NASA made arrangements with the military for the Agency to not only use Plum Brook's remaining 3,500 acres but to permanently take possession of the entire property.[455] The Plum Brook Ordnance Works (PBOW) was now officially called Plum Brook Station.

Lewis constructed or repurposed 19 facilities for space-related research during the 1960s. Several prominent aeronautical facilities, including the Icing Research Tunnel (IRT) and the Four Burner Area were effectively shuttered.[456]

One of the most dramatic alterations was the conversion of the storied AWT into two altitude chambers known as the Space Power Chambers (SPC). The center had already removed the tunnel's cooling coils and turning vanes in 1959 for the Project Mercury tests. In 1961 Lewis management decided to permanently alter the AWT by creating two vacuum chambers inside of it. NASA was initiating a host of new space missions, but there were no large vacuum chambers that researchers could use to simulate the space environment.[457] Crews removed the AWT's drive fan, inserted three bulkheads to block off portions of the tunnel, and upgraded the vacuum pumping system. The result was two large chambers— one a high vacuum and the other simulating upper altitudes. Just as those modifications were completed, Lewis required a vacuum chamber for the newly acquired Centaur Program. A vertical extension with a removable dome was added so that the rocket stage could be stood up within the vacuum chamber.[458] The facility was finally completed in 1963.

*Image 191: Three bulkheads were placed inside the AWT to create the SPC. The largest is seen here being inserted approximately where the wind tunnel fan was located (NASA C–1961–58551).*

*Image 192: Interior of the SPC's 51-foot-diameter high-altitude test area inside the former AWT (GRC–1963–C–67001).*

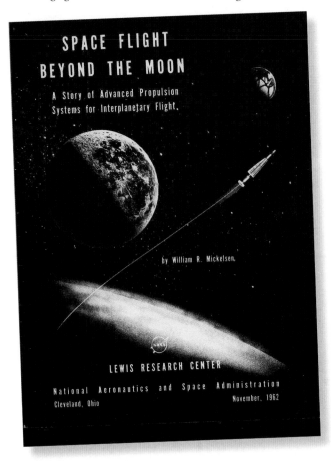

*Images 193: Lewis report from 1962 describing the history of electric propulsion and requirements for long-duration interplanetary missions.*

## Electric Propulsion

Lewis researchers were interested in virtually any form of propulsion that was applicable to aircraft, missiles, or spacecraft. In the 1950s Abe Silverstein encouraged the pursuit of two diametric experimental approaches—electric propulsion and nuclear propulsion. Nuclear propulsion yields an extremely high level of thrust that can transport large payloads for long-duration missions. Electric thrusters are more efficient than chemical or nuclear rocket engines. They operate over a long timeframe but produce extremely low amounts of thrust. In the vacuum of space, the low thrust significantly increases the spacecraft's velocity as the mission progresses. Small-scale electric engines are applicable for low-power missions such as keeping a satellite in its correct position or propelling deep space probes on lengthy journeys.[459]

The concept of electric propulsion had been proposed by rocket pioneers Konstantin Tsiolkovsky and Robert Goddard in the first years of the 20th Century.

Herman Oberth brought the obscure concept to public attention with his 1929 book, *Ways to Spaceflight*. A number of scientists studied Oberth and became interested in the feasibility of ion propulsion after World War II, including Wernher von Braun's colleague Ernst Stuhlinger.[460] Stuhlinger published a groundbreaking paper in 1954 that covered all aspects of the electrically propelled space vehicle.[461] Under Stuhlinger's supervision, the army initiated testing of ion thrusters in 1958.[462] NASA Marshall contracted with Hughes Research Laboratory to design and test the first electric propulsion engine, which Hughes first operated the thruster in September 1961.[463]

At Lewis, Wolfgang Moeckel of the Special Projects Branch discovered Stuhlinger's paper in 1956 and began calculating the theoretical capabilities of an electric thruster. Moeckel and his colleagues evaluated the ability of electric propulsion to perform a variety of space missions. The findings were presented at the Lewis Flight Propulsion Conference in November 1957. These mission studies led directly to Lewis's implementation of a new research program to study different facets of electric propulsion.[464]

In 1960 the center restructured its advanced propulsion efforts. Among the new groups were the Electromagnetic Propulsion Division led by Moeckel and the Chemistry and Energy Conversion Division led by Walter Olson. The former was responsible for the electric propulsion work. The latter studied new sources of energy and power such as solar cells, fuels cells, heat transfer, and ion energy.[465] Electric propulsion engines rely on energy conversion systems to power the thrusters, so Olson's group worked in tandem with Moeckel's.

Meanwhile Howard Childs and Bill Mickelsen had begun designing a suite of new test facilities for conducting electric propulsion research. In 1958 Lewis installed a 16-foot-long vacuum tank in the Prop House and renamed the building the Electric Propulsion Research Building (EPRB). Three more tanks were added shortly thereafter.[466] The vacuum tanks were essential since the thrusters could not operate in atmospheric air. In 1960 the center constructed a new facility with even larger and more powerful vacuum tanks, the Electric Propulsion Laboratory (EPL), and the Energy Conversion Laboratory (ECL) in the

West Area. The EPL includes 15- and 25-foot-diameter tanks, several smaller chambers, and a clean room. The ECL contained laboratory facilities used to study fundamental elements pertaining to the conversion of energy into electrical power.[467] The completion of the new facilities in 1961 coincided with NASA's decision to consolidate all of the Agency's electric propulsion at Lewis.[468] Lewis emerged as the nation's preeminent electric propulsion research organization, and remains so to this day.

❖ ❖ ❖ ❖ ❖ ❖

The Electromagnetic Propulsion Division pursued electrothermal, electromagnetic, and electrostatic propulsion systems. The most promising of the three fields was the electrostatic, or ion, engine. There were two ion engine design options—contact and electric bombardment thrusters. The first was based on a Stuhlinger concept, the latter on a design by Lewis researcher Harold Kaufman. The Kaufman engine vaporized liquid mercury, which was then bombarded by electrons to create more electrons and ions. A negatively charged electric field and a positively charged screen drew the ions rearward and out of the engine as thrust.[469] The electron bombardment thruster demonstrated 90-percent efficiency during extensive testing in vacuum tanks at the EPL. All ensuing U.S. ion propulsion systems were derived from the original Kaufman thruster.[470]

Although Lewis's vacuum tanks could simulate space conditions, it was important to demonstrate that the thruster would perform during an actual space mission. By April 1960 Lewis was planning the Space Electric Rocket Test (SERT–I). The SERT vehicle included two engines—a cesium contact thruster and the electron bombardment thruster—mounted on the outer edge of the spacecraft so that the thrust would rotate the vehicle. Lewis researchers tested SERT–I throughout 1963 and 1964 in the EPL vacuum tanks, and it was launched from Wallops Island on 20 July 1964. Although the contact thruster failed to function, the electron bombardment thruster operated for over 30 minutes. The suborbital flight lasted only 50 minutes, but it provided the first demonstration of electric propulsion technology in space.[471]

*Image 194: Space Electric Rocket Test I (SERT–I) spacecraft and thrusters tested in EPL's Tank No. 3 in June 1964 (GRC–1964–C–70258).*

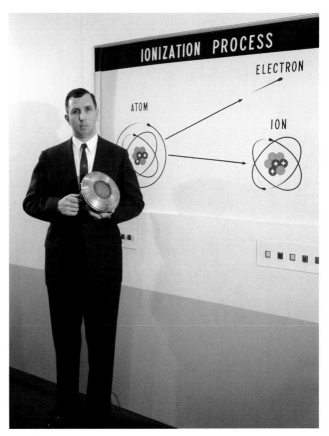

*Image 195: Kaufman with his electron bombardment thruster in the early 1960s (GRC–2001–C–01603).*

*Image 196: Technicians prepare the SERT–I spacecraft in an EPL cleanroom in February 1964 (GRC–1964–C–68553).*

## Nuclear Propulsion

Lewis researchers had been interested in nuclear propulsion since the end of World War II. Despite a variety of studies at a number of institutions, including Lewis, the nuclear aircraft program never really progressed. There were inherent design issues, such as the massive amount of shielding needed to protect the crew and the possibility of crashes, that posed major obstacles. Lewis researcher Frank Rom worked on designs that included a thick stainless steel shell for the reactor that would shield the crew and provide containment protection in the event of a crash. In the end, however, engineers could not design an aircraft large enough to carry the heavy shielding or convince the public of its safety.[472] The military lost interest in the concept in the late 1950s. The program was officially canceled in spring 1961 just as the Plum Brook Reactor Facility (PBRF) began operation.

The cancellation coincided with President Kennedy's call to expand research on nuclear rockets, which offer superior performance in comparison to chemical rockets. Unlike nuclear aircraft, these upper-stage rockets would not be activated until they were out of the Earth's atmosphere, thus reducing the danger posed by a crash. Nuclear rockets could also be designed so that the crew members were a significant distance away from the reactor. This would reduce the amount of required shielding.[473]

The design of the nuclear engine was similar in nature to the high-energy chemical rocket engines under development. Both systems used cryogenic liquid hydrogen as the fuel, turbopumps to pump the fuel from the tank to the engine, and regeneratively cooled nozzles. The main difference was that nuclear engines used thermal radiation to heat the hydrogen and thus did not require an oxidizer or combustion.[474]

In November 1955 the AEC and the Air Force instituted Project Rover to develop nuclear-powered missiles. It was the nation's first attempt to design a nuclear rocket engine. Work began almost immediately to develop engine components and systems at the Los Alamos National Laboratory. The Air Force soon lost interest in the nuclear rocket, however, and their responsibilities and funding were transferred to NASA in December 1959. At this point the proposed mission changed from a nuclear-powered missile to a nuclear rocket for long-duration space exploration. The AEC was responsible for designing the vehicle's reactor, and Lewis managed the development of the liquid-hydrogen system. The Space Nuclear Propulsion Office (SNPO), led by Lewis veteran Harold Finger, was established in August 1960 to coordinate the AEC and NASA activities.[475]

*Image 197: An unfueled Kiwi B–1–B reactor and its Aerojet Mark IX turbopump being prepared for installation in the B–3 test stand (GRC–1967–P–01289).*

Project Rover included several different and sometimes overlapping phases. The first, Kiwi, consisted of a series of increasingly powerful, but not flightworthy, reactors that researchers could use to study basic reactor designs. The second phase, NERVA, would develop a flyable nuclear engine. The third part of the program—the Reactor-In-Flight-Test—would involve an actual launch test. The AEC tested eight Kiwi reactors between 1959 and 1964 at its Los Alamos and Nevada test sites with varying results. The SNPO felt confident enough to use the Kiwi 1–B1 design as the basis for the NERVA engine design in 1960. The SNPO contracted Rocketdyne, under the supervision of Lewis, to develop the hydrogen pumping system.[476]

Meanwhile Lewis was conducting a variety of studies in the Rocket Systems Area on the use of hydrogen in nuclear engine systems. The center conducted nearly all of its nuclear rocket testing at the new Plum Brook Station facilities. Plum Brook research included studies on specific Kiwi engines and basic research that could be applied to a variety of nuclear rocket designs. The researchers did not use any functioning reactors in their investigations, concentrating instead on the hydrogen propellant system, nozzles, and other nonnuclear components. As such, much of the research could be applied to both chemical and nuclear rockets.

It was crucial for the planned long-term missions that the nuclear rocket be able to vary its speed and restart its engine without any external power. This was accomplished by allowing a small amount of liquid hydrogen to be vaporized by the reactor. The gaseous hydrogen activated a turbine that drove a turbopump to pump additional liquid hydrogen to the reactor.[477]

The process was similar to that used in chemical rockets, but researchers needed to verify that it would apply to nuclear rocket systems. Lewis began a nearly two-year test program at the B–1 and B–3 stands in September 1964 to verify the engine's startup, test two different turbopump designs, and obtain data to establish a preprogrammed startup system.[478-480] The propellant would be pumped through the rocket system as during a normal startup, but the engine would not be fired. Researchers used other Plum Brook sites to study the heat transfer properties of a nuclear rocket nozzle, verify the design of a nuclear rocket heat exchanger, test a variety of pumps and turbines, and study materials in the PBRF.

*Image 198: The B–1 and B–3 test stands, 135 and 210 feet tall, could test different components of high-energy rocket engines under flight conditions (GRC–1964–C–01310).*

Meanwhile the AEC conducted a series of tests of the NERVA engine in Nevada. Early problems with the reactor's fuel elements were improved. Then during 1969 a modified version was started and stopped over two dozen times in a vacuum environment. After plateauing in the mid-1960s, the program's budget began to decline as funding for extended missions to Mars or other distant destinations faded.[481]

### Taming Liquid Hydrogen

Although General Dynamics and NASA had committed to using liquid hydrogen in the Centaur and Saturn upper stages, there were still many unknowns regarding its use that had to be resolved before they could be successfully utilized in space. These included determining the optimal storage and pumping systems and learning how to control the liquid in microgravity conditions. Lewis conducted extensive research programs on all of these issues throughout much of the 1960s.

Most high-performance rocket engines use turbopumps that rotate thousands of times per minute to deliver the propellant from the tank to the engine at high rates of speed. A failure will likely end the mission. The use of liquid hydrogen posed a particular problem for pumping. It was stored just below its boiling point, so a portion of the flow was likely to vaporize before reaching the pump. Lewis researchers tested hundreds of different spiral-shaped impeller designs to improve pump performance in these conditions. They often analyzed the new designs first in the Engine Research Building's water tunnel. The pumps or components were then brought out to Plum Brook Station and run with liquid hydrogen in the Rocket Systems Area test sites.[482] Researchers performed generic studies that could be applied to a variety of pumps and tested specific pumps for the NERVA and Centaur programs.

Cryogenic propellants, such as liquid hydrogen, must be stored at extremely low temperatures. Thermos-like dewars suffice on Earth, but they are too heavy to include in rocket designs. Despite the coldness of space, the rocket engine, its electronics, and solar radiation all increase the propellant's temperature, which results in fuel evaporation. Insulation solutions for relatively short-term missions, such as Centaur and Saturn, were developed, but long-term missions to other planets required more robust designs. Lewis conducted extensive research and testing on different

*Image 199: Ron Roskilly demonstrates the testing of a hydrogen turbopump (right) at the Rocket Systems Area's A Site (left) (GRC–1962–C–61077).*

insulation systems for propellant tanks on space vehicles and for proposed fuel depots in space.[483]

K Site was one of Plum Brook Station's most active test facilities. The facility, which became operational in 1965, consisted of a 25-foot-diameter vacuum chamber in which cryogenic propellant tanks were tested in simulated space conditions. The facility included equipment that could create the coldness of space and shake the test article to simulate launch vibrations.[484] Researchers used the site extensively to study the use of different gases to expel propellant from the tank, transfer propellant from one tank to another, and test a variety tank insulation systems. The most successful insulation designs were multilayer foil and mylar wraps, particularly a self-evacuating layered system.[485]

❖ ❖ ❖ ❖ ❖ ❖

Early in the space program, engineers did not know how liquid would behave in low gravity. On Earth, gravity separates liquid from vapor, but it was unclear if that would occur in space. Lewis undertook an extensive study to determine the behavior of liquid hydrogen in microgravity so that proper fuel systems

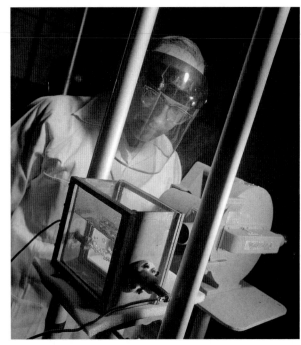

Image 200: *Bob Siegel with a test rig for high-speed filming of liquid behavior in microgravity in the 8×6 wind tunnel (GRC–1960–C–54149).*

Image 201: *Fred Haise, Lewis pilot and future Apollo astronaut, monitors the cameras and instrumentation for the experimental liquid-hydrogen container in the bomb bay of the AJ–2 aircraft (GRC–1961–C–56862).*

could be designed. Rocket designers needed to know where the propellant would be inside the fuel tank so that it could be pumped to the engine. Lewis researchers utilized sounding rockets, research aircraft, and the 2.2-Second Drop Tower for these early microgravity studies.

The 2.2-Second Drop Tower descended 100 feet into a ravine along the Lewis campus, providing researchers 2.2 seconds of freefall for simulating microgravity. Researchers used the tower as an inexpensive tool for a large number of studies. Tests were set up in experiment carts equipped with cameras to film the behavior of liquid during freefall so that researchers could investigate the wetting characteristics of liquid and

Image 202: *An Aerobee rocket being prepared for launch in January 1961 (GRC–1961–C–55686).*

liquid-vapor configurations and predict the equilibrium state in microgravity conditions. The early experimental fluid studies verified the predictions made by Lewis researchers that the total surface energy would be minimized in microgravity. Thousands of drop tower tests in the early 1960s provided an increased understanding of low-gravity processes and phenomena.[486] The tower afforded only a relatively short experiment time, but the results were sufficient for the research to be expanded with longer-duration free-fall tests on sounding rockets or aircraft.

The center acquired two North American AJ–2 Savage aircraft in the early 1960s to fly microgravity-inducing parabola flight patterns. Lewis engineers installed a 100-liter liquid-hydrogen tank, a cryogenic cooling system, and cameras in the bomb bay. A pilot would fly the aircraft on a 13,000-foot level course over western Lake Erie. Then the pilot would pull up the nose 40°. The decrease in speed would nullify both the latitudinal and longitudinal accelerations. Upon reaching 17,000 feet, the pilot would turn the aircraft into a 45° dive. As the speed reached 390 knots, the pilot would pull the aircraft up again. Each maneuver produced approximately 27 seconds of reduced gravity. A ping-pong ball was hung in the cockpit to let the pilot know when the aircraft was in microgravity.[487]

*Image 203: Orbit article on the recovery of an Aerobee telemetry unit from the Atlantic Ocean.*

*Image 204: A NASA AJ–2 Savage makes a pass for cameramen at the Cleveland Municipal Airport in November 1960. The AJ–2 was a Navy-carrier-based bomber in the 1950s (GRC–1960–C–54979).*

For longer durations of microgravity, Lewis researchers utilized sounding rockets, an Atlas booster, and Scott Carpenter's *Mercury 7* mission. The STG assigned Carpenter five somewhat basic experiments to perform during his June 1962 orbital flight, including a study of liquid behavior in microgravity for Lewis. The experiment consisted of a 3.5-inch-diameter glass sphere that was partially filled with a water mixture. Inside the sphere was a perforated tube that caused the liquid to rest at the lower portion of the tank while the vapor rose to the top. The experiment demonstrated that uneven acceleration and baffle misalignment did not alter the fluid's position.[488]

In the early 1960s researchers launched a series of microgravity fluids experiments inside a Lewis-designed 9-inch-diameter tank on Aerojet's Aerobee sounding rockets. A camera inside each suborbital Aerobee missile filmed the behavior of liquid hydrogen during its 4 to 7 minutes of freefall. Between 1961 and 1964, Lewis conducted nine Aerobee launches over the Atlantic Ocean from Wallops Island. The flights provided data on nucleate boiling and pressure increases during microgravity. Lewis's Flight Operations Branch and the Photo Lab participated in the tracking and recovery of the test package from the Atlantic.[489]

The Aerobee launches provided sufficient data for most conditions, but additional time was needed to study situations with low heat flux. Lewis researchers arranged for an experimental tank to be filled nearly half way and flown on an Atlas booster. The 25 February 1964 flight provided over 21 minutes of microgravity. The instrumentation measured temperature, pressure, vacuum, and liquid level. Temperature instrumentation indicated wall drying during the freefall. The resultant pressure increases were similar to those experienced during the normal-gravity test.[490]

## The Development Side

The Centaur upper-stage rocket was the first space vehicle to use liquid hydrogen as a propellant. General Dynamics designed the unique spacecraft to be paired with the Atlas booster.[491] The military originally intended to use Centaur to launch the Advent satellite, but Centaur development problems ultimately convinced the military to use another booster. NASA, however, selected Centaur to launch a series of *Surveyor* spacecraft that would land on the lunar surface and explore the Moon prior to the Apollo missions.

The 15,000-pound-thrust Pratt & Whitney liquid-hydrogen-fueled RL–10 engines not only powered Centaur, but were also incorporated into the early upper stages for Saturn, so it was a serious concern when combustion problems caused several RL–10 engines to explode at Pratt & Whitney's Florida facilities. NASA Headquarters assigned Lewis, the only center with extensive hydrogen experience, the responsibility for investigating the RL–10. In March 1961 Lewis's Chemical Rocket Division began a series of RL–10 tests at the Propulsion Systems Laboratory. Researchers first verified that the engines could be steered and throttled. Then they determined that the injection of a stream of helium gas into the liquid-oxygen tank immediately resolved the engine's combustion instability problems.[492]

The overall Centaur Program, managed by Marshall, continued to be beset by delays and quality issues. In May 1962 the first Atlas-Centaur (AC) launch attempt ended when the rocket exploded less than a minute after liftoff. Marshall (which did not care for the rocket's use of liquid hydrogen propellant or its

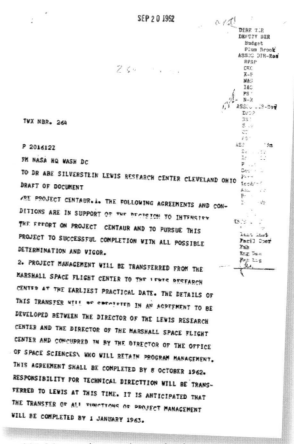

*Image 205: Memo authorizing the transfer of Centaur to Lewis in 1962.*

Image 206: Atlas booster being hoisted into the Dynamics Stand at Plum Brook Station in 1963. The Atlas and Centaur were tested individually, as a pair, and with a simulated Surveyor payload (GRC–1963–P–01700).

unique pressurized balloon-like structure) and *Surveyor's* developer (the Jet Propulsion Laboratory, JPL) called for Centaur's cancellation.[493] Instead, after congressional hearings and internal NASA deliberations, NASA Headquarters decided to transfer the Centaur Program to Lewis in September 1962.[494,495] Although Lewis had scant experience managing large development programs, they were the experts on liquid hydrogen and had experience with the RL–10 engines.

Center Director Abe Silverstein personally oversaw the intensive two-year checkout of Centaur, and numerous test facilities at Lewis and Plum Brook Station were built or modified specifically for this upper-stage vehicle. The engineers not only had to make sure that the rocket systems performed well but had to develop methods for handling the propellants in the microgravity of space, ensure that the engines could be restarted, and verify that the vehicle could withstand the vibrations of a launch.

Image 207: A Centaur stage is lowered through the dome and into the SPC vacuum tank (GRC–1964–C–68846).

Lewis conducted an extensive examination of Centaur in late 1962 and 1963. This included a full-scale test of the separation system in the new SPC and verification of the vehicle's structural integrity in Plum Brook's Dynamics Stand. The former led to a redesign of the retrorocket system. The latter verified that the Atlas could handle the unprecedented weight of the Centaur during a simulated launch. The test allowed NASA to lift the flight restraint for the upcoming launch attempt.[496] Lewis engineers were at the controls as AC–2 successfully blasted off from Cape Canaveral on 23 November 1963.

The AC Program included five developmental flights that would be flown with simulated payloads before the actual *Surveyor* missions began. There were still many questions to answer before the valuable lunar landers were risked on the new rocket. The failure of the next launch attempt, AC–3, to reach orbit in June 1964 raised concerns across the Agency again.[497] Lewis stepped up its efforts during the next two years to ensure that Centaur remained on schedule and did not delay the *Apollo* landings. The SPC became the hub for the Centaur test program.

Image 208: AC–2 on a launch pad at Cape Canaveral in November 1963. It was the first successful launch of a liquid-hydrogen rocket. The Centaur upper stage from the AC–2 launch remains in orbit today (GRC–2015–C–06539).

Centaur was designed to fire its engine multiple times during flight. For simplicity's sake, however, the engines were only fired once during each of the first three test launches. The AC–3 launch failed when the jettisoning of the vehicle's nose cone disrupted its guidance system. The manufacturer had tested the cone repeatedly prior to the mission in atmospheric conditions, but Lewis researchers wanted to verify its performance in a simulated space environment. The shroud was mounted vertically on a platform in the SPC vacuum tank, and its jettison system was activated. The shroud suffered significant damage during one of the runs, so Lewis researchers redesigned the shroud components and retested the system during fall 1964.[498] The AC–4 launch in November 1964 proved to be the first Centaur mission to have a completely error-free shroud jettison. Further modifications to the shroud were made, and it was requalified at the SPC in summer 1965.

AC–4 was also the first attempt at restarting the engines in space. When the first engine burn ended, the liquid hydrogen sloshed forward, causing some of the propellant to be lost through the tank vents. This threw off the vehicle's balance and caused enough hydrogen spillage to prevent the engines from restarting for the second burn. Afterward, Lewis engineers developed and tested in the SPC a new balanced venting system that expelled gas in an even and nonpropulsive manner. The new design also included a baffle to prevent sloshing inside the tank.[499,500] The next attempt at a two-burn mission in April 1966 was a success.

Lewis researchers considered electronic malfunctions to be one of the most likely causes of failures in space so they initiated a test program to operate a complete Centaur vehicle in the SPC vacuum chamber. The facility permitted all elements of a Centaur mission to be simulated except the firing of the engines. The engineers operated the vehicle's electronics and controls systems as they would during all phases of the missions.

The tests revealed that the pressurization of the canisters that housed the electronics caused problems, that the minimization of the necessary power level reduced equipment temperatures, and that the electronics did not transfer heat to the propellant tanks. The study of these systems for long periods of time in the SPC space tank facilitated the calibration of the electronics and improved inflight monitoring of the spacecraft.[501]

Lewis resolved issues with Centaur's pumping, electronics, shroud, and propellant storage systems without impacting the Centaur schedule. The massive effort paid off on 30 May 1966 with the successful launch of *Surveyor 1*. Three days later, *Surveyor* became the first U.S. spacecraft to land on another celestial body. Centaur rockets sent six more *Surveyors* to the Moon over the next year and a half to scout landing sites for the *Apollo* missions and perform geological research. Lewis had not only directly contributed to the Apollo Program but had proven the capability of a hydrogen-powered rocket and demonstrated the center's proficiency at program

*Image 209: Researchers prepare a Centaur-Surveyor nose cone shroud for a separation test in the SPC vacuum tank (GRC–1964–C–71091).*

management. Henceforth, Lewis was recognized across the Agency for its expertise in launch vehicle operations.[502]

❖ ❖ ❖ ❖ ❖ ❖

Lewis acquired two additional development programs in fall 1962—the Agena upper-stage rocket and the M–1 engine. Originally there was a dearth of open office space to house the personnel for these new programs, so a blockhouse with offices was quickly constructed inside the hangar, similar to what had been done for the Langley transfers in 1942. Meanwhile

*Image 210: Lewis's new DEB.*

Lewis acquired property across Brookpark Road to construct the new DEB to permanently house the staff. The three-story, K-shaped DEB opened in 1964, and the staff transferred from the hangar.[503]

Agena was an upper-stage rocket used in tandem with Thor or Atlas boosters. Unlike Centaur, the Agena was fairly well established before being transferred to Lewis, so it did not require extensive modifications.[504] For each mission, Lewis's Agena Office, led by Seymour Himmel, was responsible for determining the launch vehicle requirements, integrating the payload into the vehicle, and calculating the proper trajectories.[505]

The 17 Agena launches that took place before the program was transferred from Marshall to Lewis resulted in five failures and two partial failures. Lewis oversaw 28 successful Agena missions between 1962 and 1968, including several Ranger and Lunar Orbiter missions to the Moon, a number of Earth observation satellites, and *Mariner* missions to photograph Mars and Venus in 1964 and 1967. *Mariner-D*, launched on 28 November 1964, became the first successful mission to Mars and the first spacecraft to photograph another planet up close.[506] Future Agena launches would be assigned to the more powerful Centaur stage once Centaur's obligations to the Surveyor Program were complete.

The 1.5-million-pound-thrust M–1 engine was the largest hydrogen-fueled rocket engine ever created. The M–1 Program began in 1962 as a joint project between Marshall and Aerojet-General. When NASA transferred the program to Lewis in 1962, the M–1 was being considered for upper-stage vehicles that would be paired with the Saturn rocket for a human mission to Mars in the 1970s. Although much larger in size, the M–1 used liquid-hydrogen technology developed for the RL–10 and J–2.[507] In August 1965, soon after Lewis reanalyzed the configuration to improve thrust, NASA canceled the M–1 program because of budget cuts and the lack of a post-*Apollo* mission. Although the full M–1 engine was never completed, its thrust chamber was the largest ever tested.[508]

*Image 211: Lewis launch team monitors the Thor-Agena launch of an Orbiting Geophysical Observatory satellite in 1965 (GRC–2015–C–60541).*

❖ ❖ ❖ ❖ ❖ ❖

The early 1960s were one of the most productive periods in Lewis's history. The center leadership was critical to the formation of NASA in 1958 and the planning of the human missions of the 1960s. During this period, Lewis had scrambled to quickly refocus its activities on spaceflight, adapting many of the skills and technologies that it had developed for aircraft and missile propulsion to space applications.

The center's fundamental work with liquid hydrogen directly impacted not only Centaur, but also the upper stages for Saturn. Lewis provided key testing for Project Mercury, Centaur, and Saturn. Lewis's electric and nuclear propulsion work also paved the way for more ambitious space missions that would not occur for decades—even some that have not yet taken place. Just as space accomplishments were finally beginning to pay off with the Apollo Program, Lewis began to change course once again.

*Image 212: Bill Harrison films a test analyzing the effect of a lander's jets on simulated Moon dust. This experimental tank was located in the 8×6 complex (GRC–1960–C–53768).*

## Endnotes for Chapter 5

391. Alex Roland, *Model Research* (Washington, DC: NASA SP–4103, 1985).

392. David Portree, *NASA's Origins and the Dawn of the Space Age* (Washington, DC: NASA, 1998).

393. Walter Olson, "A Suggested Policy and Course of Action for NACA on Space Flight," 2 December 1957, NASA Glenn History Collection, Directors Collection, Cleveland, OH.

394. Irving Pinkel, Minutes of meeting of Research Planning Council, 5 December 1957, NASA Glenn History Collection, Directors Collection, Cleveland, OH.

395. Bruce Lundin interview, Cleveland, OH, by John Sloop, 30 May 1974, NASA Glenn History Collection, Oral History Collection, Cleveland, OH.

396. Bruce Lundin interview, Cleveland, OH, by Eugene Emme, 5 June 1974, NASA Glenn History Collection, Oral History Collection, Cleveland, OH.

397. "Doolittle Visits Lewis Lab," *Wing Tips* (3 July 1958).

398. Bruce Lundin, "Some Remarks on a Future Policy and Course of Action for the NACA," 1957, NASA Glenn History Collection, Bruce Lundin Collection, Cleveland, OH.

399. J. Howard Childs, Minutes of December 20, 1957, Meeting of Committee on Space Flight Laboratory, 26 December 1957, NASA Glenn History Collection, Directors Collection, Cleveland, OH.

400. Virginia Dawson, *Engines and Innovation: Lewis Laboratory and American Propulsion Technology* (Washington, DC: NASA SP–4306, 1991).

401. Childs, Minutes of December 20, 1957, Meeting.

402. Hugh Dryden, "A National Research Program for Space Technology," A staff study of the NACA (14 January 1958), *http://history.nasa.gov/SP-4103/app-h4.htm* (accessed 15 April 2015).

403. Roland, *Model Research*.

404. Portree, *NASA's Origins*.

405. Joseph Chambers and Mark Chambers, *Emblems of Exploration: Logos of the NACA and NASA* (Washington, DC: NASA/SP—2015-4556, 2015).

406. "Staff Invited To Submit Designs for NASA Insignia," *Orbit* (30 September 1958), p. 1.

407. Chambers, *Emblems of Exploration*.

408. "This is NASA Insignia," *Orbit* (31 July 1959), p. 1.

409. Chambers, *Emblems of Exploration*.

410. "Lewis Men Now at NASA Headquarters," *Orbit* (9 January 1959).

411. "Biographical Sketch of Abe Silverstein," August 1973, NASA Glenn History Collection, Silverstein Collection, Cleveland, OH.

412. Arnold S. Levine, *Managing NASA in the Apollo Era* (Washington, DC: NASA SP–4102, 1982) chap. 2.

413. Lloyd Swenson, James Grimwood, and Charles Alexander, *This New Ocean: A History of Project Mercury* (Washington, DC: NASA SP–4201, 1966).

414. James Grimwood, *Project Mercury: A Chronology* (Washington, DC: NASA SP–4001, 1963), Part II(A). [Maxime A. Faget, Chief, Flight Systems Division, memo to Project Director, "Status of Test Work Being Conducted at the Lewis Research Center in Conjunction With Project Mercury," 22 October 1959.]

415. Space Task Group, *Project Mercury Preliminary Flight Test Results of the "Big Joe," Mercury R and D Capsule* (Washington, DC: NASA Project Mercury Working Paper No. 107 and NASA TM–X–73017, 1959).

416. Lou Corpas, Frank Stenger, and Robert Miller interview, Cleveland, OH, by Bonita Smith, 17 September 2001, NASA Glenn History Collection, Oral History Collection, Cleveland, OH.

417. Swenson, *This New Ocean*.

418. "It Was Like This at Canaveral," *Orbit* (25 September 1959).

419. George C. Marshall Space Flight Center, *The Mercury-Redstone Project* (Washington, DC: NASA TM–X–53107, 1963), pp. 4–7.

420. Grimwood, *Project Mercury*, Part II(A). [Faget memo to Project Director.]

421. McDonnell Aircraft Corporation, *Project Mercury Familiarization Manual* (St. Louis, MO: McDonnell SEDR–104, 1959), pp. 6–13 and 5–12.

422. James Grimwood, *Project Mercury: A Chronology* (Washington, DC: NASA SP–4001, 1963), Part II(A). [George Low memo to NASA Administrator, "Status Report No. 14, Project Mercury," 22 May 1959.]

423. NASA Space Task Group, "Project Mercury (Quarterly) Status Report No. 6 for Period Ending April 30, 1960," 1960.

424. James Grimwood, *Project Mercury: A Chronology* (Washington, DC: NASA SP–4001, 1963), Part II(B). [NASA Space Task Group, Project Mercury [Quarterly] Status Report No. 6 for Period Ending April 30, 1960.]

425. Swenson, *This New Ocean*.

426. Swenson, *This New Ocean*.

427. James Useller and Joseph Algranti, "Pilot Reaction to High Speed Rotation," *Aerospace Medicine* 34, no. 6 (June 1963).

428. Corpas, Stenger, and Miller interview by Smith.

429. Neal Thompson, *Light This Candle: The Life and Times of Alan Shepard America's First Spaceman* (New York, NY: Crown Publishers, 2004).

430. James Useller and Joseph Algranti, "Pilot Reaction to High Speed Rotation," *Aerospace Medicine* 34, no. 6 (June 1963).

431. Kim McQuaid, "The Space Age at the Grass Roots: NASA in Cleveland, 1958–1990," 2007, NASA Glenn History Collection, Cleveland, OH, pp. 12–14.

432. McQuaid, "The Space Age." November to December 1962.

433. McQuaid, "The Space Age at the Grass Roots."

434. "Spacemobiles Tell NASA Story to Many," *Lewis News* (24 November 1967).

435. Dr. Walter T. Olson, Vitae, (Glenn History Collection, Walter Olson Collection, Cleveland, OH).

436. Calvin Weiss interview by Rebecca Wright, 6 June 2014, NASA Glenn History Collection, Oral History Collection, Cleveland, OH.

437. Billy Harrison and Nicki Harrison interview, Cleveland, OH, 14 October 2005, NASA Glenn History Collection, Oral History Collection, Cleveland, OH.

438. John Sloop, *Liquid Hydrogen as a Propulsion Fuel, 1945–1959* (Washington, DC: NASA SP–4404, 1977).

439. A. O. Tischler interview, Cleveland, OH, by John Sloop, 25 January 1974, NASA Glenn History Collection, Oral History Collection, Cleveland, OH.

440. David Akens, *Saturn Illustrated Chronology* (Huntsville, AL: NASA MHR–5, 1971).

441. Saturn Vehicle Team, "Report to the Administrator, NASA, on Saturn Development Plan," December 15, 1959, pp. 1–4, 7–9.

442. Akens, *Saturn Illustrated Chronology.*

443. Akens, *Saturn Illustrated Chronology.*

444. Robert Wasko, *Wind Tunnel Investigation of Thermal and Pressure Environments in the Base of the Saturn S-IC Booster from Mach 0.1 to 3.5* (Washington, DC: NASA TN D–3612, 1966).

445. Berhard J. Blaha, Robert A. Wasko, and Donald L. Bresnahan, *Wind Tunnel Investigation of Saturn S-IC Aerodynamic Engine Gimbal Forces and Base Pressures Using a Cold-Flow Jet Simulation Technique* (Washington, DC: NASA TM X–1470: 1967).

446. John F. Kennedy, "Special Message to the Congress on Urgent National Needs," 25 May 1961.

447. Billy Harrison and Nicki Harrison interview, Cleveland, OH, 14 October 2005, NASA Glenn History Collection, Oral History Collection, Cleveland, OH.

448. "Personnel Summary," *1997 NASA Pocket Statistics* (Washington, DC: NASA History Division, 2007).

449. Raymond Bisplinhoff, "Summary of Remarks by R. L. Bisplinghoff, NASA-OART Management Meeting," 9 December 1964, NASA Glenn History Collection, John Sloop Collection, Cleveland, OH.

450. Eugene Manganiello to William Woodward, "Justification for Energy Conversion Building," 28 October 1959, NASA Glenn History Collection, Directors Collection, Cleveland, OH.

451. Robert Sessions memorandum for those concerned, 1 July 1959, NASA Glenn History Collection, Buildings and Grounds Collection, Cleveland, OH.

452. "Ohio Historic Preservation Office: Resource Protection and Review" Section 106 Review, Project Summary Form (Columbus, OH: Ohio Historic Preservation Office, 2007).

453. "Missile Lab Here Getting 2 Buildings," *Cleveland Plain Dealer* (4 September 1957).

454. Joy Owens, "NASA Asks $12 Million for Plum Brook Expansion," *Elyria Chronicle-Telegram* (26 May 1962).

455. Department of the Army to the National Aeronautics and Space Administration, "Transfer and Acceptance of Military Real Property," 15 March 1963, NASA Glenn History Office, Plum Brook Ordnance Works Collection, Cleveland, OH.

456. Eugene Manganiello memo to I. H. Abbott and E. W. Conlon at NASA Headquarters, "Curtailment of Research Areas and Facilities for Fiscal Year 1962," 20 January 1960, NASA Lewis Research Center, Cleveland, OH.

457. R. T. Hollingsworth, *A Survey of Large Space Chambers* (Washington, DC: NASA TN D–1673, 1963), p. 12.

458. John Povolny, *Space Simulation and Full-Scale Testing in a Converted Facility* (Washington, DC: NASA N66 35913, 1970), p. 2.

459. Wolfgang Moeckel, *Electric Propulsion for Spacecraft* (Washington, DC: NASA SP–22, 1962).

460. Edgar Choueri, *A Critical History of Electric Propulsion: The First Fifty Years (1906–1956)* (Reston, VA: AIAA–2004–3334, 2004).

461. Ernst Stuhlinger, "Possibilities of Electrical Space Ship Propulsion," 5th International Astronautical Congress (Innsbruck: 1954), pp. 100–119.

462. Mike Wright, "Ion Propulsion: Over 50 Years in the Making," Science@NASA (6 April 1999), *http://science.nasa.gov/science-news/science-at-nasa/1999/prop06apr99_2/* (accessed 1 April 2015).

463. Wright, "Ion Propulsion," *http://science.nasa.gov/ science-news/science-at-nasa/1999/prop06apr99_2/*

464. Wolfgang Moeckel to Ernst Stuhlinger, 2 June 1964, NASA Glenn History Collection, Electric Propulsion Collection, Cleveland, OH.

465. "Organizational Changes and Personnel Appointments," *Orbit* (12 August 1960).

466. Wolfgang Moeckel, "Lewis Research Center Electric Propulsion Program," 5 June 1961, NASA Glenn History Collection, Electric Propulsion Collection, Cleveland, OH.

467. *Glenn Research Center Test Facilities* (Cleveland, OH: NASA B–1128, August 2005).

468. Wright, "Ion Propulsion," *http://science.nasa.gov/ science-news/science-at-nasa/1999/prop06apr99_2/*

469. Hugh Harris, NASA Lewis Research Center press release 70–74, 16 September 1970, NASA Glenn History Collection, Cleveland, OH.

470. Michael Patterson and James Sovey, "History of Electric Propulsion at NASA Glenn Research Center: 1956 to Present," *Journal of Aerospace Engineering* 26 (2013): 300–316.

471. Radio Corporation of America, *Summary Report on the Development of the SERI I Spacecraft* (Washington, DC: NASA CR–54243, 1966).

472. Frank Rom interview, Cleveland, OH, by Virginia Dawson, 14 June 2002, NASA Glenn History Collection, Oral History Collection, Cleveland, OH.

473. William Corliss and Francis Schwenk, *Nuclear Propulsion for Space* (Oak Ridge, TN: Atomic Energy Commission, 1971).

474. Corliss, *Nuclear Propulsion*.

475. James Dewar, To the End of the Solar System: *The Story of the Nuclear Rocket* (Lexington, KY: University Press of Kentucky, 2004).

476. "NERVA," *Flight International* (9 August 1962).

477. Herbert Heppler, et al., *Startup Dynamics and Control* (Washington, DC: NASA X66–51410, 1966).

478. Wojciech Rostafinski, *An Analytical Method for Predicting the Performance of Centrifugal Pumps During Pressurized Startup* (Washington, DC: NASA TN D–4967, 1969).

479. John Clark, *Analytical and Experimental Study of Startup Characteristics of a Full-Scale Unfueled Nuclear Rocket Core Assembly* (Washington, DC: NASA TM X–1231, 1966).

480. David Robinson, et al., *Flow Initiation with Turbine Accelerated Tank Pressure 35 PSIA; Runs 12, 14–17* (Washington, DC: NASA TM X–52089, 1965).

481. James Dewar, *To the End of the Solar System: The Story of the Nuclear Rocket* (Ontario, Canada: Apogee Books, 2007).

482. Lewis Research Center, *Facilities and Techniques Employed at Lewis Research Center in Experimental Investigations of Cavitation in Pump* (Cleveland, OH: NASA TP 19–63, 1964).

483. Richard Priem, *Liquid Rocket Technology for the Chemical Engineer* (Washington, DC: NASA TM X–52695, 1969).

484. Lewis Research Center, "Capabilities and Facilities of the Plum Brook Station," c1970, NASA Glenn History Collection, Plum Brook Collection, Cleveland, OH.

485. Lightweight, *Self-Evacuated Insulation Panels* (Washington, DC: NASA Tech Brief 70–10646, 1970).

486. Edward Otto, *Static and Dynamic Behavior of the Liquid-Vapor Interface During Weightlessness* (Washington, DC: NASA TM X–52016, 1964).

487. James Useller and Fred Haise, Jr., *Use of Aircraft for Zero Gravity Environment* (Washington, DC: NASA TN D–3380, 1966).

488. Donald Petrash, Thomas Nelson, and Edward Otto, *Effect of Surface Energy on the Liquid-Vapor Interface Configuration During Weightlessness* (Washington, DC: NASA TN D–1582, 1963).

489. David Chato, *The Role of Flight Experiments in the Development of Cryogenic Fluid Management Technologies* (Washington, DC: NASA/TM—2006-214261, 2006).

490. Kaleel L. Abdalla, Thomas C. Frysinger, and Charles R. Andraechio, *Pressure-Rise Characteristics for a Liquid-Hydrogen Dewar for Homogeneous Normal Graivty Quiescent, and Zero Gravity Tests* (Washington, DC: NASA TM X–1134, 1965).

491. Virginia Dawson and Mark Bowles, *Taming Liquid Hydrogen: The Centaur Upper Stage Rocket 1958–2002* (Washington, DC: NASA SP 2004–4230, 2004).

492. William E. Conrad, Ned P. Hannum, and Harry E. Bloomer, *Photographic Study of Liquid-Oxygen Boiling and Gas Injection in the Injector of a Chugging Rocket Engine* (Washington, DC: NASA TM X–948, 1964).

493. Brian Sparks to Homer E. Newell, 21 September 1962, NASA Glenn History Collection, Centaur Collection, Cleveland, OH.

494. Edgar Cortright to Abe Silverstein, "Re: Project Centaur," 20 September 1962, NASA Glenn History Collection, Centaur Collection, Cleveland, OH.

495. Edgar M. Cortright interview, Houston, TX, by Rich Dinkel, 20 August 1998, NASA Johnson Space Center Oral History Project, Houston, TX.

496. Rocket Systems Division Reports, Dynamics Laboratory, October 1963.

497. Rocket Systems Division Reports.

498. Jack Humphreys, *Centaur AC-4 Nose Jettison Tests* (Washington, DC: NASA TM–X–52154, 1966).

499. Raymond Lacovic, et al., *Management of Cryogenic Propellants in a Full-Scale Orbiting Space Vehicle* (Cleveland, OH: NASA TN D–4571, 1968).

500. William Groesbeck, *Design of Coast-Phase Propellant Management System for Two-Burn Atlas-Centaur Flight AC-8* (Washington, DC: NASA TM–X–1318, 1966), p. 2.

501. Ralph Schmiedlin, et al., *Flight Simulation Tests of a Centaur Vehicle in a Space Chamber* (Washington, DC: NASA TM–X–1929, 1970).

502. Robert C. Seamons to Abe Silverstein, "Atlas Centaur Success. The Perfection of the Centaur AC-10," 22 October 1962, NASA Glenn History Collection, Cleveland, OH.

503. "Moving Days at DEB," *Lewis News* (5 June 1964).

504. "Agena: Fact Sheet for FY 1963 Congressional Presentations," 1963, NASA Glenn History Collection, Agena Collection, Cleveland, OH.

505. Public Information Office, "The Agena Project," February 1967, NASA Glenn History Collection, Agena Collection, Cleveland, OH.

506. Public Information Office, "The Agena Project."

507. Walter Dankhoff, *The M-1 Rocket Engine Project* (Washington, DC: NASA TM X–20854, 1963).

508. Aerojet-General Corporation, *Development of a 1,500,000-Lb-Thrust Liquid Hydrogen/Liquid Oxygen Engine* (Washington, DC: NASA N68–15861, 30 August 1967).

Image 213: The new Space Power Facility opened in late 1969. It was one of three world-class facilities brought online at Plum Brook Station (GRC–1969–C–03156).

# 6. Limitless Opportunities

*"I am firmly convinced that without the staff that NACA had in its three research centers that we would never have been able to put a man on the moon."*

—Jesse Hall

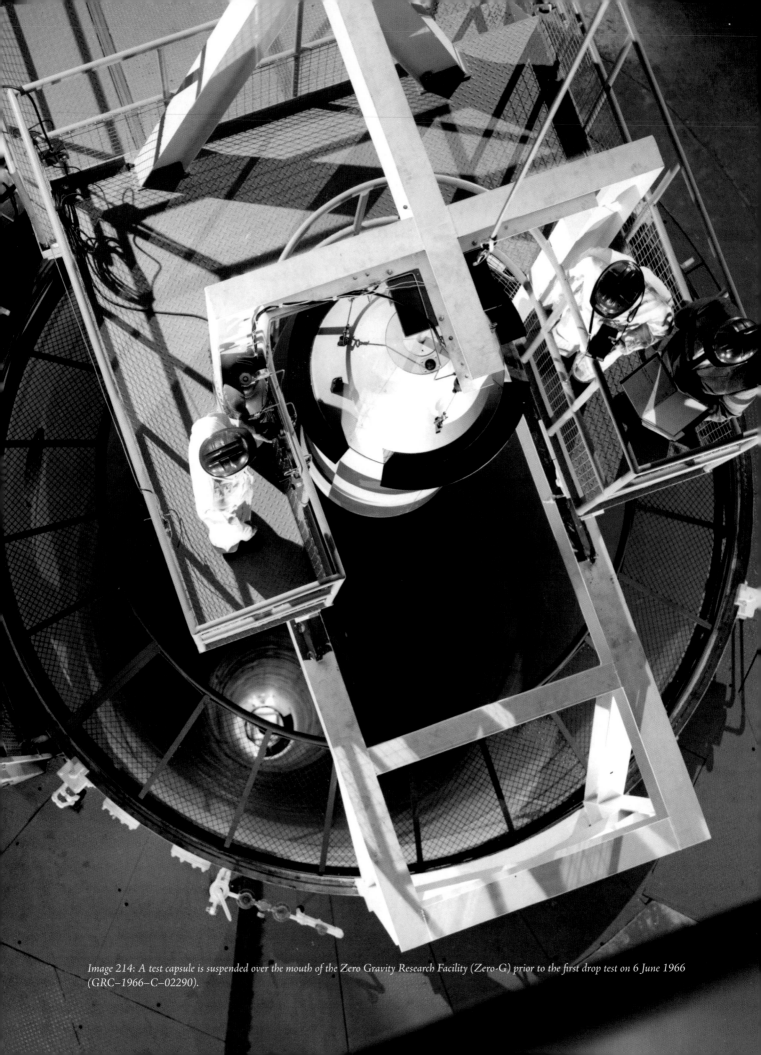

*Image 214: A test capsule is suspended over the mouth of the Zero Gravity Research Facility (Zero-G) prior to the first drop test on 6 June 1966 (GRC–1966–C–02290).*

# Limitless Opportunities

The successful launch of *Surveyor 1* on an Atlas-Centaur in May 1966 brought a sense of closure to the Lewis Research Center's pressure-filled first decade of space activities. The landing coincided with the center's 25th anniversary. The nation was experiencing growing civil rights unrest, escalation of the Vietnam War, and increasing poverty. With *Apollo's* final sprint to the Moon, NASA distanced itself both literally and figuratively from the nation's turmoil. With the exception of the shocking *Apollo 1* accident in January 1967, NASA's achievements in the late-1960s provided the nation with an unrivaled source of accomplishment. Although the Soviet Union had achieved another round of space firsts in 1966, the untimely death of program chief Sergei Korolev in January essentially ended any Soviet race to the Moon. *Apollo 8's* successful orbiting of the Moon in December 1968 quickly led to the *Apollo 11* landing the following July. NASA's achievements were a unifying source of pride and excitement across the nation and the globe.

With the initial successes of the *Surveyor* Program in 1966, Lewis's contributions to the *Apollo* Program were mostly complete. Just as the public was beginning to witness the *Apollo* milestones, Lewis redirected a large portion of its efforts and plunged determinedly into aviation. A host of new aeronautical challenges had arisen during the previous decade, including attempts at the long-awaited supersonic transport (SST) vehicle, the need for quieter and more efficient engines, and an increased desire for vertical and short takeoff aircraft. Although Lewis continued to manage Centaur and pursue space power and propulsion alternatives, the center increasingly focused on aeronautics.

Lewis had a diversified research portfolio, an experienced 5,000-person staff, and a battery of facilities at its disposal.[509] Nothing seemed impossible. While the nation was beset by turmoil, Lewis was flush with a sense of confidence that had not been present in years. This manifested itself in the construction of several of the center's marquee facilities: the Zero-G, the Spacecraft Research Propulsion Facility, the Space Power Facility (SPF), the Propulsion Systems Laboratory (PSL) No. 3 and 4, and the 9- by 15-Foot Subsonic Wind Tunnel.

## Zero Gravity Demonstration

Events were held throughout 1966 to commemorate Lewis's 25 years of research—including a city-sponsored Lewis Day on 23 January 1966 and a three-day Inspection in October. The Inspection, Lewis's first since 1957, drew 2,000 business, industry, and government executives and included an employee open house. The visitors witnessed presentations at the major facilities; viewed the *Gemini VII* spacecraft, a Centaur rocket, and other displays in the hangar; and witnessed what a press release referred to as "a dramatic demonstration in Lewis's 500-foot-deep Zero-G during the first day...."[510,511]

The Zero-G was a new drop tower that provided researchers with 5.2 seconds of microgravity. It had over twice the capability of the center's 2.2-Second Drop Tower and was the largest facility of its kind in the world. Lewis engineers began planning the facility

*Image 215: Zero-G drop preparations in September 1966 (GRC–1966–C–03685).*

in the early 1960s as researchers struggled to understand the effects of microgravity on liquid hydrogen. The construction process, which began in April 1964, was similar to that used to build missile silos: A hole was excavated in 25-foot segments, then lined with concrete. After that, a steel vacuum chamber was built inside the completed shaft to prevent any air resistance from impeding the freefall.[512,513]

The experiments were enclosed in a 2,500-pound experiment package that was suspended over the hole then released. Special test equipment and high-speed cameras allowed researchers to film and take measurements of the experiment as it fell, and a 15-foot-deep canister of special polystyrene pellets at the base of the shaft slowed and stopped the test package. Determining the proper type of pellets to use proved to be one of the most troubling aspects of the facility's design. Engineers tested different samples right up until the facility's completion in 1966. They conducted the tests at the Space Power Chambers (SPC), the Nuclear Rocket Dynamics and Control Facility (B–3) test stand at Plum Brook, and even in the elevator shaft of the Terminal Tower, Cleveland's tallest building at the time.[514]

The center sought to show off the brand new Zero-G at the 1966 Inspection. During the first day, the operators realized that they did not have enough of the foam pellets to continue the planned demonstrations, so the center ordered a new load from the manufacturer and installed them for the second day of the Inspection. A group of visitors gathered around the shaft the next morning and watched the staff drop the experiment package down the shaft. Black smoke began pouring out of the hole and soon covered all the equipment in the facility. The engineers had not given the pellets enough time to go through the required off-gassing period. The fire was easily contained, but embarrassed Lewis officials had to quickly usher the tour group out and get all available hands to help wipe down the sooty surfaces in the facility. The tours resumed shortly thereafter without any additional demonstration drops.[515]

Although the issues concerning the behavior of liquid hydrogen in space had largely been resolved by the time that the Zero-G was completed, researchers used the facility to study fluid pumping, heat transfer, and sloshing.[516] The Zero-G enabled researchers to study fire dynamics and fluid motion much more closely than previously possible.

## Reappearance of Aeronautics

After reaching its peak in 1966, NASA's funding began subsiding. The rush toward the Moon was well underway, and it was becoming evident that that level of support would not be forthcoming in the future. Most of the urgent problems regarding the early space program had been resolved, and plans for more advanced missions began fading as fiscal realities set in. NASA was just beginning to analyze design concepts for what would become the space shuttle, and Lewis was not directly involved.

Meanwhile the Federal Aviation Administration (FAA) was seeking assistance with a host of aeronautical issues brought on by the advent of the jet airliner in the late 1950s and early 1960s. The jet engine transformed commercial aviation—enabling transcontinental flights and lowering passenger expenses. The jet also introduced concerns about engine noise and

*Image 216: Construction of the 50-foot-diameter cooler for PSL No. 3 and 4. The cooler contained 5,500 water-fed cooling tubes and three banks of spray nozzles to reduce the engine exhaust temperature from 3,500 to 120 degrees Fahrenheit (GRC–1969–C–03898).*

*Image 217: Pilots Clifford Crabs and Byron Batthaer with Lewis's F–106B Delta Dart. The modified fighter aircraft could record 480 pressure measurements and nearly 100 other flight characteristics every 11 seconds (GRC–1971–C–00847).*

pollution as airport congestion increased. There also was interest in new SST and vertical or short-takeoff and landing (V/STOL) aircraft, which required more efficient engines than those currently available. In addition, engine manufacturers were developing engines that would create much higher turbine temperatures than previous models.[517] In the mid-1960s the FAA asked NASA's three former NACA centers—Ames, Langley, and Lewis—to address these and other related problems.

Abe Silverstein's redirection of the center in 1966 included the addition of the Airbreathing Engine Division, the acquisition of a F–106B Delta Dart aircraft, the shift from rocket to aircraft engine testing in the PSL, and the initiation of design work for two new, more powerful PSL test chambers.[518] Silverstein added three new aeronautics divisions in 1968.[519]

Unlike the center's groundbreaking military-driven engine work in the 1940s and 1950s, these new studies also included civilian applications. Lewis would never again forsake its aeronautics base.

❖ ❖ ❖ ❖ ❖ ❖

The major breakthroughs in propulsion during the 1940s and 1950s led many to believe that supersonic airlines were just over the horizon. In response to new programs by a French/British partnership and the Soviet Union, President John F. Kennedy initiated the National Supersonic Transport Program in 1963. The government selected Boeing to design the airframe and General Electric to develop the engines for a vehicle that would surpass the European efforts.[520] Lewis engineers served as advisors to the U.S. program for several years.

*Image 218: Lewis researchers take ground-based noise measurements from the F–106B (GRC–1971–C–00775).*

The two critical issues were drag and noise. Supersonic aircraft burn a large quantity of fuel while accelerating to their cruising speed, so decreasing aerodynamic drag was imperative; and significant noise reductions would have to be demonstrated before the FAA would certify the vehicle.[521] Lewis partnered with General Electric under the Department of Transportation's Supersonic Transport Program to conduct aerodynamics and noise investigations. The former sought to investigate how changing pressure levels affected inlet performance and airflow over the engine nacelle.[522] The latter analyzed different types of noise suppressors, compared static noise levels with those in flight, and assessed the effect of velocity on nozzle shapes.[523]

Lewis acquired the F–106B in October 1966 to facilitate this research. Then mechanics and engineers went through an extensive effort to remove the weapons systems and install research equipment. The most obvious modification was the addition of two General Electric J85 turbojets underneath the aircraft's wings—one was a standard model and the other incorporated a series of experimental inlets and nozzles.[524] The delta-winged F–106B with the additional engines underneath roughly simulated the shape of the SSTs. Flight testing was critical because wind tunnel investigations needed to use small models to avoid wall interference.[525]

Lewis and General Electric developed different advanced inlet and nozzle configurations and installed them on the J85 engine. They tested the J85 extensively in the PSL and the 8- by 6-Foot Supersonic Wind Tunnel (8×6). The modified engine was then installed underneath the F–106B and flight-tested in a 200-mile flight corridor over Lake Erie.[526]

Over the course of several years, Fred Wilcox and his colleagues used the aircraft to analyze noise and drag levels for the various configurations. They investigated various types of nozzles and found that mounting the nozzle directly underneath the wing reduced drag.[527] For the noise investigations, microphones were set up on the tarmac. The aircraft first performed static engine tests on the ground, then flew low-altitude passes to determine which nozzles yielded the least noise.[528] Wilcox and his team found that the experimental nozzles did not prove as effective as hoped for.[529]

The National Supersonic Transport Program was not progressing well, and public opposition to the aircraft's noise and pollution levels grew. The program was canceled in 1971, but Lewis completed the planned tests. Meanwhile the Soviets and Europeans flew their SSTs.[530] The F–106B executed 300 test flights during its almost nine-year detail to Lewis and led to 45 technical reports.[531] A second F–106 aircraft that was acquired during this period to serve as a chase plane went on to perform a variety of other research flights for the center.

❖ ❖ ❖ ❖ ❖ ❖

The sharp increase in commercial flights during the late 1950s and early 1960s led to airport congestion, which exacerbated aircraft pollution and noise. The noise issue was such a concern that President Lyndon Johnson personally called on federal agencies to develop abatement programs.[532] During this period of airliner expansion, engine manufacturers developed a new type of engine: the turbofan. The turbofan's exhaust speed was lower than that of turbojets. This provided better fuel efficiency and reduced exhaust noise. The fan and the other internal machinery of turbofans, however, generated higher levels of mechanical noise than turbojets did. Lewis researchers, who had conducted engine noise-reduction studies in the mid-1950s, took up the effort again in the late 1960s. They developed several experimental options for redesigning turbofans to reduce the fan noise.[533]

NASA initiated the Quiet Engine Program in 1969 to demonstrate that it was possible to incorporate noise-reduction technologies into the types of engines typically used on Boeing 707 or McDonnell-Douglas DC–8 airliners without diminishing engine performance. The effort was aided by the new low-bypass-ratio turbofan engines, the emergence of new acoustic technologies, and improved understanding of how an engine fan creates noise.[534]

Also in 1969 Lewis contracted with General Electric to build and aerodynamically test three experimental engines.[535] Engineers reduced the speed and pressure of the fan, and incorporated acoustic treatments in the engine cover, or nacelle.[536] The engines were then brought to Lewis in December 1971 for testing. To accommodate the engine, Lewis engineers built an outdoor engine stand, the Fan Noise Test Facility, off of the Main Compressor and Drive Building of the 10- by 10-Foot Supersonic Wind Tunnel.

Carl Ciepluch and his colleagues ran the engine at different speeds with microphones set up 10 feet away. The studies revealed that not only did the untreated version of the General Electric engines generate less noise than was anticipated, but the modified nacelle reduced engine noise substantially. Lewis researchers then verified the engine performance in simulated altitude conditions in the PSL. The Quiet Engine Program engines proved to be significantly quieter than any contemporary commercial engine.[537] Lewis hosted an Aircraft Noise Reduction Conference in May 1972 to share the results of the Quiet Engine Program with industry,[538] and NASA and industry used the resulting technology in future noise reduction programs.

*Image 219: A General Electric Quiet Engine tested with Lewis's acoustically treated nacelle on the hangar apron (GRC–1972–C–01486).*

## New Materials

Engines are complex machines that operate in a variety of environments, and it is essential to develop specialized materials for specific components or applications. Temperature, strength, and weight are the primary concerns, but reliability, durability, flammability, maintainability, and cost-effectiveness are also important. Lewis's Materials and Structures Division increased its efforts to address these concerns in the 1960s. The center built several new facilities to accommodate this research, including the Materials Processing Laboratory and the Materials and Structures Laboratory.[539] Lewis researchers investigated the reliability of materials based on both their likelihood to fracture under certain conditions and their physical breakdown over time. The researchers subjected materials to stresses or temperature fluctuations and then analyzed their physical properties to predict behavior.

In the 1960s Lewis developed several key steel alloys for advanced jet engines. These materials had to be heat resistant, formable, and affordable. Lewis alloys from this period include Tungsten RHC, which had higher strength at temperatures over 3,500 degrees Fahrenheit than any other metal, a cobalt-tungsten alloy that combined strength at high temperatures and magnetic properties, and a nickel-tungsten alloy that could withstand high temperatures and repeated temperature fluctuations.[540,541] The center also began research in ceramic materials, which were lighter than steel and could withstand higher temperatures.[542]

❖ ❖ ❖ ❖ ❖ ❖

During this time, the researchers also were using the Plum Brook Reactor Facility (PBRF), NASA's only nuclear reactor, to study the effects of radiation on materials. Although this was primarily in support of the nuclear rocket effort, there were other applications as well. The 60-megawatt test reactor went critical for the first time in 1961 and began its full-power research operations in 1963. Over the next decade, the reactor performed some of the nation's most advanced nuclear research and proved itself as both a safe and efficient test facility.[543]

The reactor core, where the chain reaction occurred, sat at the bottom of a tubular pressure vessel. The core and its fuel rods with uranium isotopes were surrounded by deep pools of water that cooled the reactor and blocked any escaping radiation. The test articles were generally tiny samples of materials that were encapsulated in small shuttles referred to as "rabbits." The rabbits were pneumatically pushed into one of the reactor core's 44 test locations. After a predetermined amount of time, the samples were removed and remotely analyzed in thickly shielded test cells to determine changes in their physical properties.[544,545]

*Image 220: Pouring of a nickel alloy at Lewis's Technical Services Building in April 1966 (GRC–1966–C–01563).*

The PBRF was unique in its ability to subject test specimens to radiation and cryogenic temperatures simultaneously. This was essential for the study of materials for nuclear rockets that would use liquid hydrogen. The reactor's cryogenic test facility was large enough to test small mechanical components and instrumentation devices. Researchers used the PBRF to support studies from Lewis, Westinghouse, and Lockheed for the Nuclear Engine for Rocket Vehicle Application (NERVA), Systems for Nuclear Auxiliary Power–8 (SNAP–8), and other programs.[546,547]

The reactor also was used for nonaerospace applications. Dean Schiebley led an effort for the Environmental Protection Agency in 1972 to use neutron activation to determine the presence of pollutants in fuel, air, and vegetation. The studies were part of a larger effort to identify and catalog local pollutants. The reactor also tested corn samples for the Department of the Interior to help identify the nutritional value of crow-pecked corn, which seemed to grow larger than unpecked corn.[548]

## Interactive Computing

Lewis's computer systems made significant improvements in the 1960s through the introduction of transistors and interactive computing. During the center's 1959 reorganization, Lewis merged the computing personnel that had been assigned to the research divisions into the new Instrument and Computing Division. This group included data processing, mechanical computations and analysis, machine computing, and instrument systems.

The development of the transistor in the 1950s revolutionized computer systems. The initial electronic computers used vacuum tubes—which were

Image 221: *Plum Brook Station manager Alan "Hap" Johnson and reactor chief Brock Barkley examine Moon dust that had been irradiated in the PBRF to identify its composition. The 25 milligrams of lunar soil had been retrieved by the Soviet Luna XVI spacecraft in September 1970 (GRC–1970–C–00950).*[549]

Image 222: *The reactor was submerged in deionized water to assist with the cooling, and its core emitted a blue glow known as Cherenkov radiation, c1962 (GRC–1996–C–03983).*

Image 223: Betty Jo Armstead [nee Moore] monitors an IBM 1403 high-speed printer in February 1964. This was linked to the IBM 7094/7044 Direct Couple System (GRC–1964–C–68508).

larger, hotter, more expensive, and less reliable than transistors. IBM introduced the first line of affordable transistor computers, the 1401, in late 1959 to an overwhelming response. In 1962 Lewis purchased two IBM 1401s for business applications.[550]

In 1962, Lewis also obtained the transistor-based IBM 7090 for analytical processing. The performance of the 7090 was on par with its vacuum-tube predecessor, but by 1964 Lewis had purchased the 7094, which handled twice as much data. The 7094 was coupled with a 7044, permitting flexible job scheduling for the first time. Lewis acquired a second 7094 in 1968.[551]

The next big phase was interactive computing, which allowed a user to make changes to the computer program as it was running. Lewis acquired an IBM 360 in 1966 for interactive and traditional analytical and data processing applications. The system had two central

processing units (CPUs) that could handle 79 interactive, graphic, and communication lines simultaneously. Researchers also could use the 360 for computer modeling work.[552]

Lewis upgraded its centralized data recording capabilities in 1968 with the addition of the CADDE II system. The staff used the system to record data from 63 different Lewis facilities. Unlike the original CADDE, CADDE II could record test data from up to three facilities simultaneously. The new system was connected to the IBM 360, resulting in almost real time computational performance. The center utilized the 1401s, 7094s, and CADDE II until the mid-1970s.[553]

## Lewis Reaches Out

In the mid-1960s the Lewis staff was larger than at any other time in Lewis's history. The core group of NACA veterans and their families had literally grown up together at the lab. Many had relocated to Cleveland, and coworkers became friends and, in some instances, spouses. This cycle repeated itself with the influx of new personnel in the early 1960s. The new hires were primarily young people recently graduated from universities. Many moved to Cleveland and were beginning families. The veterans recognized the importance of the tight-knit Lewis community to the overall performance of the center and sought to include the younger generation.

Image 224: Patricia Coles, center, is named Miss NASA Lewis at the 1971 annual picnic. Coles, the daughter of renowned Lewis magneticism researcher Willard Coles, worked in the center's Personnel Division (GRC–2015–C–06554).

Image 225: *Lewis employees enjoying themselves at the 1971 annual picnic at the picnic grounds (GRC–2015–C–06567).*

The center created the Lewis Social Activities Committee (LeSAC) in September 1963 to bring the two demographic groups together. Director Abe Silverstein appointed key personnel from each division to ensure wide participation in LeSAC. The committee devised a variety of events, including mixers, nights out on the town, picnics, formal dances, and theme parties. All assistance—from photographers and graphic artists, to carpenters and fabricators, to those procuring bands or food and beverage—was volunteered. The *Lewis News* quoted one volunteer in 1964 as saying

that the Friday night preparations were sometimes more fun than the Saturday event.[554]

LeSAC instituted the Miss NASA Lewis competition at the 1964 annual picnic. Contestants from various offices around the center vied for trophy and crown. Duties primarily involved participation at the center's charitable functions and other LeSAC events. Over the years, winners emerged from the Chemical Rocket, Personnel, and Procurement divisions and the library. The LeSAC activities and Miss NASA Lewis competition ceased in the mid-1970s.

The center also channeled its abundant resources into efforts for the local community. Lewis hosted programs that provided an opportunity for underprivileged youths to learn science and engineering skills, Christmas events for needy children, and regular activities at a newly established Boy Scout Explorer Post.

Beginning in 1963 Lewis and Case Institute of Technology cosponsored an annual 10-week fellowship for approximately two dozen science and engineering faculty members. The faculty attended lectures by the Lewis staff and pursued research in their fields of interest. By 1967 five other NASA centers had taken Lewis's lead and extended their own fellowships.[555] Lewis also hosted a Youth Science Congress

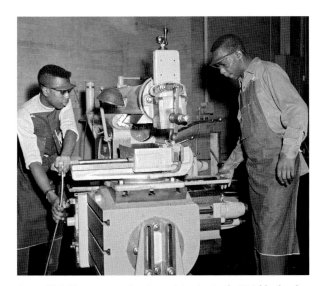

Image 226: *Young men at Lewis participating in the Neighborhood Youth Corps in January 1967 (GRC–1967–C–00640).*

Image 227: *Members of the Boy Scout Explorer Post in the cockpit of Lewis's North American AJ–2 Savage in 1968 (GRC–2015–C–06569).*

*Image 228: Over 500 people attended Lewis's Conference on Aerospace Related Technology for Industry in Commerce in May 1967 (GRC–1967–C–01813).*

for 20 exceptional Midwest high-school students. The students were able to talk with researchers and tour facilities while gaining experience in writing and presenting research reports.[556] The Lewis Speakers Bureau remained active throughout this period with new topics relating to civilian air travel and interplanetary space missions.

Lewis instituted a technology utilization program in the mid-1960s to demonstrate possible terrestrial applications of NASA's technological breakthroughs. As a civilian agency, NASA considered the transfer of technology to industry to be one of its primary missions. The transmission of technology had to be rapid in order for it to be relevant to the private sector. Lewis issued Special Publications and Tech Briefs regarding specific innovations. To facilitate the dissemination of information, Walter Olson, Lewis Director of Public Affairs, organized a series of annual conferences in the mid-1960s to share NASA research with particular fields—including the steel industry in 1964, the petroleum industry in 1965, small businesses in 1967, the electric power industry in 1968, and aerospace materials manufacturers in 1969. These events drew hundreds of guests and often included talks by NASA administrators.

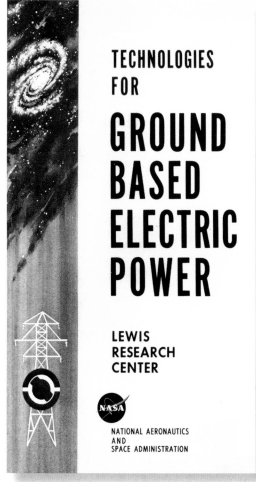

**TECHNOLOGIES FOR GROUND BASED ELECTRIC POWER**

**LEWIS RESEARCH CENTER**

NATIONAL AERONAUTICS AND SPACE ADMINISTRATION

*Image 229: Pamphlet from a Lewis conference for the power industry.*

## A New Generation of Space Missions

Lewis's space propulsion work in the late 1960s was dominated by two very different efforts—the liquid-hydrogen-fueled Centaur rocket and the electron bombardment thruster. Both had already demonstrated their performance in space but were now embarking on new, more complex endeavors. Atlas-Centaur began a new career orbiting large satellites and sending probes across the universe; and the ion thruster powered a new *Space Electric Rocket Test II (SERT–II)* vehicle, which produced unprecedented electric propulsion performance.

Lewis's launch vehicles team and the Centaur vehicle had proven themselves with the *Surveyor* missions in the mid-1960s. Lewis personnel were recognized as experts in the launch business, and Centaur was now the nation's most powerful space tug.[557] As such, NASA phased out its use of the Agena rocket. Scheduled Agena missions such as the Advanced Test Satellite, Orbiting Astronomical Observatory (OAO), Intelsat, *Mariner*, and *Pioneer* were assigned to Centaur.[558]

Lewis created the Launch Vehicles Division in 1969 to address all Atlas-Centaur issues, including integration of the payload into the launch vehicle, establishment of the correct flight trajectories, and preparation of the vehicle for launch. The staff worked closely with the launch crews at the NASA Kennedy Space Center, the spacecraft designers from other NASA centers, and eventually commercial and military organizations. Lewis also worked with General Dynamics to update Centaur with a new computer system and an equipment module to house the new electronics.[559]

In August 1968 Centaur's first post-*Surveyor* attempt failed when the vehicle's engines failed to restart.[560] This added to the pressure regarding the launch of OAO–2 four months later. OAO–2 was a space telescope designed to operate above atmospheric distractions. At 4,436-pounds, it was Centaur's largest payload to date, and its unprecedented size forced Lewis engineers to use a much longer Agena shroud on the Atlas-Centaur. The Agena shroud and its jettison system were combined with the cylindrical section of the Centaur shroud and a new interstage adapter. Lewis verified the performance of the new system with full-scale tests in the SPC's altitude tank. The 7 December 1968 launch was a success, and OAO–2's

space telescopes provided a wealth of astral information during their four years of operation.[561,562] The Lewis launch team received an Agency Group Achievement Award for their efforts.[563]

Centaur then launched the *Mariner 6* and *7* spacecraft in February and March 1969. After a six-month journey, the twin probes provided the most accurate photography of the Martian surface yet and revealed the makeup of the planet's polar ice cap. The findings disputed NASA's belief, based on the earlier *Mariner 4* images, that the Mars landscape was similar to that of the Moon.[564,565]

The ensuing May 1971 launch of *Mariner 8* failed when Centaur's guidance system malfunctioned, but less than one month later, Atlas-Centaur sent *Mariner 9* into Mars orbit. It was the first spacecraft

*Image 230: Atlas-Centaur launch of the OAO–2 satellite (GRC–2015–C–06558).*

Image 231: *Three successful SPC tests of the unique OAO–2 shroud jettison system, half of which lies on the net in the foreground, verified its flight performance and led to Centaur's first post-Surveyor success (GRC–1968–C–01258).*

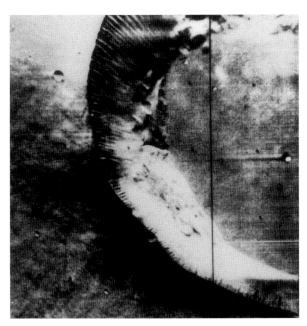

Image 232: A 25-mile-diameter Martian volcano photographed by Mariner 9 in November 1971. Mariner 9 revealed that the complex Martian terrain included channels, polar-layered materials, mammoth volcanos, sand dunes, and a vast canyon system (JPL–PIA04003).

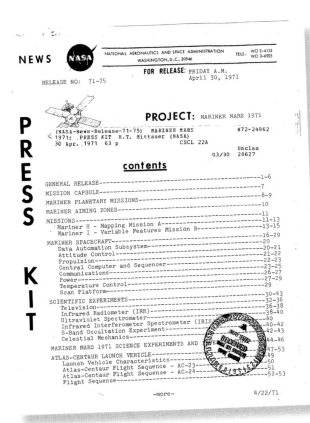

Image 233: Press kit for the Atlas-Centaur launch of the Mariner Mars spacecraft.

to orbit another planet. *Mariner 9* yielded over 7,000 images, which for the first time revealed the full extent of the Martian terrain. In March 1972 Centaur launched *Pioneer 10*, which became the first spacecraft to visit Jupiter and the first to exit the solar system. In addition, Atlas-Centaur began launching a series of Intelsat communications satellites for COMSAT Corporation. These early communications satellites enabled global television broadcast and dramatically reduced the cost of intercontinental telephone transmissions.[566]

❖ ❖ ❖ ❖ ❖ ❖

The *SERT–I* mission in 1964 successfully demonstrated that the electron-bombardment ion thruster could perform in space. Lewis researchers immediately began taking steps to conduct a longer space demonstration. NASA approved the *SERT–II* mission in August 1966. Lewis was responsible for the entire launch vehicle, the *SERT–II* spacecraft, and its propulsion system. Raymond Rulis was project manager, and Harold Kaufman was responsible for the thruster fabrication. Over the next three and a half years, the Lewis staff subjected the *SERT–II* thrusters to extensive qualification testing, including six months of continuous operation in the Electric Propulsion Laboratory (EPL) vacuum tanks. Lewis then subjected the entire spacecraft to vacuum testing for another 50 hours.[567]

*SERT–II* employed two solar-powered mercury ion bombardment thrusters. The spacecraft used 1 kilowatt of electric power to ionize mercury molecules and accelerate them in 6-inch-diameter exhaust beams at up to 100,000 mph. It was important to determine if the thruster would interfere with the spacecraft's electronics or communications systems or damage the solar arrays.[568]

A Thor-Agena rocket launched *SERT–II* into orbit on 3 February 1970, and the first thruster was activated a week later. Lewis researchers calibrated the propellant flows, verified the thruster's flight power conditioner, and tested the overall thruster performance on the spacecraft.[569] A piece of space debris however, caused an electrical short that caused the thrusters to shut down before *SERT–II* completed its six-month goal.

Lewis researchers were able to dislodge the debris and restart one of the engines in 1973.[570] Over the next eight years, they were able to repeatedly restart the thrusters until the mercury propellant finally ran out. During this period, researchers restarted the thruster 300 times without issue, and reactivated one thruster after 10 years of dormancy.[571]

*SERT–II* came to life once again in 1989 when the Lewis team was able to reboot the vehicle without the thrusters for a two-year analysis of solar array erosion. Although initially dubbed as a failure, *SERT–II* set a duration record for in-space operation of a solar electric propulsion system, demonstrated the thrusters' starting capabilities, and revealed the durability of the solar arrays after extended time in space.[572]

*Image 234: SERT–II spacecraft in the foreground after testing in the EPL in December 1969 (GRC–1970–C–00112).*

*Image 235: Termination of SERT–II thruster operations in June 1981 from the control room in the 8×6 building (GRC–1981–C–02861).*

## Emergence of Space Power

Long-duration spacecraft require self-sustaining on-board power to operate their electronics, communications, and navigation systems, and in some instances, to drive electric propulsion systems. Space power systems have two tasks: obtaining energy and converting it to electricity. The two primary sources used to supply energy are solar energy and nuclear energy. Rankine, Brayton, and Stirling systems have been used to convert this energy into electricity.[573] In 1963 Lewis created the Space Power System Division, led by Bernie Lubarsky, to advance these space power and related technologies. Decades later, space power remains one of the center's core programs.

Solar cells were developed in the mid-19th century and improved on over the next century. The mid-1950s saw the emergence and improvement of silicon solar cells and thin-film solar cells.[574] The first significant demonstration of the potential of solar cell technology was in space. Vanguard, the first U.S. satellite, employed a small solar array to power the satellite's radio systems in 1958. Virtually all succeeding space vehicles and satellites have employed solar cells to generate power. Joseph Mandelkorn and several of the researchers who designed the Vanguard system joined NASA Lewis in 1961. The radiation in space severely shortened solar cell life. To address this problem, Mandelkorn and a colleague developed a new type of radiation-resistant silicon solar cell. It became an industry standard.[575]

Lewis's High Energy Fuels Laboratory included equipment that allowed researchers to create their own silicon crystals and subject them to a multistep manufacturing process to create solar cells.[576] At Lewis, Mandelkorn continued to perfect the solar cell. His radiation-resistant cells lasted 25 times longer than existing solar cells. NASA mandated that only this "superblue" type of cell be used on its spacecraft and presented Mandelkorn with a unique $5,000 bonus for his contributions.[577] In general Lewis photovoltaic research in the 1960s sought to determine and address the causes of solar cell inefficiency. Key developments included shallow junction cells and back surface fields.[578]

The Chemistry and Energy Conversion Division began investigating the high-power and low-cost thin-film solar cell concept in 1961. The expensive conventional low-power solar cells that most space vehicles used at

*Image 236: High-voltage solar array test in the Engine Research Building during July 1965 (GRC–1965–C–01839).*

the time were both delicate and heavy. *The Mariner IV* spacecraft had required 28,000 solar cells for its flyby of Mars in 1964.

Thin-film cells were made by heating semiconductor material until it evaporated. The vapor was then condensed onto an electricity-producing film only 1000th of an inch thick. The physical flexibility of the new thin-film cells allowed them to be furled, or rolled up, during launch. Lewis researchers tested the new thin-film solar cells for 18 months in an EPL space simulation chamber with rotating periods of light and found no degradation.[579,580]

Early thin-film solar cells degraded while in storage, however, if moisture was present. By the late 1960s researchers improved the shelf life, but temperature fluctuations continued to impact performance.[581] When Lewis researchers began investigating this problem in 1967, they found that the cause stemmed from flaws in the manufacturing of the grid that joined the cells.[582] Research continues today, and thin-film solar cells are still considered to be a less expensive option for the future.[583]

During the mid-1960s Lewis researchers discovered that solar cell behavior on Earth varied at different altitude levels because of the difference in the amount of atmospheric particulates. Solar cell performance needed to be predicted for different altitudes before the cells could be standardized.[584] In the early 1960s Henry Brandhorst developed a method to calibrate solar cells using Lewis's B–57B aircraft. The pilots would take the aircraft up into the troposphere and expose the solar cell to the sunlight. The aircraft would steadily descend while instruments recorded how much energy was being captured by the solar cell. From these data, Brandhorst could estimate the power for a particular solar cell at any altitude. The aircraft-based system permitted measurements above the ground haze and weather and provided a multitude of plot points.[585] These calibration flights have been a critical component of Lewis's solar cell work for decades.

*Image 237: A model of the SNAP–8 power system created by the Lewis Fabrication Shop prior to a September 1966 Space Power Conference at the center (GRC–1966–C–03304).*[586]

❖ ❖ ❖ ❖ ❖ ❖

Lewis also examined thermonuclear methods of acquiring energy for spacecraft. The most significant work went into the SNAP program, which was initiated by the military and the Atomic Energy Commission (AEC) in the 1950s to develop a series of radioisotope and reactor-powered systems for a variety of applications, including space.[587] NASA became involved with the effort in 1960 while seeking a way to power its own space efforts. In March 1960 NASA contracted with Aerojet-General to develop what was called the SNAP–8 power conversion system. Henry Sloan's Space Power Division provided oversight of the project.

SNAP–8 integrated a high-temperature Rankine conversion system with the AEC reactor. It was the first attempt to develop a space power system that utilized liquid metals. In these systems, the liquid metal flows through the reactor and carries heat to the heat exchanger; there the metal is vaporized and used to operate the turbine.[588] Because metals can operate at higher temperatures than water can, such reactors require less heat-transfer material.[589]

Developmental problems and budget reductions led to a reorganization of the SNAP–8 program in 1963. The new effort concentrated more on subsystems and component technology than on a complete power system.[590,591] In 1964 Lewis technicians set up a SNAP–8 test rig with a mercury boiler and condenser in Cell W–1 of the Engine Research Building. Researchers used the rig to study the heat transients in the system's three loops. In 1967 Lewis operated a complete Rankine system for 60 days in W–1 to verify the integrity of the Lewis-developed mercury boiler.[592,593] Further tests in 1969 confirmed the shutdown and startup of the system under normal and emergency conditions.[594]

Meanwhile Aerojet had operated the first complete Rankine system in June 1966 and completed a 2,500-hour endurance test in early 1969. The SNAP–8

*Image 238: Brayton system setup in the SPF prior to the installation of the space radiator. The tests, which began in September 1969, were the first ever conducted in the new vacuum tank at Plum Brook Station (GRC–1970–C–01966).*

Rankine system had reached a state of mission readiness, but the AEC struggled to develop a reactor. The fuel elements repeatedly cracked during operation. Reductions in government funding led to the cancellation of the program in the early 1970s. Although SNAP–8 never flew in space, NASA recognized the Lewis and Aerojet success with the Rankine system with a NASA Group Achievement Award in November 1970.[595,596]

❖ ❖ ❖ ❖ ❖ ❖

Lewis researchers were also investigating the potentially more efficient and powerful Brayton power conversion system in the late 1960s and 1970s.[597] Brayton engines are similar to turbojets except that the high-temperature gas is recycled instead of exhausted. The system circulates a helium-xenon liquid that can be heated by either nuclear or solar energy. The liquid passes through turbines that drive the compressor and alternator, is cooled, and is then pressurized by the compressor. Excess heat is dissipated through a radiator.[598]

Lewis's testing of the individual components, subsystems, and complete systems revealed that the technology performed as good as or better than predicted. The system's main component, the Brayton Rotating Unit, included the drive shaft with a turbine, alternator, and compressor.[599] In the late 1960s Lewis tested each of the components and then the entire system. In 1968 Lewis researchers subjected the rotating units to up to 21,000 hours of endurance testing to verify their ability to meet the proposed five-year mission goals. The units exhibited a high rate of energy conversion. The Brayton Rotating Unit program demonstrated that the materials, manufacturing processes, and startup were reliable.[600] The success of these tests ultimately led Lewis researchers to develop a solar dynamic power system for the space station in the 1990s.[601]

Lewis researchers designed a space radiator and tested it with a 15- to 20-kilowatt Brayton cycle power system in Plum Brook's new SPF vacuum chamber. They sought to determine the radiator's effect on the overall power system in a simulated space environment.

After subjecting the Brayton system to simulated orbits with 62 minutes of Sun and 34 minutes of shadow, the researchers found a 4-percent power variation during the transitions.[602] Lewis continued its Brayton system and overall space power research in the following decades.

**Aerospace Safety**

Despite NASA's overall success in the late-1960s, the dangers of aerospace research were never far. During 1966 and 1967 three astronauts perished in T–38 fighter crashes and another in an X–15. On 8 June 1966 the F–104 flown by former Lewis pilot Joe Walker came into contact with an XB–70 during a formation flight over Muroc Lake and fatally plummeted onto the desert floor. The most shocking incident was the January 1967 launch pad fire that claimed Gus Grissom, Ed White, and Roger Chaffee during preparation for the Apollo 204 (renamed *Apollo 1*) mission.[603]

NASA immediately launched an urgent accident investigation into *Apollo 1* and requested that Irving Pinkel apply his experience from the Crash Fire Test program to the investigation. Pinkel arrived at Kennedy within days of the incident to help inspect the damaged capsule, develop an event timeline, and offer

*Image 240: Pinkel, foreground, along with Homer Carhart, Alan Krupnick, and Robert Van Dolan, inspect the Apollo 1 review capsule on 10 March 1967 (NASA 265–577C–4).*

insight into the origin and propagation of the fire.[604] The investigation board was under intense pressure to identify the cause quickly so that the investigation would not impact the overall Apollo schedule. The board issued its 400-page report two months later. It concluded that a host of manufacturing and oversight errors led to the deadly spark in the capsule wiring that ignited the oxygen-rich cabin air.[605] Pinkel spent the next year helping to redesign the *Apollo* capsule and spacesuits with flame-resistant materials.[606]

Following the investigation, NASA Administrator James Webb asked Pinkel to serve as director of the new Aerospace Safety Research and Data Institute (ASRDI). This Lewis-based group collected information about all of NASA's safety problems in a single dynamic database and sponsored a limited amount of safety research.[607]

The Safety Data Bank contained over 10,000 entries associated with fire, explosions, lightning strikes, cryogenic fuel hazards, rotor bursts,

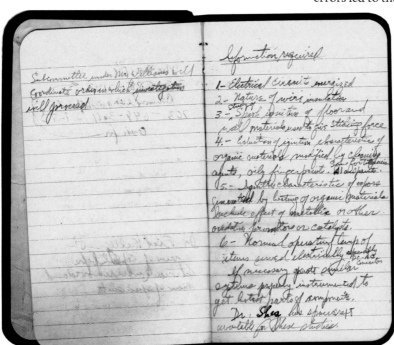

*Image 239: Notebook kept by Irv Pinkel during the Apollo 1 investigation.*

*Image 241: Damage to the PSL Equipment Building (GRC–1971–C–01422).*

*Image 242: Wrecked F–8 with Pinkel, seen to the right, who led the ensuing investigation (GRC–1969–C–02422).*

hazardous materials, structural failures, and aircraft systems and operation. Solomon Weiss took over the ASRDI in 1972, with George Mandel handling the publications and information services. Budget reductions ultimately forced the incorporation of the data bank into NASA's centralized database at the Scientific and Technical Information Office in 1976.[608]

In April 1970 Pinkel was summoned again, this time to serve as observer for the *Apollo 13* review board. Lewis utilized its Zero-G in June 1970 to help determine the likelihood that a short circuit ignited Teflon wire insulation and caused the *Apollo 13* oxygen tank failure. The experiments demonstrated that the capsule's oxygen-rich environment permitted the insulation to burn in a microgravity setting.[609,610] The review board concluded that an electrical fire started in one of the service module oxygen tanks.[611]

Board Chairman Edgar Cortright thanked Pinkel afterward, "It was a long and concentrated effort and required a considerable personal sacrifice on your part to be so long away from home and office. However, your dedicated participation was instrumental in bringing our work to a satisfactory conclusion in a relatively brief time span."[612] Pinkel retired two years later, but was called on again and again to assist with accident investigations of all types.

❖ ❖ ❖ ❖ ❖ ❖

Lewis also experienced some serious, but fortunately nonlethal, accidents during the late 1960s, including the crash of a Vought F–8 Crusader that the center had used as a chase plane. On 14 July 1969 the F–8 clipped the exposed raised end of the runway while it was performing some practice landings. It skidded off the pavement onto its side and burst into flames. The center's emergency crews responded rapidly and were able to rescue the pilot.[613]

In the early morning hours of 7 April 1971, a massive explosion ripped through the floor of PSL's Equipment Building. The blast seriously injured two employees and caused major damage to the facility. The building's I-beams, windows, heavy equipment, and 6-inch-thick concrete floor were damaged. It was later determined that the sealing of a 96-inch valve caused the exhauster equipment to operate like compressors, quickly over-pressurizing the system to failure.[614]

Although fire and security forces had been at the lab since the 1940s, Lewis formally established a Safety Office in 1952 to proactively protect employees and property.[615] This new office established Area Safety Committees to review potential hazards posed by all research within a particular zone. The committee worked with the engineers and researchers to mitigate hazards, then issued the required safety permits. An Executive Safety Board of upper-level management handled any disputes between the committees and the researchers.[616] This system has functioned well for over 60 years. Despite the presence of large quantities of fuels, heavy machinery, radiation, and other hazards, Lewis has maintained an excellent safety record.

### Giants at Plum Brook Station

Plum Brook Station's twin research facility titans, the SPF and the Spacecraft Propulsion Research Facility (B–2), officially opened on 7 October 1969 after nearly six years of construction. SPF's 100-foot-diameter, 150-foot-tall vacuum tank and B–2's ability to fire large rocket stages in a space environment were unrivaled in the United States. SPF was designed to test full-scale nuclear power sources, such as the SNAP–8, at simulated altitudes of 300 miles. The chamber is accessible on each side by massive 50-by 50-foot doors that open into large assembly and disassembly rooms.[617]

The first series of tests at SPF involved the Brayton Cycle space power system.[618] Over the next several years, researchers used SPF to test the large shroud-jettison systems for the *Apollo*-Skylab and Titan-Centaur launch vehicles.[619,620]

Lewis created B–2 to test complete rocket propulsion systems with up to 100,000 pounds of thrust in a simulated space environment. The facility has the unique ability to maintain a vacuum at the rocket's nozzle while the engine is firing. Giant diffusion pumps are employed to reduce chamber pressure at both B–2 and SPF to simulate the vacuum of space. The rocket fires into a 120-foot-deep spray chamber that cools the exhaust before it is ejected outside the B–2. The facility also utilizes a cryogenic cold wall to create the temperatures of space and quartz lamps to replicate the radiation of the Sun.[621]

Image 243: A Centaur D1–A rocket is readied for a test firing inside the B–2 vacuum chamber. The test chamber, 55 feet high by 33 feet in diameter, can handle rockets up to 22 feet long (GRC–1969–C–02596).

Image 244: The shroud for Skylab installed inside the SPF vacuum chamber for a jettison test. The shroud enclosed the multiple docking adapter, the top of the airlock, and the Apollo telescope mount. Problems with the ejection system were found during two tests in winter 1970. The issues were remedied, and the shroud was successfully jettisoned at a simulated 330,000-foot altitude in June 1971 (GRC–1969–C–03690).

The first test at B–2 was a hot firing of a Centaur D1–A vehicle on 18 December 1969. The D1–A was a next-generation Centaur that included improved computer and avionics systems. Engineers had been trying to determine if the propellant system's boost pumps were necessary for the engines to start. Tests of the pump system in Plum Brook's High Energy Rocket Engine Research Facility (B–1) test stand indicated that the pumps were superfluous. The B–2 test program confirmed these findings by testing a full-scale Centaur system without the pumps. The eventual elimination of the boost pumps simplified the system and reduced vehicle cost by over $500,000.[622]

### Departures

The late 1960s brought the first wave of retirements of the original staff from the early 1940s. Some—like upper-level managers Charles Herrmann, Oscar Schey, Irv Pinkel, James Braig, and Newell Sanders—were prominent figures at the center. Others—like assistant Mary Lou Gosney, nursing supervisor Ruth Elder, and waiter Josephus Webb—had become beloved for their personalities and service. Still others—like illustrator Richard Buchwald, architect Amuil Berger, and mechanical supervisor Melvin Harrison—let their skills speak for them. Many others who had begun their careers with the NACA during World War II retired in the late 1960s and early 1970s, including Center Director Abe Silverstein.

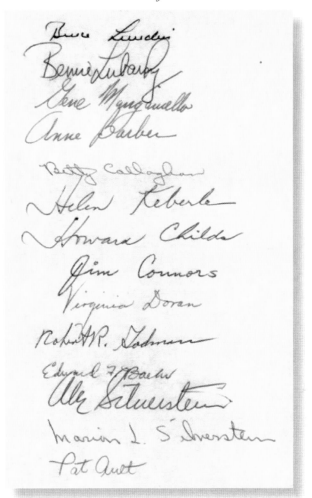

*Image 245: Irving Pinkel's 1974 retirement mat signed by the Silversteins, Bruce Lundin, Bernie Lubarsky, and others (NASA Pinkel Collection).*

*Image 246: Silverstein bids Christine Truax farewell as she retires as head of the Computing Section in July 1967. She had joined the NACA in the early 1940s (GRC–2015–C–06573).*[623]

*Image 247: Silverstein badge for Apollo 11 launch.*

Although he had 40 years of service with the NACA and NASA, the 61-year-old Silverstein was still relatively young when he announced his retirement in October 1969. The Apollo Program was at its apex, and NASA was already planning shuttle activities for the 1980s. He explained at the time, "It may be important that the men whose decisions initiate the new long-range projects be available to complete them. Since I do not think I can stretch my 40 years of service into 50, it is perhaps best for me and for the Lewis Center if I bow out now."[624]

As Center Director, Silverstein had personally ensured the success of the Centaur Program in the early 1960s and initiated the 1966 transition back to aeronautics. The center was healthy and had an experienced management staff. Silverstein's final years with NASA were relatively uneventful. In February 1968 he gave the Theodore Von Karman Lecture in Tel Aviv and became active in community and civic affairs. He served as a trustee at several institutions, established a successful Boy Scout Explorer Post at Lewis, and created a local Soviet Anti-Semitism Council. The awards and honors only increased during his final years, including a NASA Distinguished Service Medal and a Rockefeller Public Service Award in 1968.[625]

## Endnotes for Chapter 6

509. Lynn Manley, NASA Lewis Research Center press release 66–54, 16 September 1966, NASA Glenn History Collection, Cleveland, OH.
510. Manley, NASA Lewis Research Center press release 66–54.
511. Lynn Manley, NASA Lewis Research Center press release 66–61, 4 October 1966, NASA Glenn History Collection, Cleveland, OH.
512. Jack Lekan, Eric S. Neumann, and Raymond G. Sotos, *Capabilities and Constraints of NASA's Ground-Based Reduced Gravity Facilities* (Washington, DC: NASA N93–20184, 1993).
513. "515-Foot Zero-G Facility Ready for Major Mission," *Lewis News* (25 November 1966).
514. "515-Foot Zero-G Facility."
515. Howard Wine interview, Cleveland, OH, by Sandra Johnson, 3 June 2014, NASA Glenn History Collection, Oral History Collection, Cleveland, OH.
516. "Zero Gravity Research Facility at Lewis Research Center," c1966, NASA E–882, NASA Glenn History Collection, Facilities Collection, Cleveland, OH.
517. Abe Silverstein, *Progress in Aircraft Gas Turbine Engine Development* (Washington, DC: NASA TM X–52240, 1966).
518. Lynn Manley, NASA Lewis Research Center press release 66–2, 14 January 1966, NASA Glenn History Collection, Cleveland, OH.
519. Charles Kelsey, NASA Lewis Research Center press release 68–29, 3 May 1968, NASA Glenn History Collection, Cleveland, OH.
520. Erik Conway, *High-Speed Dreams: NASA and the Technopolitics of Supersonic Transportation, 1945–1999* (Baltimore, MD: Johns Hopkins University Press, 2008).
521. "Staff Men are Part of SST Study Group," *Lewis News* (10 November 1966).
522. Fred Wilcox to the Record, "Definition of F-106 Research Program and Results of Study of Accuracy," 8 December 1967, NASA Glenn History Collection, Flight Research Collection, Cleveland, OH.
523. J. F. Brausch, *Flight Velocity Influence on Jet Noise of Conical Ejector, Annular Plug, and Segmented Suppressor Nozzles* (Washington, DC: NASA CR–120961, 1972).
524. Fred Wilcox, "F106B Retirement, Comments on Its Tour of Duty at Lewis," 17 May 1991, NASA Glenn History Collection, Facilities Collection, Cleveland, OH.
525. Wilcox to the Record.
526. Wilcox, "F106B Retirement."
527. Wilcox, "F106B Retirement."
528. Brausch, *Flight Velocity Influence.*

529. Wilcox, "F106B Retirement."

530. *High-Speed Research—The Tu-144LL: A Supersonic Flying Laboratory* (Washington, DC: FS-1996-09-18-LaRC, September 1996), *http://www.nasa.gov/centers/langley/news/factsheets/TU-144.html* (accessed 23 May 2011).

531. "Final Flight," *Lewis News* (4 February 1977).

532. Lyndon B. Johnson to Heads of Departments and Agencies, "Memorandum on Aircraft Noise and Land Use in the Vicinity of Airports" (22 March 1967), Online by Gerhard Peters and John T. Woolley, The American Presidency Project, *http://www.presidency.ucsb.edu/ws/?pid=28160* (accessed 8 July 2015).

533. "Quieting the Skies," 19 September 1973, NASA Glenn History Collection, Inspection Collection, Cleveland, OH.

534. James J. Kramer and Francis J. Montegani, *The NASA Quiet Engine Program* (Washington, DC: NASA TM–X–67988, 1972).

535. James Kramer, *The NASA Quiet Engine* (Washington, DC: NASA TM X–67884, 1971).

536. Dennis Huff, "NASA Glenn's Contributions to Aircraft Engine Noise Research," *Journal of Aerospace Engineering* (April 2013).

537. Carl Ciepluch, et al., "Quiet Engine Test Results," *Aircraft Engine Noise Reduction* (Washington, DC: NASA SP–311, 1972), pp. 183–214.

538. Charles Kelsey, NASA Lewis Research Center press release 72–29, 1 May 1972, NASA Glenn History Collection, Cleveland, OH.

539. "Materials for Man," 19 September 1973, NASA Glenn History Collection, Inspection Collection, Cleveland, OH.

540. Hugh Harris, NASA Lewis Research Center press release 67–60, 16 September 1967, NASA Glenn History Collection, Cleveland, OH.

541. "First Exclusive Patent License," *Lewis News* (28 February 1964).

542. "Materials for Man."

543. Lynn Manley, NASA Lewis Research Center press release 66–45, 12 August 1966, NASA Glenn History Collection, Cleveland, OH.

544. Mark Bowles, *Science in Flux: NASA's Nuclear Program at Plum Brook Station 1955–2005* (Washington, DC: NASA SP–2006–4317, 2006).

545. Mark Bowles and Robert Arrighi, *NASA's Nuclear Frontier: The Plum Brook Reactor Facility* (Washington, DC: NASA SP–2004–4533, 2004).

546. Bowles, *Science in Flux*.

547. Mark Bowles, *NASA's Nuclear Frontier*.

548. Charles Mitchell, NASA Lewis Research Center press release 72–21, 10 March 1972, NASA Glenn History Collection, Cleveland, OH.

549. "Plum Brook Irradiates Russian Moon Dust," *Lewis News* (30 July 1971).

550. Frank da Cruz, "The IBM 1401" (2009), *http://www.columbia.edu/cu/computinghistory/1401.html* (accessed 17 June 2015).

551. Mainframe Systems Branch, "History of Computing at Lewis Research Center," (NASA B-0300, September 1990) (Glenn History Collection, Computing Collection, Cleveland, OH).

552. Mainframe Systems Branch, "History of Computing."

553. Mainframe Systems Branch, "History of Computing."

554. "LeSAC Marks Its First Birthday," *Lewis News* (11 September 1964).

555. Charles Kelsey, NASA Lewis Research Center press release 67–36, 13 June 1967, NASA Glenn History Collection, Cleveland, OH.

556. Hugh Harris, NASA Lewis Research Center press release 68–24, 9 April 1968, NASA Glenn History Collection, Cleveland, OH.

557. Cary Nettles to Virginia Dawson, "Centaur Notes," 8 May 2002, NASA Glenn History Collection, Cleveland, OH.

558. "3 Atlas-Agena Launches Assigned to Atlas-Centaur," *Lewis News* (6 January 1967), p. 1.

559. Virginia Dawson and Mark Bowles, *Taming Liquid Hydrogen: The Centaur Upper Stage Rocket 1958–2002* (Washington, DC: NASA SP 2004–4230, 2004).

560. Lewis Research Center, *Atlas-Centaur AC-17 Performance for Applications Technology Satellite ATS-D Mission* (Washington, DC: NASA TM X–2525, 1972).

561. John J. Nieberding, *Atlas-Centaur AC-16 Flight Performance Evaluation for the Orbiting Astronomical Observatory OAO-II Mission* (Washington, DC: NASA TM X–1989, 1970).

562. "OAO Meets Milestone," *Lewis News* (31 January 1969), p. 2.

563. "NASA Cites Atlas, Centaur for OAO II," *Lewis News* (6 June 1969).

564. KSC Centaur Operations Branch, *Atlas/Centaur-20 Mariner-6 Flash Flight Report* (Washington, DC: NASA TR–954, 5 March 1969).

565. S. A. Collins, *Mariner 6 and 7 Pictures of Mars* (Washington, DC: NASA SP–263, 1971).

566. "Big Boost From Rockets," 13 September 1973, NASA Glenn History Collection, Inspections Collection, Cleveland, OH.

567. William Kerslake and Louis Ignaczak, *Development and Flight History of SERT II Spacecraft* (Washington, DC: NASA TM–105636, 1992).

568. Joann Temple, NASA Lewis Research Center press release 66–19, 5 May 1966, NASA Glenn History Collection, Cleveland, OH.

569. W. R. Kerslake, et al., *Flight and Ground Performance of the SERT II Thruster* (Washington, DC: NASA TM X–52848, 1970).

570. Louis R. Ignaczak, N. John Stevens, and Bruce E. LeRoy, *Performance of the SERT II Spacecraft After 4.5 Years in Space* (Washington, DC: NASA TM X–71632, 1974).

571. William Kerslake, *SERT-II Thrusters—Still Ticking After Eleven Years* (Reston, VA: AIAA 81–1539, 27 July 1981).

572. Kerslake, *Development and Flight History.*

573. Ronald Sovie, *Power Systems for Production, Construction, Life Support, and Operations in Space* (Washington, DC: NASA TM–100838, 1988).

574. Sheila Bailey, Larry Viterna, and K. R. Rao, "Role of NASA in Photovoltaic and Wind Energy," *Energy and Power Generation Handbook: Established and Emerging Technologies* (New York, NY: ASME, 2011).

575. Valerie Lyons, "Power and Propulsion at NASA Glenn Research Center: Historic Perspective of Major Accomplishments," *Journal of Aerospace Engineering* 26 (April 2013): 288–299.

576. "Solar Cell Fabrication Is Shown," *Lewis News* (9 June 1967).

577. "Mandelkorn Presented Largest Contributions Award at Lewis," *Lewis News* (14 August 1964).

578. Bailey, "Role of NASA."

579. "New Thin-Film Solar Cells May Provide Space Power," *Lewis News* (5 August 1966).

580. Daniel Bernatowic, *Cadmium Sulfide Thin-Film Solar Cell Review Introduction* (Washington, DC: NASA TM X–52579, 11 June 1968).

581. F. A. Shirland, Americo Forestieri, and A. E. Spakowski, *Status of the Cadmium Sulfide Thin-Film Solar Cell* (Washington, DC: NASA TM X–52436, 1968).

582. Bernatowic, *Cadmium Sulfide Thin-Film Solar Cell Review.*

583. K. Zweibel, *Thin Film Photovoltaics* (Washington, DC: NREL/CP–520–25262, 1988).

584. "Preparing for Solar Cell Tests," *Lewis News* (10 April 1964).

585. Henry W. Brandhorst, *Airplane Testing of Solar Cells* (Washington, DC: NASA TM–X–51946, 2–3 June 1964).

586. "SNAP-8 (System for Nuclear Auxiliary Power)," press release, January 1967, NASA Glenn History Collection, Cleveland, OH.

587. Susan Voss, *SNAP Reactor Overview* (Kirtland Air Force Base, NM: Air Force Weapons Laboratory, AFWL–TN–84–14, 1984).

588. "SNAP-8 (System for Nuclear Auxiliary Power)."

589. "SNAP-8 Passes Milestone," *Lewis News* (13 December 1964).

590. James A. Albers, Ronald H. Soeder, and Pierre A. Thollot, *Design-Point Performance of a Double-Containment Tantalum and Stainless Steel Mercury Boiler for SNAP-8* (Washington, DC: NASA TN D–4926, 1968).

591. "SNAP-8 Clocks 2500 Hours," *Lewis News* (14 March 1969).

592. Robert P. Macosko, William T. Hanna, and Sol H. Gorland, *Performance of an Experimental SNAP-8 Power Conversion System* (Washington, DC: NASA TM X–1732, 1969).

593. Aerojet-General Corporation, *SNAP-8 Electrical Generating System Development Program* (Washington, DC: NASA CR–1970, 1971).

594. Voss, *SNAP Reactor Overview.*

595. "NASA Awards Laud Efforts," *Lewis News* (20 November 1970).

596. "Lewis Holds Conference on Space Power Systems," *Lewis News* (2 September 1966).

597. "Looks to 1971 Accomplishments," *Lewis News* (29 January 1971).

598. Lee S. Mason and Jeffrey G. Schreiber, *A Historical Review of Brayton and Stirling Power Conversion Technologies for Space Applications* (Washington, DC: NASA/TM—2007-214976, 2007).

599. Donald G. Beremand, David Namkoong, and Robert Y. Wong, *Experimental Performance Characteristics of Three Identical Brayton Rotating Units* (Washington, DC: NASA TM X–52826, 1970).

600. Lee Mason, "Dynamic Energy Conversion: Vital Technology for Space Nuclear Power," *Journal of Aerospace Engineering* 26 (2013): 352–360.

601. Lyons, "Power and Propulsion."

602. Ralph C. Nussle, George M. Prok, and David Fenn, *Performance of a Brayton Power System With a Space Type Radiator* (Washington, DC: NASA TN D–7708, 1974).

603. Courtney G. Brooks, James M. Grimwood, and Lloyd S. Swenson, *Chariots for Apollo: A History of Manned Lunar Spacecraft* (Washington, DC: NASA SP-4205, 1979).

604. Floyd Thompson to Irving Pinkel, 3 February 1967, NASA Glenn History Collection, Irving Pinkel Collection, Cleveland, OH.

605. Brooks, *Chariots for Apollo.*

606. Irving Pinkel interview, Cleveland, OH, by Virginia Dawson, 30 January 1985, NASA Glenn History Collection, Oral History Collection, Cleveland, OH.

607. Pinkel interview by Dawson.

608. Aerospace Safety Research and Data Institute, *The Safety Data Bank*, March 1975, NASA Glenn History Collection, Cleveland, OH.

609. Thomas Cochran, et al., *Burning of Teflon-Insulated Wires in Supercritical Oxygen at Normal and Zero Gravities* (Washington, DC: NASA TM X–2174, 1971).

610. Hugh Harris, NASA Lewis Research Center press release 70–22, 12 June 1970, NASA Glenn History Collection, Cleveland, OH.

611. Hearing Before the Committee on Aeronautical and Space Sciences, *Apollo 13 Mission Review*, United States Senate, 30 June 1970.

612. Edgar Cortright to Irving Pinkel, 11 June 1970, NASA Glenn History Collection, Irving Pinkel Collection, Cleveland, OH.

613. "NASA Aircraft Accident Report," 1969, NASA Glenn History Collection, Flight Research Collection, Cleveland, OH.

614. Jack Esgar, "NASA Industrial Mishap Report," 27 April 1971, NASA Glenn History Collection, Test Facilities Collection, Cleveland, OH.

615. Edward Sharp, "Establishment of Safety Office," 18 August 1952, NASA Glenn History Collection, Directors Collection, Cleveland, OH.

616. William Brown, *Safety Management of Complex Research Operations* (Washington, DC: NASA TM–81772, 1981).

617. Hugh Harris, NASA Lewis Research Center press release 69–54, 7 October 1969, NASA Glenn History Collection, Cleveland, OH.

618. Nussle, *Performance of a Brayton Power System.*

619. Charles Daye, *Skylab Payload Shroud Jettison Tests* (Washington, DC: NASA TN D–36913, 1972).

620. Lewis Research Center Staff, *Centaur Standard Shroud Heated Altitude Jettison Tests* (Washington, DC: NASA TM X–71814, 1975).

621. Lewis Research Center, "Capabilities and Facilities of the Plum Brook Station," c1970, NASA Glenn History Collection, Plum Brook Collection, Cleveland, OH.

622. Lewis Research Center, "Centaur Space Vehicle Pressurized Propellant Feed System Tests" (Washington, DC: NASA TN D–6876, 1972).

623. "Three Lewis Staffers to Retire," *Lewis News* (7 July 1967).

624. "Director Announces Retirement," *Lewis News* (24 October 1969).

625. Charles Kelsey, NASA Lewis Research Center press release 69–1, 11 January 1969, NASA Glenn History Collection, Cleveland, OH.

*Image 248: Lewis's front entrance reflects the center's new focus (GRC–1980–C–04980).*

*"Today there is much more to be done,
our country's needs are more manifold and our
capabilities far broader than was the situation in 1958."*

—Bruce Lundin

Image 249: A 5,000-pound-thrust rocket engine is fired at the Rocket Engine Test Facility as part of a thermal fatigue investigation in September 1975 (GRC–1975–C–03125).

# Crisis

ASA Lewis Research Center's heady days of the 1960s came to a sobering end at noon on 5 January 1973. Center Director Bruce Lundin informed the staff assembled in the Plum Brook cafeteria that the Office of Management and Budget had canceled all research programs without near-term applications. This included NASA's nuclear propulsion and power efforts. The Plum Brook Reactor Facility would cease operations immediately, and the station's other sites would be shut down over the next 18 months.[626] The reactor crew filed out of the hall and gathered in the reactor control room to watch Bill Fecych terminate the chain reaction for the final time.[627]

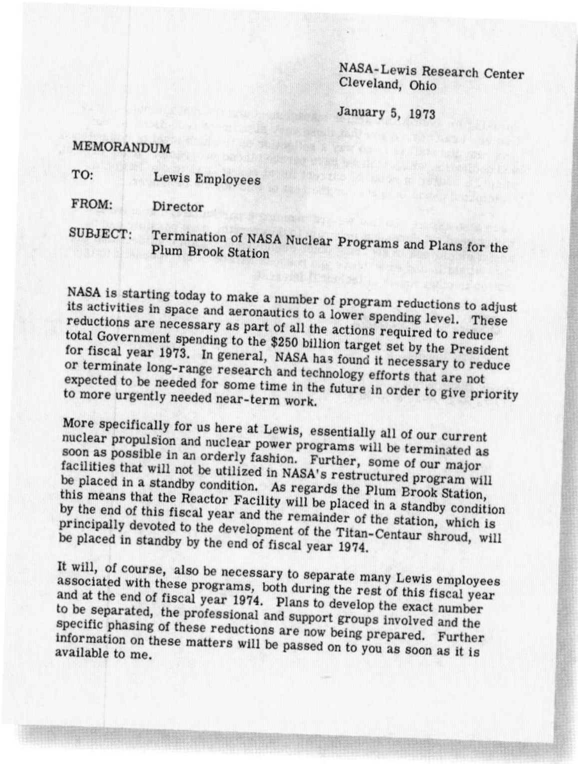

The cuts were the latest in a series of reductions coinciding with the end of the Apollo Program. The nation's priorities were changing, and Congress was reigning in NASA's once generous budget. The Agency was transitioning, as well, with the development of a new reusable space shuttle. Although Lewis found itself largely outside of the shuttle development, the center soldiered on with its successful aircraft

NASA-Lewis Research Center
Cleveland, Ohio

January 5, 1973

MEMORANDUM

TO:       Lewis Employees

FROM:     Director

SUBJECT:  Termination of NASA Nuclear Programs and Plans for the Plum Brook Station

NASA is starting today to make a number of program reductions to adjust its activities in space and aeronautics to a lower spending level. These reductions are necessary as part of all the actions required to reduce total Government spending to the $250 billion target set by the President for fiscal year 1973. In general, NASA has found it necessary to reduce or terminate long-range research and technology efforts that are not expected to be needed for some time in the future in order to give priority to more urgently needed near-term work.

More specifically for us here at Lewis, essentially all of our current nuclear propulsion and nuclear power programs will be terminated as soon as possible in an orderly fashion. Further, some of our major facilities that will not be utilized in NASA's restructured program will be placed in a standby condition. As regards the Plum Brook Station, this means that the Reactor Facility will be placed in a standby condition by the end of this fiscal year and the remainder of the station, which is principally devoted to the development of the Titan-Centaur shroud, will be placed in standby by the end of fiscal year 1974.

It will, of course, also be necessary to separate many Lewis employees associated with these programs, both during the rest of this fiscal year and at the end of fiscal year 1974. Plans to develop the exact number to be separated, the professional and support groups involved and the specific phasing of these reductions are now being prepared. Further information on these matters will be passed on to you as soon as it is available to me.

*Image 250: Lundin memo announcing the termination of the nuclear program.*

*Image 251: The staff gathers in the control room on 5 January 1973 as the Plum Brook Reactor is shut down one final time (GRC–2003–C–00847).*

*Image 252: Silverstein and Lundin talk at a 1968 reception for Lundin at the Guerin House (GRC–2015–C–06553).*

propulsion, space communications, and Centaur programs. In addition, Lundin and his staff began applying the center's aerospace skills to the resolution of newly emerging problems on Earth—particularly pollution, alternative energy resources, and energy efficiency.

Despite these efforts, Lewis's budget and staffing levels continued to decline throughout the 1970s, and several of the large test facilities were operating below their capacity. John McCarthy, who succeeded Lundin as Center Director in 1978, launched a campaign to improve Lewis's image and increase congressional support. These efforts, however, were overshadowed by the Reagan Administration's attempt to downsize the federal government in 1980 and 1981.[628] Although the center's 40-year existence was in jeopardy for a time, Lewis managed to weather the storm and emerge more resilient than ever.

### Turbulent Skies

NASA's budget had been declining steadily since its peak in 1966, resulting in increasingly severe cuts at the centers, particularly the former NACA laboratories—Langley, Ames, and Lewis. Between 1966 and 1973 Lewis's 5,000-person workforce declined by 20 percent. The cancellation of the nuclear programs caused the elimination of another 700 positions.[629-634]

There was a real danger of losing what Walter Olson referred to as "the fine problem-solving machine known as Lewis."[635] Although management attempted to ameliorate the reductions by encouraging those who had begun their careers in the 1940s to retire, large numbers of layoffs were inevitable.

As NASA's overall budget shrank, the Agency increased its spending on the space shuttle development. The shuttle was considered the first step to the creation of a large orbiting space station. The center did not play a primary role in the overall development of the space shuttle. Lewis's independence—first under Abe Silverstein and then Bruce Lundin—kept Lewis out of NASA's mainstream space programs in the 1970s. Lundin's unfavorable technical opinions on the shuttle and his unhappiness with the Lewis reductions strained the center's relationship with headquarters at a time when NASA's structure was evolving.[636,637]

Nearly all of the center's funding came from the Office of Aeronautics and Space Technology (OAST). OAST served other programs and institutions and did not depend on large projects. Thus OAST was only a fraction of NASA's overall budget.[638] NASA was beginning to divide up elements of major programs among several centers instead of allocating them entirely to a single center. This increased the competition among the centers, and the larger institutions like Johnson Space Center and Marshall Space Flight Center did not look favorably toward the former NACA centers now under OAST. Lewis decided to seek additional work outside the Agency.

### Pollution Monitoring

One unexpected result of the space program was a new appreciation for the uniqueness and fragility of Earth. Spacecraft had photographed the inhospitable surfaces of Mars and Venus and men had trod over the desolate lunar landscape. These images contrasted with Apollo 8's celebrated earthrise photograph, which vividly captured Earth's beauty and vulnerability. The nation began taking measures to protect natural resources, conserve energy, and reduce pollution. This new interest in the environment led to the creation of the Environmental Protection Agency (EPA) in 1970.

Lewis staff members developed an in-house school to learn about a variety of environmental problems and initiated efforts to monitor and reduce pollution. At the time, Cleveland was infamous for its collection of steel mills and for its fires on the Cuyahoga River. In the early 1970s Lewis agreed to assist the city with its efforts to control its air pollution. Lewis designed inexpensive and easy-to-use pollution-monitoring equipment and developed methods for cataloguing pollution particles by size, shape, and type. Lewis personnel also set up a series of monitoring stations around the city to identify the sources of particular contaminants and the effect of weather on pollution. They placed some of the stations at schools so that students could participate in the effort.[639] The researchers were able to construct a mathematical model of the city based on the data.[640] Lewis announced the results of this two-year study in November 1972. The air quality in the main industrial valley did improve over the course of the study.[641]

Lewis also initiated a broader effort to monitor pollution in the skies and waterways. The center added the Combustion and Pollution Research Branch to the Airbreathing Engines Division to investigate aircraft emission levels. During normal conditions jet engines operate at nearly 100-percent combustion efficiency, but when the engines are idling and during takeoff,

not all of the fuel burns. The unburned elements mix with the air and other pollution, and the Sun transforms the mixture into smog. Nitrogen emitted during takeoff contributes the most to smog. Lewis set out to measure pollutants at high altitudes and stratosphere to determine the extent of the problem.[642]

Lewis created the Global Air Sampling Program (GASP) in 1972 to measure pollution levels along the air lanes frequently traversed by the airline industry.[643] Porter Perkins, a prominent icing researcher from the NACA era, served as project manager. Participating airline companies installed commercial air-sampling equipment on Boeing 747 aircraft to automatically measure a variety of pollutants and collect dust particles to analyze their chemical composition.[644,645]

After several years of preliminary tests, the research flights commenced in March 1975. Four airline companies installed the sampling equipment on their aircraft and flew routes all over the world. By July 1978 nearly 7,000 GASP flights had taken measurements. These included samples from both poles during a special round-the-world flight to mark Pan Am's 50th anniversary.[646] Lewis also installed the GASP equipment on their F–106B Delta Dart. The F–106B not only augmented the Boeing 747 sampling but measured the amount and distribution of ash and gases

*Image 253: NASA's F–106B Delta Dart was acquired as the chase plane for the center's first F–106B. After that program ended, the chase plane was equipped with air-sampling and ocean-scanning equipment and performed remote sensing throughout the 1970s. The ocean-scanning equipment was stored in the nose section of the F–106B (GRC–1979–C–02423).*

*Image 254: A NASA OV–10 aircraft participates in the Project Icewarn program during March 1973 (GRC–1973–C–00948).*

factors indicated the presence of algae and pollution. The satellite's scanning system, however, developed problems just months after deployment. In an effort to resolve the issue, Lewis agreed to install the scanning equipment on the F–106B, and Lewis pilots flew the aircraft along the same flight path as Nimbus 7 to gather the desired data. NOAA engineers were able to use the Lewis data to recalibrate the satellite system.[648] The NOAA/NASA scanning effort was then extended to the Great Lakes and Midwest river systems.

❖ ❖ ❖ ❖ ❖ ❖

in the stratosphere following Mount St. Helen's eruption in 1980. GASP significantly increased existing climate information, providing the first climate information for some geographical areas and significantly increasing the existing data for others. The data helped researchers determine the effect of air traffic, seasons, and weather on pollution concentrations.[647]

Lewis also utilized its F–106B to support the National Oceanic and Atmospheric Administration's (NOAA) water-monitoring efforts. NOAA's Nimbus 7 satellite included the Coastal Zone Color Scanner to measure salinity, color, and temperature in coastal waters. These

NASA also partnered with NOAA and the Coast Guard for another water-monitoring effort—not for pollution, but for ice. The freezing of large portions of the Great Lakes regularly halted the region's shipping industry during the winter months. Even with satellite imagery, ship captains had difficulty identifying areas of open water or thin ice. Lewis researchers developed the Side Looking Airborne Radar (SLAR), which employed microwaves to measure ice distribution. When used in conjunction with electromagnetic systems, SLAR enabled researchers to determine the thickness of the ice and provided accurate data on the ever-changing ice flows.[649,650]

*Image 255: Two-way ship traffic through Neebish Channel, Michigan, in January 1976 (GRC–1976–C–00365).*

Project Icewarn was a multi-year feasibility study to determine if the monitoring system could be implemented in a way that was practical for ship captains to use. During the 1972–73 winter, Lewis and Coast Guard aircraft worked in conjunction with researchers in small boats to calibrate and verify the systems.[651] An aircraft with this dual system made a single pass over an area and relayed the information via a NOAA satellite to a ground station. NOAA personnel converted the information into maps and transmitted them to ships. That winter, shipping continued throughout the season for the first time in memory. Lewis participated in Project Icewarn for several years until the Coast Guard took full control. The ice monitoring was eventually performed by satellite technology, which improved dramatically over the years.[652]

Image 256: A J–58 engine in a new test chamber in the Propulsion Systems Laboratory (PSL) with a swirl-can combustor on display (GRC–1973–C–03376).

❖ ❖ ❖ ❖ ❖ ❖

In 1973 Congress added significant amendments to the national air pollution standards set out in the 1963 Clean Air Act. These new standards led to comprehensive emissions regulations across the nation.[653] The 1973 statutes required the airline industry to reduce their emissions of carbon monoxide, unburned hydrocarbons, nitrogen oxides, and smoke by two-thirds by 1979.[654] Lewis initiated in-house efforts and large programs in conjunction with engine manufacturers to develop new lower-emission combustors. Combustors mix compressed airflow with fuel, then ignite it as it flows toward the exhaust nozzle to provide thrust.[655]

In the 1940s and 1950s, Lewis combustion researchers focused their efforts on reliability, particularly at high altitudes. By the mid-1960s they were seeking to improve combustion efficiency and reduce emissions. Many of the engines on civilian aircraft during that period were originally created for military applications. These engines used fuel-rich mixtures that emitted high levels of pollution. One of Lewis's first steps was to reduce the fuel levels in the combustion chamber.[656] Many fuel injectors at the time did not properly atomize the fuel, which led to the emission of carbon dioxide while the aircraft was idling. The researchers found that injecting a small amount of air into the fuel nozzle improved the atomization.[657]

Nitrogen, the other main pollutant, was more difficult to control. A new type of combustor would have to be designed. Traditional combustors use fuel-rich mixtures that produce carbon dioxide and high temperatures. The high temperatures generate excessive levels of nitrogen oxide pollutants and require a longer combustor to reduce the temperature of the fuel-air mixture before it comes in contact with the turbine. Lewis began this effort in the early 1970s by creating the swirl-can combustor, which employed a lean direct-injection concept. Such systems inject fuel directly into the flame and employ additional small injectors.[658,659] The design spreads the flame zone uniformly and quickly mixes hot gases with combustor air to reduce flame temperature.[660]

In the 1972 Experimental Clean Combustor Program, Lewis contracted with Pratt & Whitney and General Electric to develop two complete combustors incorporating the swirl-can design and compare it to a standard model.[661] General Electric developed the dual-annular combustor, and Pratt & Whitney the Vorbix combustor. The two companies continued to mature their designs in the ensuing years, and they are now available on the CFM56, GE 90, and V2500 engines.[662] Combustors would continue to be a key component of the center's aeropropulsion research.

## AWARENESS

Bruce Lundin introduced the Acquainting Wage Board, Administrative and Research Employees with New Endeavors of Special Significance (AWARENESS) program in fall 1971 to periodically brief the staff on relevant topics. Nearly 800 employees and guests turned out for the inaugural event in November 1971, a talk by Wernher von Braun. Center management renamed the program Alerting Lewis Employees on Relevant Topics (ALERT) in November 1973.[663] Hundreds of staff members attended the monthly sessions to hear lectures by Lewis researchers, NASA officials, scientists, local politicians, and academics from outside the Agency.[664]

Meanwhile the AWARENESS title was applied to a new program that sought to improve staff morale through discourse and recognition. The center adopted the "Lewis Means Teamwork" slogan and strove to demonstrate how each Lewis position contributed to the goals of both Lewis and the nation. AWARENESS sought to acknowledge group accomplishments and highlight some of the center's lower-profile activities.[665] The recognition aspect had two components—events like the one honoring the extended team that helped build the new Propulsion Systems Laboratory (PSL) and promotional efforts like multipage *Lewis News* features highlighting the work of a specific program or branch. The committee

*Image 257: Lundin with Gerald Soffen, Cleveland native and Chief Scientist for the Viking missions, at an ALERT event in the Developmental Engineering Building auditorium in February 1977 (GRC–1977–C–02740).*

also organized a series of informal meetings between members of the upper management and the staff to improve communication. Later in the 1970s the Lewis Center Director periodically invited individual staff members to sit on stage with him and have a dialogue. The AWARENESS and ALERT programs were very successful and continued into the mid-1990s.

*Image 258: Acting Director Bernard Lubarsky takes questions from staff members during a 13 April 1978 AWARENESS forum (GRC–1978–C–01278).*

*Image 259: Display for the Big Boost From Rockets presentation at the 1973 inspection, which discussed the applications of rocket technology for everyday life (GRC–1973–C–3372).*

❖ ❖ ❖ ❖ ❖ ❖

In mid-December 1972, just as NASA was deciding to cancel its nuclear program, Deputy Administrator George Low asked OAST to initiate a series of Inspections similar to those held by the NACA.[666] Although the stated goal was similar to the NACA events—"displaying our programs and accomplishments and rapidly disseminating technology to the user"—these new Inspections were much more politically motivated. The public's interest in NASA was waning after Apollo, and the three research centers were struggling. It was imperative to demonstrate their contributions to life on Earth. The theme, which echoed a similar campaign introduced by President Richard Nixon, was "Technology in the Service of Man."

Walter Olson served as chairman of the planning committee. The eight tour stops focused primarily on terrestrial applications of NASA research, including aircraft noise reduction, pollution control, and clean energy. Even the launch vehicle presentation emphasized spinoff benefits such as the semiconductors, materials, insulation, and cryogenic storage devices that had found their way into a myriad of everyday applications.[667]

The September 1973 event was a tremendous success in nearly every way. It received rave reviews. Nearly 900 invited very important persons (VIPs) participated in the Inspection, and another 22,000 people attended the employees' day and public open house held afterward.[668] Perhaps more importantly, the Inspection focused Lewis's attention and boosted employee morale in the midst of major reductions. The Inspection failed, however, in the most consequential matter—generating congressional interest. John Donnelly, head of NASA's public affairs programs, commented afterward, "I'm not convinced that we got our money's worth. We just didn't get the people who count."[669] The Lewis event was NASA's last attempt to revive the fabled Inspections.

*Image 260: Bruce Lundin watches as NASA Administrator James Fletcher and ERDA Administrator Robert Seamans start the Plum Brook wind turbine for the first time (GRC–1975–C–03866).*

## Energy Systems for Earth

The dramatic increase in U.S. energy consumption in the 1960s and 1970s coupled with growing demands for cleaner energy began taxing the energy production industry. Both conservation and the development of alternative energy sources would be needed to improve the situation, which was exacerbated by the Arab Oil Embargo in October 1973.[670] For the first time, Americans experienced fuel shortages and price spikes. President Nixon challenged the nation to achieve energy self-efficiency in the same unified manner that had driven the successes of the Manhattan Project and the Apollo Program. He and his successor, Gerald Ford, created the Federal Energy Administration and the Energy Research and Development Administration (ERDA) to handle current energy allocation and long-term energy research, respectively. In 1977 President Jimmy Carter combined these two groups into the new Department of Energy (DOE).[671]

Lewis researchers had been urging center management to utilize their expertise in areas such as materials, bearings, lubrication, seals, combustion, aerodynamics, and turbomachinery to address these new energy problems. Bruce Lundin was able to convince headquarters that Lewis's aerospace experience could be applied to new fields opened up by the Energy Crisis and the environmental movement—fields that were previously outside of NASA's scope.[672] The 1974 NASA Authorization Act gave the center approval to pursue energy-related research according to its own judgment.[673]

Lundin spearheaded an effort to transform Lewis into the Agency's lead energy conversion center. Lewis considered aeropropulsion, space power and propulsion, communication satellites, and terrestrial propulsion and power to be part of this effort.[674] Soon Lewis was partnering with DOE to develop terrestrial power sources using renewable energy sources like wind and solar, as well as new technologies like the Stirling engine and nickel batteries.

❖ ❖ ❖ ❖ ❖ ❖

The Wind Energy Program was perhaps Lewis's most well-known alternative energy endeavor. Although the windmill had a long tradition of use around the world, it had primarily been applied to small, private water-pumping machines. Beginning in the 1930s there had been limited efforts to develop larger wind turbines to generate electrical power. The Smith Putnam wind turbine, which operated briefly in the United States during 1941, was the largest of these.[675] These early devices were technically feasible but were unable to compete with the low cost of traditional fossil fuels. Engineers revisited the wind turbine concept in the 1970s as energy prices escalated and concern grew over emissions from fossil fuels.[676,677] Although the technology existed, there were questions regarding wind energy's reliability, public acceptance, integration with the utility grid, and most importantly, cost.[678]

While visiting Puerto Rico for an unrelated project, Lewis researchers were approached by local officials interested in utilizing wind energy to create electrical power. Joe Savino, Ron Thomas, and Richard Puthoff proposed applying the center's experience with aerodynamics, powerplants, and energy conversion to wind energy. At the time, the National Science Foundation (NSF) was studying improvements to the energy field as a whole, but it did not have any research facilities of its own. The NSF partnered with Lewis in 1973 to conduct a symposium that assembled wind turbine experts from around the world, including some who had pioneered the field in the 1930s. As a result, the

Image 261: *Lewis engineers set up a Mod–0A–2 wind turbine in Culebra, Puerto Rico (GRC–1978–C–02389).*

NSF and NASA announced a joint effort with ERDA to pursue large-scale wind energy technology.[679] ERDA partnered with other institutions on smaller, more experimental designs.

ERDA assigned Lewis the responsibility for developing a series of large and increasingly powerful horizontal-axis wind turbines to be integrated into power grids. Lewis began by translating all known wind energy documents. The engineers then began designing three generations of turbines—one to obtain basic design and operation information, one to demonstrate that a working turbine and energy conversion system was feasible, and a third to create a cost-effective, repeatable system that would make the concept appealing to commercial groups.[680]

The first step was the construction of an experimental 100-kilowatt (kW) wind turbine at the recently mothballed Plum Brook Station in 1975. The device was large enough to obtain useful data, but small enough to permit frequent updates and modifications.[681] Plum Brook veteran Henry Pfanner was responsible for the operation of the 125-foot-diameter turbine atop a 100-foot tower. The researchers first used the wind turbine to validate analytical predictions, then to study the effectiveness of various configurations. It was necessary to maintain the predetermined power output so as to not overpower the power grid.[682]

The Plum Brook wind turbine efficiently produced power, but problems arose with the structural integrity

Image 262: *Wind energy program pamphlet.*

of the aluminum blades. The rotor design, unexpected load levels, and airflow blockage by the steel truss and stairway leading to the top of the machine all contributed to the problem. Lewis engineers were able to successfully redesign the blades,[683] and the turbine became NASA's workhorse for testing new technologies before they were transferred to industry. These technologies included new low-cost towers, new control systems, more efficient airfoils, and rotor hubs with teeter mechanisms to reduce structural loads. During its 10 years of operation, the Plum Brook wind turbine was reconfigured over a dozen times.[684]

Meanwhile NASA and ERDA engineers expanded the Plum Brook design into new 200-kW wind turbines. Westinghouse used the NASA design to manufacture these turbines and install them in Puerto Rico, New Mexico, Rhode Island, and Hawaii.

*Image 263: Installation of 2-MW wind turbine in Goldendale, Washington, in November 1980 (GRC-1980-C-05886).*

Although these early machines demonstrated that large turbines could be integrated into the grid, they could not compete with the price of fossil fuels. However, the 200-kW turbines did demonstrate an increasing reliability that spurred Westinghouse to privately develop a 600-kW version. Fifteen of these more powerful models were sold and successfully operated in Hawaii and Colorado.[685]

In 1979 the significantly more powerful 2-megawatt (MW) turbine was then installed in the mountains of North Carolina near the town of Boone. Although still considered a first-generation device, the machine was twice the size of the earlier models and generated the highest power levels to date. Local residents complained that the machine interfered with local television signals and generated low-frequency noise.[686] NASA engineers were able to resolve the noise problem by reducing the rotational speed of the blades, but the television interference proved more difficult to remedy, so the nearby homes were upgraded to cable television.[687] The first-generation multimegawatt design, which was removed after 2 years, did not demonstrate the long life necessary for commercial development. Nonetheless, the North Carolina machine provided important lessons for the future industry, including the need to locate turbines an appropriate distance from populated areas.[688]

The first second-generation turbine was a 2.5-MW model developed under contract with Boeing. The design was a major wind turbine advance that replaced the steel truss tower with a hollow steel tube, rotated the blade tips instead of the entire blade for control, and reduced loads by allowing the blades to teeter at the hub. Three of the turbines were built at same site in Washington State to study the interaction between units.[689] The researchers, however, found that a manufacturing error in the rotor caused greater stresses than anticipated, which required modifications to the turbines. Operation ceased in 1986 as the federal government cut funding, and the three turbines were eventually sold for scrap.[690,691] In the mid-1980s Boeing built another 2.5-MW turbine that became the first U.S. wind turbine to be sold to a commercial customer.[692]

Image 264: *The Boeing 7.2-MW wind turbine began operation in in Oahu in July 1987 (GRC–1987–C–05991).*

During this period Lewis also provided the technical management for the development of a large wind turbine in Wyoming by Hamilton Standard and the Department of the Interior. The 4-MW turbine included lightweight, flexible towers and fiberglass blades manufactured on automated winding machines.[693] It proved to be the nation's most powerful turbine for over 20 years. A comparable 3-MW machine operated for many years in Europe with similar success.[694]

In 1980 NASA issued both General Electric and Boeing contracts to develop the third generation of large-horizontal-axis wind turbines. As federal funding for wind energy decreased, General Electric decided not to pursue the development. Its model was completed under the Federal Wind Energy Program. Boeing proceeded with the development of its turbine at Oahu, Hawaii, using a design that expanded on NASA's second-generation models. Completed in 1987, the turbine had the largest blade diameter in the world and was the first turbine that could operate at variable speeds—a feature that increased efficiency and output.[695] The Boeing turbine was the most successful of all the NASA/ERDA machines in terms of performance and commercial market potential. The device was sold to the local utility in 1988 and set a one-month record for wind turbine production in March 1991.[696]

The Oahu machine brought a close to Lewis's wind energy research. The center's efforts peaked in 1979 and 1980 with the involvement of nearly 100 staff members. Much of the research was aimed at supporting technology—rotors, structural dynamics, electrical systems, controls, utility integration, economics, and environmental effects.[697] As Lewis moved away from terrestrial energy research in the early 1980s, NASA decided to transfer the Plum Brook wind turbine to DOE's Solar Energy Research Institute. In late 1985 over 100 people attended a ceremony at the site to mark the conclusion of NASA's 10-year operation of the device.[698]

By this time the Energy Crisis had passed, and the government stopped funding most wind energy projects. As the price of oil plummeted, the numerous requests to install the Boeing wind turbine in Hawaii were canceled, and the manufacturers soon ceased their efforts.[699] The vast majority of U.S. wind turbine and other renewable energy manufacturers went out of the business during this period. The few manufacturers that made it through, developed small turbines that required less capital expense during the tight market.[700]

Increased funding by several European nations in the 1980s and 1990s, particularly Denmark and Germany, expanded wind energy technology and provided Europe with most of the existing market. During this period the size of the turbines increased, eventually rivaling and exceeding the size of the large NASA turbines. Today the vast majority of wind turbines are large devices that offer increased reliability and lower maintenance costs in comparison to the use of many small machines.[701] It was not until the 2000s that the United States resumed its efforts in the wind energy field. In 2008 and 2009, the United States increased its installed capability for wind energy by 45 percent and achieved the largest installed wind energy capacity of any country in the world. General Electric is now the largest U.S. producer of wind turbines.[702]

Image 265: Robert Ragsdale briefs Senator Howard Metzenbaum on the Solar Simulation Laboratory during a visit to Lewis on 13 February 1974 (GRC–1974–C–00599).

The legacy of the NASA/DOE program can be found throughout the wind industry today. NASA demonstrated that two blades provide the optimal efficiency, but the European manufacturers who had dominated the market in the 1990s felt that the three-bladed design was more aesthetically pleasing. Nonetheless NASA-derived technologies in use today include steel tube towers, variable-speed generators, high-performance laminar airfoils, and fiberglass blade manufacturing.[703] In addition, engineering models, such as the Viterna Method for wind turbine aerodynamics developed by NASA engineer Larry Viterna are used throughout the world.[704]

❖ ❖ ❖ ❖ ❖ ❖

Lewis researchers used their experience with photovoltaic power generation for space vehicles to develop terrestrial applications for solar energy during the 1970s. Except for higher levels of radiation in space, the development was essentially the same. ERDA sought to spur the development of low-cost solar cell arrays within 20 years, and the Jet Propulsion Laboratory (JPL) sponsored a program that analyzed different commercial solar cells. This program included testing high-output solar panels in atmospheric conditions at Lewis's outdoor Photovoltaic Systems Test Facility in 1976. Lewis was able to obtain the first data on high-power solar arrays in the environment.[705]

Lewis's Solar Simulator in the High Temperature Composites Laboratory was one of the nation's largest facilities for testing terrestrial solar cells. In this indoor facility, solar cells were arranged in a collector that was aligned with a bright solar simulator. The setup allowed researchers to study a number of solar collectors and high-efficiency single-junction solar cells. The researchers could control the radiation levels, temperature, airflow, and fluid flow, as well as modify any aspect of the collector. Susan Johnson and Frederick Simon developed a single mathematical equation to correlate the efficiency data from 35 unique solar collectors. These collectors had a variety of applications, including temperature control systems in buildings.[706]

Lewis researchers also developed power-conversion systems to transform solar energy into electrical power for remote areas. In May 1978 NASA and the DOE signed an agreement with the Papago Native American tribe in Schuchuli, Arizona, to supply the village with solar-generated electricity within the year. Lewis provided all of the equipment and technical assistance, and the tribe's construction team built the arrays and support equipment. The 3.5-kW system, activated on 16 December 1978, was modest in scope but resulted in the first solar electric village. The system provided power to operate a refrigerator, freezer, washing machine, and water pump for the village and lights in each of the 16 homes.[707]

Lewis engineers improved the photovoltaic system for the DOE and the Agency for International Development and oversaw its installation in a number of remote villages without utilities in South America, India, and Africa. Lewis researchers consulted with local users regarding the amount of power needed and the solar panel design. Lewis solar power systems were installed to provide electricity for remote villages, medical clinics, and refrigeration machines.[708] Later, Lewis engineers doubled the initial power system at

*Image 266: Residents of Schuchuli, Arizona, attend the dedication ceremony for the solar village project (GRC–1978–C–05107).*

Tangaye, Upper Volta, by adding additional arrays.[709] Between 1975 and 1985 Lewis designed, fabricated, and installed 58 photovoltaic systems in 28 countries.[710] In addition to the larger projects, Lewis installed many smaller systems at fire lookout stations, weather stations, highways, and other locations.[711]

Lewis also continued its research into solar cells for space applications. These efforts included developing easy fabrication techniques, increasing efficiency, reducing cost, reducing weight, and developing strategies for limiting damage from radiation.[712]

❖ ❖ ❖ ❖ ❖ ❖

One of the more unique applications of Lewis's aerospace skills to benefit life on Earth was the use of the cyclotron to treat cancer patients. General Electric built the device beneath the Materials and Stresses Building in the early 1950s to test the effects of radiation on different materials being considered for nuclear aircraft engines. The cyclotron could split beryllium atoms to cause neutrons to be released.[713] In the early 1970s Lewis physicist James Blue developed a technique to stream these neutrons directly at an object such as a tumor.

In 1975 the Cleveland Clinic Foundation partnered with NASA Lewis to employ the center's cyclotron to study fast neutron radiation treatments for certain cancer patients. There had been debates in the medical

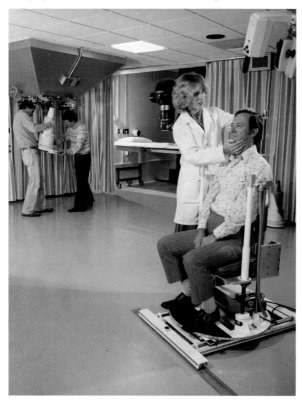

*Image 267: Medical personnel treat patients in the cyclotron area (GRC–1978–C–04709).*

community for years comparing the effectiveness of fast neutron therapy with that of x-ray radiation. The use of fast neutrons allowed better targeting of the tumor, but only a limited number of facilities could produce sufficient neutrons to carry out this treatment.[714] Lewis's cyclotron had a dual-beam system that could target the tumor with fast neutrons both vertically and horizontally.[715] In 1979 the National Cancer Institute incorporated the Lewis treatments in a larger 10-year study of the effectiveness of fast neutron treatments. Blue considered the cyclotron particularly effective for salivary gland and prostate tumors, but less so for those affecting the central nervous system. Over the years, the clinic staff treated 1,200 patients at the cyclotron.

The clinic terminated the program in 1990 as it began concentrating its efforts on treatments that did not involve radiation. Substantial improvements to x-ray technology and the cost of transporting patients and equipment to the center had made the Lewis program, which had always been considered temporary, less imperative. The end of the Lewis program coincided with the cessation of cyclotron cancer treatment activities around the nation.[716]

## Space Technology Hits the Road

Congressional studies during the 1970s determined that current automotive engines did not comply with the nation's new long-term emissions and efficiency objectives. There were a number of alternative propulsion concepts that had the potential to meet these goals if sufficiently supported. Congress enacted legislation in February 1978 that instructed the DOE to pursue these advanced automotive engines—including a Stirling-based system and a gas turbine engine.[717] The DOE partnered with NASA Lewis on the Automotive Stirling Engine (ASE) program. Lewis had investigated Stirling systems for space power applications in the 1960s.

The ASE was originally a five-year study, but budget cutbacks in the early 1980s resulted in a lower intensity 10-year effort. The effort included research on materials, combustion, controls, and other technologies, but the main thrust was the development of a working engine. The engine was to improve fuel economy by 30 percent, be able to use a variety of fuels, and reduce emissions.[718] These objectives had to be reached without significantly impacting vehicle performance, reliability, and cost.[719]

Like the Brayton Cycle engine, the Stirling engine is a closed-cycle engine that recirculates a working gas. In the automotive Stirling engine, hydrogen is used as the working gas. Continuous burning of the air and fuel heats the hydrogen to power the pistons. By design, Stirling engines are quiet and highly efficient, have low emissions, and can use different types of fuel.[720]

In March 1978 NASA signed contracts with Mechanical Technology Incorporated (MTI), United Stirling of Sweden, and AM General (AMG) to develop the Stirling automotive engine. MTI managed the effort, United Stirling developed the engine, and AMG integrated the system in its vehicles.[721]

ASE began by improving the net efficiency of a baseline engine from 3 to 37 percent. The researchers then developed two generations of Stirling engines. The first was installed in an AMC Spirit, which operated for over 9,000 hours and was driven over 1,000 miles. The engine was tested in a variety of AMC vehicles.[722] The second-generation engine—a substantial improvement over the early designs—was both

*Image 268: American Motors Pacer vehicle modified to run on twenty 6-volt batteries (GRC–1977–C–02096).*

reasonably priced and compact enough to be installed in the AMC Celebrity. The engine demonstrated a high level of efficiency during its test program in 1986 and 1987.[723]

The ASE program concluded in 1989. The technology met or exceeded all of its original goals and produced the first demonstration that the Stirling system could be applied and operated. By this time, however, the energy crisis had passed, and automobile manufacturers were not ready to overhaul their production plants and train their employees on Stirling technology.[724]

❖ ❖ ❖ ❖ ❖ ❖

Lewis created a Transportation Propulsion Division in 1977 to coordinate a number of automotive projects, including research on new drivetrains, improved powertrains, and overall engine improvements. Lewis also contributed to a renewed interest in electric automobiles,[725] which had been popular in the United States until the Ford Model T made gas-powered vehicles affordable.[726] Lewis focused on battery performance—the most critical component of electric vehicle operation. The researchers applied their experience with batteries for satellites to the automobile issue and developed a nickel-zinc battery that offered almost twice the range of traditional automobile batteries.[727]

Lewis also analyzed the performance of all 16 commercially available electric vehicles to determine the current state of electric vehicle technology. These were gas-powered vehicles that had been modified to run on batteries. Some of the vehicles had analog data-recording systems to measure the battery during operation and sensors to determine speed and distance. The vehicles were run on a local test track to analyze their range, acceleration, coasting, braking, and energy consumption. Lewis researchers found that the performance of the different vehicles varied significantly but that the range, acceleration, and speed were generally lower than those found on conventional vehicles. They concluded that traditional gasoline-powered vehicles were as efficient as the electric vehicles, but advances in battery technology and electric drive systems would significantly improve efficiency and performance in the coming years.[728]

## Aeronautics Expanding

The airline industry continued to grow exponentially, increasing air pollution, noise, and airport congestion. The resurgence of Lewis's aeronautics programs, which had begun in the mid-1960s, continued unabated throughout the 1970s. Unlike the research of the NACA era, Lewis now worked in tandem with industry on research and technology. Industry was responsible for developing the resulting concepts into commercial products. Lewis efforts included work on new technologies like vertical or short-takeoff and landing (V/STOL) aircraft, noise reduction, energy efficiency, and control systems. The research was aided by the two new altitude chambers in the PSL complex—No. 3 and 4—and the new 9- by 15-Foot Low-Speed Wind Tunnel (9×15).

The design of cost-effective and functional V/STOL aircraft has proven to be one of the aircraft industry's most elusive endeavors. Since the 1950s aeronautical engineers have been trying to perfect vehicles that combine the helicopter's ability to elevate vertically with the horizontal cruising speeds of traditional aircraft. V/STOL aircraft not only had military applications, but also had the potential to relieve airport congestion by requiring shorter runways so smaller airports could handle more traffic.[729] Tilt engines, tilt wings, and vectoring engine thrust are the primary technologies necessary for V/STOL vehicles.[730]

*Image 269: A V/STOL engine with a zero-length inlet installed in the 9×15 (GRC–1980–C–04513).*

NASA's three research centers—Langley, Ames, and Lewis—began pursuing the V/STOL technology in the 1960s as interest increased globally. Lewis specialized in the engines and their integration into the airframe. The researchers initially sought to identify operating problems, develop basic technology that could be applied to future designs, and analyze different aircraft configurations based on a particular mission.[731]

Lewis designed the 9×15 to study V/STOL engine models at the low speeds encountered during takeoff and landing. Built inside the return leg of Lewis's 8- by 6-Foot Supersonic Wind Tunnel (8×6) to minimize the time and cost of its construction, the 9×15 was used to study lift fans, a multifan wing design, and V/STOL inlets for Boeing and Grumman.[732] The Lewis V/STOL effort in the 1970s also included studies of inlet guide vanes and fan stages in the Engine Research Building, propulsion control simulation and

engine tests in the PSL altitude chambers, and an array of inlet and deflector studies in the 10- by 10-Foot Supersonic Wind Tunnel (10×10).[733] These systems are very complex, and until recently there was only one successful V/STOL aircraft operating in the United States—the British-designed Hawker Harrier.[734]

Lewis's Quiet Clean Short Haul Experimental Engine (QCSEE) program actively pursued powered lift (tilting down of engine or wing for extra lift during takeoff) technology for transport aircraft with a limited flight range operating from small airports.[735] QCSEE sought to develop economical and environmentally beneficial high-speed, short-distance aircraft technologies. Regulators could use the data to create effective regulations, and industry could develop the technology.

The basic engine design was a high-bypass turbofan that could deflect its exhaust downward by using wing flaps. Lewis contributions included variable-pitch compressor blades that could reverse airflow through the engine during landing to reduce the required runway length. Lewis's improvements to turbine staging resulted in a four-stage turbine that performed as well as a normal eight-stage turbine.[736] Lewis researchers investigated two powered-lift designs and an array of new technologies in a specially designed test stand, the Engine Noise Test Facility, on the hangar apron. The QCSEE engines met the researchers' performance goals without increasing noise or pollution levels. Although the short-distance transport aircraft for which the engines were intended never materialized, different technological elements of the engine were applied to some future General Electric engines.[737]

❖ ❖ ❖ ❖ ❖ ❖

The addition of two new test chambers, No. 3 and 4, to the PSL complex was Lewis's most significant aeronautics investment since the mid-1950s. The two new 40-foot-long and 24-foot-diameter altitude chambers could test engines twice as powerful as any then in

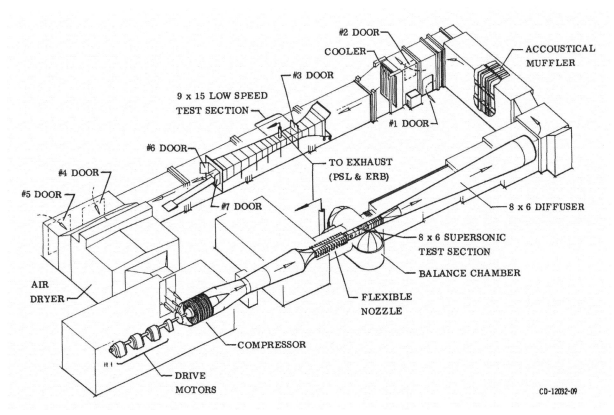

*Image 270: Diagram showing the 9×15 section in the return leg of the 8- by 6-Foot Supersonic Wind Tunnel (CD–1202–09).*

existence. The new chambers were located on the other side of the original PSL Equipment Building and used the same air-handling equipment. Nearly 5,000 tons of steel were used during the 1967 to 1972 construction.[738] The new test chambers, which began operation in 1973, were significantly more powerful than the original PSL chambers. Researchers used all four of the PSL chambers extensively during the 1970s. The center continues to upgrade the facility's capabilities. Today the PSL is NASA's only facility that can operate full-scale engines in simulated flight conditions.

Lewis collaborated with the Air Force on two comprehensive programs in the 1970s that utilized surplus Pratt & Whitney F100 and General Electric J85–21 engines as testbeds to study engine design issues such as flutter, inlet airflow distortion, and electronic controls. As engine designs became more complex, problems like flutter (compressor blade vibration due to distortions in airflow) were more pronounced, and sophisticated control systems were required. The researchers used the PSL chambers to validate predictive computer codes that were being used to design compressor blades and other components.

After mapping the engine performance under normal conditions, the researchers destabilized the airflow to determine the effects of flutter.[739] They then compared the data with results from computer models. The efforts led to technological improvements and identified areas requiring future research.[740]

❖ ❖ ❖ ❖ ❖ ❖

The PSL was also instrumental in the development of digital "fly-by-wire" engine control systems in the 1970s. Engine control systems determine the amount of fuel, the nozzle area, and other parameters required to produce specific levels of thrust in a variety of flight conditions. Engine controls had traditionally been mechanical systems, but the newer, more sophisticated engines required improved computer-based control systems that could manage multiple facets at the same time.

Researchers at the NASA Dryden Flight Research Center created the first computer-based fly-by-wire system in the late 1960s. Then Lewis researchers began to develop a single system that would integrate the control of the engine inlet, afterburner, and nozzle. Lewis researchers tested the system on

*Image 271: A Pratt & Whitney F100 engine in PSL No. 4 during September 1981 (GRC–1981–C–04382).*

*Image 272: A Refan engine installed on a DC–9 for a ground test in January 1975 (GRC–1975–C–00104).*

a Pratt & Whitney TF30 engine in the PSL during 1975, and Dryden flight-tested it soon thereafter.[741] Lewis then developed a series of equations that converted the pilot's input into optimal engine performance. In 1977 the system was verified on an F100 engine in the PSL.[742] The next step was the Full Authority Digital Engine Control (FADEC) system, which managed all engine components to maximize efficiency. The system was tested in the PSL during 1979 before being flight-tested on an F–15 at Dryden. The use of these FADEC systems is pervasive throughout the engine industry today.[743]

❖ ❖ ❖ ❖ ❖ ❖

Although noise-reduction technology was increasingly available in the 1970s, the airline industry balked at installing it on existing aircraft without a legal mandate. The Federal Aviation Administration (FAA), however, could not create regulations until the technology was demonstrated to be both effective and economically feasible to incorporate.[744] Lewis's Refan Program was an effort to demonstrate that noise-reduction modifications could be applied to existing aircraft engines at a reasonable cost and without diminishing the engine's performance or integrity. The program focused on the Pratt & Whitney JT8D turbofan—which, in the early 1970s, was one of the airline industry's most widely used engines.[745]

The Refan redesign replaced the engine's two-stage fan with a larger single-stage fan. This slowed the engine's exhaust flow and significantly reduced the amount of noise it generated. Booster stages were added to maintain the proper level of airflow through the engine. Acoustic treatments were also installed on the fan duct walls, around the exhaust, and on the inlet guide vanes.[746] These modifications added about 250 pounds to the engine, but this was compensated for by the new design's extra thrust.[747]

Lewis researchers first tested a modified JT8D in the PSL for 200 hours under simulated altitude conditions. Then the Refan engine was ground-tested on an actual aircraft before it was used for a series of flight tests in early 1975. The Refan Program reduced the JT8D's noise by 50 percent while increasing the fuel efficiency.[748] McDonnell Douglas announced that the refanned engines would be added to their DC–9s in the 1980s, but both the DC–9 and the Boeing 727 became outdated before Refan technology

was incorporated. The advancements were applied instead to the next-generation Boeing 737 and MD–80.[749,750]

❖ ❖ ❖ ❖ ❖ ❖

The 1973 Oil Embargo triggered fuel shortages and precipitous rises in energy prices that impacted the airline industry exceptionally hard. As a result NASA instituted the Aircraft Energy Efficiency Program (ACEE) in 1975 to develop new technologies to improve aircraft efficiency.[751] The ACEE effort included three airframe programs based at Langley and three propulsion programs based at Lewis—Engine Component Improvement (ECI), Energy Efficient Engine (E³), and the Advanced Turboprop (ATP). NASA's primary partners for the ACEE propulsion studies were General Electric and Pratt & Whitney.

The ECI program sought to quickly improve fuel efficiency by modifying the components on existing engines that received the most wear and by developing methods to detect engine deterioration. The Lewis staff first identified 150 prospective short-term improvements that would increase efficiency by five percent on the two most prevalent engine models for airliners. The 16 most plausible concepts were then evaluated and developed in the late 1970s and early 1980s.[752] General Electric emerged from the ECI program with enhanced components such as a redesigned fan, a single-shank blade, and active clearance control for turbines, as well as significantly improved engine diagnostics. At the time, retrofitting the existing engines was not considered cost effective, but the industry was pleased with the Lewis research results and incorporated the improvements into new engine models.[753]

The E³ Program sought to develop new turbofan engines that would reduce fuel consumption and operating costs while adhering to noise and pollution regulations. General Electric and Pratt & Whitney based the designs for their experimental engines on the high-bypass turbofan engine. The companies created preliminary designs, worked on component improvements, and then tested the new hardware on engine cores. Budget cuts in 1982 forced NASA to cancel its full-scale testing of the engines, but General Electric went on to test their model the following year, demonstrating that it had achieved the program's goals.[754] General Electric integrated its E³ advances into the design of its CF6–80E and GE90 engines. Pratt & Whitney continued to develop their E³ technologies and eventually incorporated them into their Geared Turbofan engine in the mid-2000s.[755]

The ATP program was the most significant ACEE propulsion effort. It sought to resurrect and greatly enhance the turboprop (propeller driven by a gas turbine) technology that emerged in the 1940s and 1950s. The new engines were intended to perform as well as the turbofan engines used by airliners while providing fuel savings of 20 to 30 percent. The turboprops could move large quantities of air, so they required less

engine speed and thus less fuel. Lewis had tested several turboprop models in the Altitude Wind Tunnel during the NACA period, and several airliners had employed the engines in the 1950s. The new ATP engines, however, were designed for efficient high-speed aircraft. These modern engines had at least eight blades and were swept back for better performance and noise reduction.

Daniel Mikkelson of Lewis and Carl Rohrbach of propeller manufacturer Hamilton Standard began revisiting the turboprop concept in the early 1970s before ACEE began. Soon a team of Lewis engineers

*Image 273: Salvatore Grisaffe (left) and Oral Mehmed with a Hamilton Standard SR–6 propfan in the 8×6 test section (GRC–1980–C–04620).*

was working on the project.[756] After the Lewis researchers developed the advanced turboprop theory and established its potential performance capabilities, they began an almost decade-long partnership with Hamilton Standard to develop, verify, and improve the concept.

Despite its promise, the ATP was initially left out of the 1975 ACEE program because of the perceived social stigma regarding propellers and the reluctance of the airline industry to completely overhaul its collective fleet.[757] Lewis and Hamilton Standard continued their efforts, however, and tested a series of small-scale models with different blade shapes and angles in the 8×6 wind tunnel. The performance of the single-rotation propfan proved successful enough to convince NASA to incorporate the ATP into the ACEE program in 1978. NASA would significantly expand the ATP program in the 1980s.[758]

*Image 274: General Electric E$^3$ (GRC–1983–C–03109).*

*Image 275: Display for November 1982 Propfan Acoustics Workshop at Lewis (GRC–1982–C–6372).*

## Titan-Centaur

The Centaur Program, particularly the new Titan-Centaur vehicle, produced the center's highest-profile accomplishments of the 1970s. Centaur staff members, who had a full launch schedule and offices outside the main Lewis campus, were almost impervious to the center's funding trials during this period. Lewis worked with General Dynamics to update Centaur with a new computer system and an equipment module.

Lewis managed the Atlas-Centaur launches of communications satellites and space telescopes, as well as 13 planetary launches over a 10-year span—a feat that had not been equaled before or since. These began with three *Mariner* missions to Mars and the *Pioneer 10* visit to Jupiter.

In 1973, the new Centaur D1–A sent *Pioneer 11* by Jupiter and Saturn, before this spacecraft also exited the solar system, and launched *Mariner 10*, the first spacecraft to use the gravitational pull of one planet (Venus) to reach another (Mercury).[759,760]

Centaur's greatest achievements during this busy decade were accomplished with an alternative booster vehicle, the Titan III. In the late 1960s NASA engineers began planning two ambitious space exploration initiatives—*Voyager* and *Viking. Voyager* would take advantage of a rare planetary alignment

that permitted a single spacecraft to visit multiple outer planets. *Viking* was the first attempt to send an orbiter and lander to Mars. NASA scheduled six Titan-Centaur missions— two *Vikings*, two *Voyagers*, and two German *Helios* vehicles to capture close-up data from the Sun. The *Viking* and *Voyager* missions were among NASA's most important efforts during the 1970s, and there was tremendous pressure to meet the deadlines and limited launch windows.

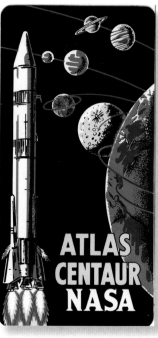

Image 276: Atlas-Centaur decal highlighting its interplanetary missions.

Because the *Viking* vehicles were over three times the weight of Atlas-Centaur's previous heaviest payload, NASA engineers decided to mate Centaur with the more powerful Titan III. General Dynamics introduced a new Centaur model for Titan—the D–1T.

Image 277: Centaur rocket control room in the Development Engineering Building during the preparations for the Titan-Centaur-Helios launch on 10 December 1974. From here the Lewis staff in Cleveland could monitor and back up the Lewis launch team in the actual control room at Cape Canaveral (GRC–1974–C–04007).

The D1–T's biggest modification was its use of a completely new fairing: the Centaur Standard Shroud. Lewis tested a scale-model of the shroud extensively in the 8×6 and evaluated the full-size fairing at Plum Brook Station's Nuclear Rocket Dynamics and Control Facility (B–3) test stand and Space Power Facility in the early 1970s.[761,762]

The first Titan-Centaur launch was a proof test of the launch vehicle. Since the proof test did not have a scheduled payload, Lewis researchers designed the small Space Plasma, High Voltage Interaction Experiment (SPHINX) research satellite to take advantage of the free ride. Researchers designed it to study the electrical interaction of SPHINX's experimental surfaces with space plasma to determine if higher orbits would improve the transmission quality of communications satellites.[763,764]

❖ ❖ ❖ ❖ ❖ ❖

The 11 February 1974 launch, however, proved to be one of the Launch Vehicle Division's darkest days. The launch pad safety officer destroyed the vehicle

*Image 278: Saturn and its moon Titan as seen from Pioneer 11 on 26 August 1979 (NASA GPN–2002–0006).*

after Centaur's RL–10 engines failed to operate. Lewis and General Dynamics investigators concluded that the failure of Centaur's liquid-oxygen boost pumps was not unique to the Titan-Centaur and could have occurred on any of the previous Atlas-Centaur missions.[765] The successful *Helios-1* launch in December 1974 restored confidence in Centaur.

Nonetheless, tensions were high as work proceeded toward the scheduled spacecraft launches, including

*Image 279: Titan-Centaur carrying Viking 1 (1975–L–07178).*

those for the $1 billion Viking Program. The marquee twin *Viking* launches occurred on 20 August and 7 September 1975. The Titan, Centaur, and Centaur Standard Shroud performed perfectly on both launches. The two orbiters and landers arrived at Mars in July and September 1976, respectively. The landings, the first ever on Mars, were one of NASA's biggest achievements in the post-*Apollo* years. *Lewis News* described the scene the following summer when *Viking* made its first transmissions from Mars, "All the problems associated with the Viking program seemed to evaporate when [Titan-Centaur Director Andy] Stofan, together with other scientists who had spent their entire careers dedicated to the program, watched the martianscape appear on the screens—line by line— a year later at the Jet Propulsion Lab in California."[766]

The successes were followed by another *Helios* launch in January 1976. The two *Voyager* missions to explore several outer planets, referred to as the Grand Tour, were next. The successful 20 August 1977 *Voyager 1* launch was followed 16 days later by *Voyager 2*. The complex missions required the Titan-Centaur to perform six different engine burns to achieve the proper trajectories.[767] The two *Voyager* spacecraft visited Jupiter in 1979 and Saturn in late 1980 and 1981

before going their separate ways. *Voyager 1* then headed toward the edge of the solar system. Meanwhile *Voyager 2* became the first spacecraft to approach Uranus and Neptune in 1986 and 1989, respectively. Then *Voyager 2* also began to exit the solar system. The two space probes succeeded in leaving the solar system in 2004 and 2007.[768]

Centaur closed out its eventful decade with the 1978 Atlas-Centaur launches that sent *Pioneers 12* and *13* toward Venus.[769] The mighty Titan-Centaur was retired, as it prepared to transfer its launch services to the space shuttle. An updated version of the vehicle was resurrected in the 1990s.

❖ ❖ ❖ ❖ ❖ ❖

Although Lewis's work on liquid-hydrogen injectors, turbopumps, and nozzles in the 1960s influenced the design of the Space Shuttle Main Engines, the center did not play a direct role in the primary development of the actual shuttle. Lewis did, however, provide critical test data in support of the shuttle design and contributed to improvements of the shuttle insulation system and fuel cells. The former was essential to protect the vehicle during reentry, and the latter powered all of the shuttle's on-board equipment.[770]

*Image 280: Ken Baskin checks a complete 2.25-scale model of the shuttle in the 10×10 test section during July 1975 (GRC–1975–C–02011).*

In July 1975 Lewis engineer Ken Baskin conducted a series of tests with a subscale model of the shuttle launch vehicle in the 10×10.[771] Baskin sought to determine the effects of the recirculation of the rocket exhaust on the shuttle's external fuel tank and solid rocket boosters. He simulated launch conditions by firing the rockets in the same manner as during a shuttle launch. Rockwell International engineers utilized the data from the 10×10 runs to facilitate the design of the shuttle's insulation systems.[772] Then Lewis and Rockwell followed up these tests in 1977 with additional heating studies on a slightly larger version of the shuttle model. Technicians installed over 100 high-temperature transducers on the model to read heat and pressure distributions as it was fired in the 10×10. The 10×10 tests were augmented by similar studies in the 8×6 and Ames's 11- by 11-Foot Transonic Wind Tunnel. Rockwell used the information to evaluate a new reusable blanket-type insulation for the shuttle.[773]

The shuttle's electric system was powered by three hydrogen and oxygen fuel cells, and NASA engineers were concerned about the longevity of these devices on extended missions. Fuel cells rely on powdered catalysts to expedite the reaction of the hydrogen with the oxygen. The effectiveness of these powders, however, declined eventually, causing short circuits and inefficiencies. In 1974 Lewis researchers determined that a new gold and platinum alloy was more efficient than the existing platinum and palladium combination. The new catalyst reduced the required fuel cells from three to two, resulting in both cost and weight savings.[774]

On 12 April 1981 Columbia rose up from Launch Pad A into the Florida sky and became the first space shuttle to enter orbit and land on a runway. The launch signaled NASA's return of humans to space and the beginning of a new era of space research. Although Lewis did not have a large role in the shuttle development, it would be a primary contributor of experiments to be performed during shuttle missions.

## Communications

Communications satellites were one of the space program's most tangible products for the public. Until NASA launched a series of new commercial satellites in the early 1960s, all telephone and television transmission required cables. The initial communications satellites were basically mirrors that reflected transmissions from one site to another.

Image 281: Aircraft models hang in the Visitor Center foyer (GRC–1977–C–04170).

*A number of test facilities, including the Space Power Chambers (SPC), began closing their doors in the mid-1970s because of staff and program reductions. The SPC included a whole complex of support buildings, including one that housed the large compressors and exhausters used to simulate altitudes within the facility. In July 1970 Lewis converted a section of this Exhauster Building into the Aerospace Information Display. The room featured models, hardware, and exhibits that had been used at various outreach events in the 1960s.*

*By 1975 all of the exhauster equipment had been removed from the building. Lewis soon expanded the display area throughout the building, which was first renamed the Visitors Information Center and then the Visitors Center. It not only housed an array of exhibit displays, but provided educational supplies for teachers, staff members, and the general public. It also served as home for Lewis's Speakers Bureau. The effort was a success, and the Visitors Center became a popular attraction for the community, particularly local schools.*[776]

Telstar, launched in 1962, was the first satellite to amplify the signal and retransmit it, providing the first transatlantic television signal. The Intelsats of the mid-1960s could handle hundreds of different channels, providing instantaneous television and telephone contact with other continents.[775]

As innovative as these devices were, they still required relatively large transmission and receiving equipment on the ground. NASA developed two programs in the early 1970s to test experimental satellites capable of significantly amplifying the signal and, therefore, reducing the size of ground equipment. The Applications Technology Satellite–6 (ATS), launched in May 1974, employed a narrow high-power beam that reduced the equipment size and allowed reception in remote areas. ATS–6 broadcast a variety of educational programs to locations around the globe, but only over an area of roughly 40,000 square miles at a time.[777]

The Communications Technology Satellite (CTS), a joint venture between Lewis and the Canadian Department of Communications, used a more powerful transmitter that could provide the same quality signal over a larger area—about one-third of the country. This capability stemmed from a new traveling-wave-tube technology developed at Lewis.[778,779]

Image 282: *Lewis News article about CTS receiving an Emmy.*
Image 283: *Lewis-designed CTS transmitting vehicle (GRC–1977–C–01038).*

Image 284: *The CTS satellite is prepped for testing in the Electric Propulsion Laboratory vacuum tanks during November 1974 (GRC–1974–C–03902).*

Henry Kosmahl's Multistage Depressed Collector was the key element of the CTS.[780] It was developed in the late 1960s to increase the efficiency of the microwave tubes used to amplify signals on communications satellites. By adding the multistage collector to the tube, Kosmahl was able to capture significantly more electrons from the signal than with a standard tube. The resulting efficiency was over twice that of contemporary tubes, and the improved traveling-wave tube enabled smaller, less-expensive ground receiving stations to be used.

Lewis tested the collector and a model of the CTS satellite in the Electric Propulsion Laboratory's vacuum tanks, and managed the Delta launch vehicle that placed CTS in orbit on 17 January 1976. In addition, the satellite's primary control center with eight reflector antennas was based at Lewis. Lewis also designed a mobile CTS station that consisted of a bus and a trailer that were equipped with a large antenna, recording equipment, and cameras for live television interviews.[781,782]

Lewis controlled and coordinated CTS's U.S.-sponsored experiments. Throughout 1977 and 1978 NASA permitted qualified groups to utilize CTS from one of the three transmission centers. The primary users came from the education, medical, special services, and technology fields, but CTS also was utilized to address specific needs, like enabling flood victims to contact out-of-state relatives.

As planned, NASA terminated CTS activities in 1979 after three successful years of operation.[783] CTS transmitted power levels that were 10 to 20 times higher than those of contemporary satellites.[784] This reduced the cost of satellite transmission and expanded reception to areas that had smaller, less sophisticated ground receiving equipment.[785]

CTS technology was successfully applied to a number of communication satellites, providing access to a variety of users who had previously been priced-out of the field, many of whom went on to use commercial communication satellites.[786]

In 1987 the CTS team received the first Emmy ever presented to a federal agency.

## Research Analysis Center

Data recording was an essential element for testing. There was no use running a test if the data could not be recorded and analyzed. As testing became more sophisticated, the costs of data recording and storage escalated.[787] Although Lewis researchers continually designed more efficient methods of data collection, it had become apparent by the mid-1970s that the center's computing capabilities were lacking. This was particularly true in the rapidly expanding new field of computational fluid dynamics and structures.[788] Despite the Agency's budget woes, NASA funded a new centralized computer center at Lewis. Ground was broken for the new Research Analysis Center (RAC) in January 1979 at the former site of the center's picnic grounds. The RAC building would be the first new building at the center since the late 1960s.

The RAC building opened in 1980. It featured a large, open area for the mainframe computers and two floors of office space. Complying with President Jimmy Carter's energy efficiency directives, the Lewis design utilized the heat from the computer systems to warm the entire building, employed energy-efficient lighting, and reduced its window areas.[789]

The $12 million new Cray S/2200 computer system went into operation in October 1982. Linked to all of Lewis's test facilities via coaxial cable, the Cray was a number cruncher that could be used for computational flow dynamics and structural analyses. It allowed researchers to use visualizations to analyze test data and predictive models. The Cray was 10 times faster and one-quarter the physical size of the IBM 370 that it was replacing.[790]

*Image 285: Work stations in the RAC building (GRC–1981–C–03752).*

*Image 286: Cray computer system in the RAC building (GRC–1982–C–05464).*

❖ ❖ ❖ ❖ ❖ ❖

Lewis had many personal computers and word processing units scattered throughout the center, but they were not always compatible or connected with one another. In the late 1970s, the Computer Services Division (CSD) began analyzing the center's future network requirements. As a result, the center sought to synchronize its disparate systems with the new Cray supercomputer in the RAC Building and to create a more interactive computing culture. Technicians installed nearly 22 miles of coaxial cables to connect the various systems across the campus.[791,792]

In August 1982 Lewis became the first NASA center with an operational integrated network.[793] CSD introduced the Lewis Information Network (LINK) to take advantage of this new cable backbone. The LINK network handled high-speed telecommunications, internet, and video transmissions. It also allowed real-time transmission of data from the test facilities to the main IBM 370 processor and back to the facilities. Lewis could connect computer systems at other centers to LINK via a satellite dish located behind the RAC.[794]

With LINK in place, the CSD began taking steps to provide desktop computing for every employee and established the Lewis Information Management System (LIMS) project office in 1984 to develop and implement the effort. In October 1987 the center held a ceremony in the Administration Building auditorium to unveil a LIMS workstation for employees.[795] Over the next three years the center installed 3,000 disk-operating-system- (DOS-) based personal computers.[796] LIMS improved word processing and other existing services, introduced new capabilities like email, and provided greater access to computational and graphical programs.[797] The benefits of computers were now literally at the fingertips of every Lewis employee.

### New Perspective

Bruce Lundin retired in August 1977 after 35 years at the center and nearly 8 as Center Director. The final years were trying, but Lundin managed to guide the center through yet another major transformation. Bernard Lubarsky, another Lewis veteran, served as Acting Director for over a year after Lundin's departure. Funding for aeronautics and energy programs improved in the mid-1970s, but support for other

Lewis mainstays like chemical rocket research was reduced by similar amounts.[798] The combination of budget cuts and Lewis's enthusiastic interest in programs that focused on Earth's resources spurred rumors that the center would be transferred from NASA to the DOE. The center's perceived lack of direction worried some staff members. Lubarsky increased the number of staff/management AWARENESS meetings to try to stem the increasing malaise at the center.[799,800]

In November 1978, NASA brought in John McCarthy, former Director of Space Research at the Massachusetts Institute of Technology, to formally succeed Lundin. Being Lewis's first Director without any

*Image 288: Director John McCarthy (left), former Acting Director Bernie Lubarsky (center), and future Director Andy Stofan (GRC–1978–C–04732).*

NASA experience, McCarthy was able to view the center's problems from a new perspective. In his first address to the staff, McCarthy did not hesitate to explain that Lewis and Cleveland had a poor reputation elsewhere in the country. In addition to the city's bankruptcy and pollution problems, which were well known, McCarthy explained that others mistakenly considered Lewis's research programs and staff as pedestrian. He continued, "I think that Lewis is plagued with the problem of spending all their time doing their job and ignoring the selling aspects and the PR aspects, including [promoting Lewis work at] headquarters."[801]

McCarthy and Lubarsky met for several days with NASA officials to discuss the Agency's long-range plans, and tasked Seymour Himmel with the development of a plan to restore Lewis's reputation. Showcasing of the center's talents was the first step.[802] McCarthy initiated a series of annual reports that described the center's competencies, goals, and accomplishments. Unlike his predecessors, McCarthy also spent a good deal of time in Washington, DC, seeking support for Lewis from Headquarters and local representatives. Lubarsky left NASA in January 1979 to serve in the Central Intelligence Agency.[803]

President Ronald Reagan's administration, which took office in 1980, implemented a series of reductions that threatened to not only derail McCarthy's efforts but shut down the center once and for all. The conservative think tank, the Heritage Foundation, issued a report advocating sharp reductions of expenditures across most federal agencies.[804] Science and research were among the areas on the chopping block, including civil aviation research. Budget Director David Stockman instructed NASA to trim its nonshuttle budget by nearly one-third. This had serious implications for not only Lewis, but Ames and Langley, as well.[805]

*Image 287: Lewis's first attempt at annual self-assessment and planning.*

McCarthy increased his efforts to drum up support locally—from industry—and in Washington, DC, where Congresswoman Mary Rose Oakar and other local representatives took up the fight. Ten upper-level Lewis managers established a Save the Center Committee to augment McCarthy's efforts.[806,807] The crisis reached its apex in December 1981 when Stockman rejected a NASA appeal to ease the cuts.[808] There was pushback from some key figures, however. Defense Secretary Caspar Weinberger stressed that the cuts would impede development of the new stealth bombers; Neil Armstrong stated that previous aeronautics cuts did not spur additional industry research; and Representative Don Fuqua warned that the cuts would "drive the nail into the coffin" of the U.S. aeronautical program, resulting in a "second rate power."[809] NASA officials met with top White House advisors on Friday, 11 December 1981. By Monday it was clear that Lewis would remain open.[810,811]

The hour of crisis had passed. Administrator James Beggs praised McCarthy for standing "strong and tall" during the budget crisis and closure talks.[812] In March 1982 McCarthy announced he was returning to academia.[813] Lewis, however, had turned the corner.

*Image 289: The Lewis hangar in November 1981 (GRC–1981–C–05731).*

## Endnotes for Chapter 7

626. Bruce Lundin, untitled talk at Plum Brook Station, 5 January 1973, NASA Glenn History Collection, Bruce Lundin Collection, Cleveland, OH.

627. Robert Didelot interview, Cleveland, OH, by Mark Bowles, 22 January 2002, NASA Glenn History Collection, Oral History Collection, Cleveland, OH.

628. "NASA Aircraft Energy Efficiency Program Marked for Elimination," *Aerospace Daily* 113, no. 36 (January 1982).

629. Bruce Lundin to Lewis Employees and Resident Personnel, "Cleveland Press News Item," 21 January 1971, NASA Glenn History Collection, Directors Collection, Cleveland, OH.

630. Bruce Lundin to Lewis Employees, "Reduction in Force," 14 March 1972, NASA Glenn History Collection, Directors Collection, Cleveland, OH.

631. William McCann, "NASA to Ax 600 Jobs at Lewis Center," *Cleveland Plain Dealer* (6 January 1973).

632. Ihor Gawdiak and Helen Fedor, *NASA Historical Databook Volume IV, NASA Resources 1969–1978* (Washington, DC: NASA SP–4012, 1994).

633. Jane Van Nimmen, Leonard Bruno, and Robert Rosholt, *NASA Historical Databook Volume I, NASA Resources 1958–1968* (Washington, DC: NASA SP–4012, 1976).

634. Lynn Manley, NASA Lewis Research Center press release 66–60, 30 September 1966, NASA Glenn History Collection, Press Release Collection, Cleveland, OH.

635. Walter T. Olson, "Lewis at 40: A Reflection," *Lewis News* (25 September 1981).

636. Virginia Dawson, *Engines and Innovation: Lewis Laboratory and American Propulsion Technology* (Washington, DC: NASA SP–4306, 1991).

637. Robert English interview, Cleveland, OH, by Virginia Dawson, 11 July 1986, NASA Glenn History Collection, Oral History Collection, Cleveland, OH.

638. Bruce Lundin, "On the Management of National Energy R&D," 4 January 1974, NASA Glenn History Collection, Directors Collection, Cleveland, OH.

639. Charles Kelsey, NASA Lewis Research Center press release 72–2, 19 April 1972.

640. Hugh Harris, NASA Lewis Research Center press release 71–1, 5 January 1971.

641. Charles Kelsey, NASA Lewis Research Center press release 72–102, 21 November 1972.

642. "Technology in the Service of Man."

643. "GASP Monitors Fluorocarbons," *Lewis News* (27 June 1975).

644. Marvin Tiefermann, *Ozone Measurement System for NASA Global Air Sampling Program* (Washington, DC: NASA TP–1451, 1979).

645. Thomas Dudzinski, *Carbon Monoxide Measurement in the Global Atmospheric Sampling Program* (Washington, DC: NASA TP–1526, 1979).

646. "Researchers Find Out What's Over the Poles," *Lewis News* (11 November 1977).

647. William Jasperson and James Holdeman, *Tabulations of Ambient Ozone Data Obtained by GASP Airliners; March 1975 to July 1979* (Washington, DC: NASA TM–82742, 1984).

648. Warren Hovis, John Knoll, and Gilbert Smith, "Aircraft Measurements for Calibration of an Orbiting Spacecraft Sensor," *Applied Optics* 24, no. 3 (1 February 1985).

649. T. D. Brennan and Richard Gedney, *Coast Guard/NOAA/NASA Great Lakes Project ICEWARN* (Washington, DC: NASA N76266555, 1976).

650. "Ice Information System May Help Shippers," *Lewis News* (6 April 1973).

651. R. J. Jirberg, R. J. Schertler, and R. T. Gedney, *Application of SLAR for Monitoring Great Lakes Total Ice Cover* (Washington, DC: NASA TM X–71473, 1973).

652. James Haggerty, "Waterway Ice Thickness Measurements," *Spinoff 1978, An Annual Report* (Washington, DC: NASA, 1978).

653. Paul Rogers, "EPA History: The Clean Air Act of 1970," *EPA Journal* (January/February 1990), *http://www2.epa.gov/aboutepa/epa-history-clean-air-act-1970* (accessed 2 March 2015).

654. C. C. Gleason and R. W. Niedzwiecki, *Results of the NASA/General Electric Experimental Clean Combustor Program* (Reston, VA: AIAA No. 76–763, 1976).

655. "Technology in the Service of Man, Lewis Research Center," 1973, NASA Glenn History Collection, Inspections Collection, Cleveland, OH.

656. C. T. Chang, et al., "NASA Glenn Combustion Research for Aeronautical Propulsion," *Journal of Aerospace Engineering* (April 2013).

657. "Technology in the Service of Man."

658. Dipanjay Dewanji, "Lean Direct Injection," *Leonardo Times* (June 2012).

659. "Low NOx, Lean Direct Wall Injection Combustor Concept Developed," *Research and Technology*, 2002.

660. "Technology in the Service of Man."

661. Lundin, "On the Management of National Energy R&D."

662. Chang, "NASA Glenn Combustion Research."

663. "ALERT Committee Offers Six Programs," *Lewis News* (5 October 1973).

664. "ALERT Committee."

665. Bruce Lundin, "Lewis Means Teamwork," pamphlet, NASA Glenn History Collection, Directors Collection, Cleveland, OH.

666. E. C. Kilgore to George Low, "NASA Inspections," 14 December 1972, NASA Glenn History Collection, Inspections Collection, Cleveland, OH.

667. "Technology in the Service of Man."

668. Walter Olson to George Low, "NASA Inspection: Technology in the Service of Man, Lewis 1973," 29 November 1973, NASA Glenn History Collection, Inspections Collection, Cleveland, OH.

669. John Donnelly to George Low, 20 December 1973, NASA Glenn History Collection, Inspections Collection, Cleveland, OH.

670. "U.S. Energy Information Administration, Energy Perspectives 2011," Table 1.1 (Washington, DC: USEIA, 2011).

671. Richard Nixon, "Address to the Nation About Policies To Deal With the Energy Shortages" (7 November 1973), Online by Gerhard Peters and John T. Woolley, The American Presidency Project, *http://www.presidency.ucsb.edu/ws/?pid=4034* (accessed 14 July 2015).

672. Olson, *Lewis at 40.*

673. Dawson, *Engines and Innovation.*

674. Andrew Stofan, "Foreword," *Technical Facilities Lewis Research Center* (Washington, DC: GPO–1982–561–508, 1982).

675. David Spera, *Wind Turbine Technology* (New York, NY: ASME Press, 2009).

676. Joseph Savino, *A Brief Summary of the Attempts to Develop Large Wind-Electric Generating Systems in the US* (Washington, DC: NASA TM X–71605, 1974).

677. "ERDA/NASA/NSF Wind Energy Program," c1973, NASA Glenn History Collection, Wind Energy Collection, Cleveland, OH.

678. "National Need: Abundant Domestic Supplies of Clean Energy," NASA CS-65027, c1972, NASA Glenn History Collection, Wind Energy Collection, Cleveland, OH.

679. Alfred Eggers and Bruce Lundin, "Memorandum of Understanding Between Research Applications Directorate of the National Science Foundation and the NASA Lewis Research Center Concerning Cooperative & Collaborative Efforts for a Program of Research in Wind Energy Conversion," 4 December 1973, NASA Glenn History Collection, Wind Energy Collection, Cleveland, OH.

680. L. V. Divone, "Evolution of Modern Wind Turbines," *Wind Turbine Technology*, D. A. Spera, editor (New York, NY: ASME Press, 1994).

681. David Spera, "Ten-Year Wind Turbine Program Sets Stage for Future," *Lewis News* (29 November 1985).

682. James Schefter, *Capturing Energy from the Wind* (Washington, DC: NASA SP–455, 1982).

683. Sheila Bailey, Larry Viterna, and K. R. Rao, "Role of NASA in Photovoltaic and Wind Energy," *Energy and Power Generation Handbook: Established and Emerging Technologies* (New York, NY: ASME, 2011).

684. Spera, "Ten-Year Wind Turbine Program."

685. Larry Viterna interview, Cleveland, OH, 4 September 2015, NASA Glenn History Collection, Oral History Collection, Cleveland, OH.

686. Divone, "Evolution of Modern Wind Turbines."

687. L. A. Viterna, "Method for Predicting Impulsive Noise Generated by Wind Turbine Rotors," *Internoise 82 Proceedings* (San Francisco, CA: International Institute of Noise Control Engineering, 1982), pp. 339–342.

688. Divone, "Evolution of Modern Wind Turbines."

689. Bradford Linscott, Joann Dennett, and Larry Gordon, *The Mod-2 Wind Turbine Development Project* (Washington, DC: NASA TM–82681, 1981).

690. Divone, "Evolution of Modern Wind Turbines."

691. Gipe, *Wind Energy Comes of Age.*

692. Larry Viterna interview.

693. Spera, *Wind Turbine Technology.*

694. Bailey, "Role of NASA."

695. Ronald Thomas, *DOE/NAA Lewis Large Wind Turbine Program* (Washington, DC: NASA TM–82991, 1982).

696. Divone, "Evolution of Modern Wind Turbines."

697. David Spera, *Bibliography of NASA-Related Publications on Wind Turbine Technology 1973–1975* (Washington, DC: NASA CR–195462, 1995).

698. Spera, "Ten-Year Wind Turbine Program."

699. Bailey, "Role of NASA."

700. Larry Viterna interview.

701. Spera, *Wind Turbine Technology.*

702. Bailey, "Role of NASA."

703. Lyons, "Power and Propulsion."

704. A. C. Hansen and C. P. Butterfield, "Aerodynamics of Horizontal-Axis Wind Turbines," *Annual Review of Fluid Mechanics* 25 (1993): 115–149.

705. Bailey, "Role of NASA."

706. Susan Johnson and Frederick Simon, *Evaluation of Flat-Plate Collector Efficiency Under Controlled Conditions in a Solar Simulator* (Washington, DC: NASA TM X–73520, 1976).

707. William Bifano, et al., *Social and Economic Impact of Solar Electricity at Schuchuli Village: A Status Report* (Washington, DC: NASA TM–79194, 1979).

708. Bailey, "Role of NASA."

709. Ken Atchison and Paul Bohn, *Solar Cell Systems to be Used in Experimental Third World Projects* (Washington, DC: NASA Release 81–10172, 1981).

710. Lyons, "Power and Propulsion."

711. Bailey, "Role of NASA."

712. "The Lewis Space Solar Cell Team," *Lewis News* (12 October 1979).

713. Del Zatroch and James Blue, "Cyclotron's Fast Neutrons Fight Cancer," *Lewis News* (14 June 1985).

714. Donald Alger and Robert Steinberg, *A High Yield Neutron Target for Cancer Therapy* (Washington, DC: NASA TM X–68179, 1973).

715. "Purer Isotope Technique May Help Medicine," *Lewis News* (23 October 1970).

716. "Lewis Cancer Treating Cyclotron Shuts Down," *Lewis News* (1 February 1991).

717. Noel Nightingale, *Automotive Stirling Engine Mod II Design Report* (Washington, DC: NASA CR–175106, 1986).

718. William Tabata, *Automotive Stirling Engine Development Program—A Success* (Washington, DC: NASA TM–89892, 1987).

719. Nightingale, *Automotive Stirling Engine.*

720. Mechanical Technology Incorporated, *Assessment of the State of Technology of Automotive Stirling Engines* (Washington, DC: NASA CR–159631, 1979).

721. Tabata, *Automotive Stirling Engine Development.*

722. William Tomazic, "Stirling Engine Technology Project Tests New Automotive and Space Power Uses," *Lewis News* (31 May 1985).

723. Nightingale, *Automotive Stirling Engine.*

724. Tabata, *Automotive Stirling Engine Development.*

725. "The ERDA–NASA Energy Activities Team," *Lewis News* (14 May 1976), p. 3.

726. Rebecca Matulka, "The History of the Electric Car" (15 September 2014), *http://energy.gov/articles/history-electric-car* (accessed 1 June 2015).

727. "New Technology for Transportation," *NASA Spinoff 1977* (Washington, DC: NASA, 1977), p. 90.

728. Miles Dustin and Robert Denington, *Test and Evaluation of 23 Electric Vehicles for State-of-the-Art Assessment* (Washington, DC: NASA TM–73850, 1978).

729. "Technology in the Service of Man."

730. Seymour Lieblein, *Problem Areas for Lift Fan Propulsion for Civil VTOL Transports* (Washington, DC: NASA TM X–52907, 1970).

731. Seymour Lieblein, "V/TOL Aircraft," *Exploring in Aeronautics* (Washington, DC: NASA EP–89, 1971).

732. Ronald Blaha, "Completed Schedules of NASA Lewis Wind Tunnels, Facilities and Aircraft 1944–1986," February 1987, NASA Glenn History Collection, Test Facilities Collection, Cleveland, OH.

733. "V/STOL Propulsion Is…," *Lewis News* (28 March 1980).

734. Seth Anderson, *Historical Overview of V/STOL Aircraft Technology* (Washington, DC: NASA TM–81280, 1981).

735. Carl Ciepluch, *A Review of the QCSEE Program* (Washington, DC: NASA TM X–71818, 1975).

736. "Technology in the Service of Man."

737. Harry Bloomer and Irvin Loeffler, *QCSEE Over-the-Wing Engine Acoustic Data* (Washington, DC: NASA TM–82708, 1982).

738. Robert Arrighi, *Pursuit of Power NASA's Propulsion Systems Laboratory No. 1 and 2* (Washington, DC: NASA SP–2012–4548, 2012).

739. Joseph Lubomski, *Characteristics of Aeroelastic Instabilities in Turbomachinery: NASA Full Scale Engine Test Results* (Washington, DC: NASA TM–79085, 1979).

740. Ross Willoh, et al., "Engine Systems Technology," *Aeronautical Propulsion* (Washington, DC: NASA SP–381, 1954).

741. Richard P. Hallion, On the Frontier: Flight Research at Dryden, 1946-1981 (Washington, DC: NASA SP–4303, 1984).

742. Link Jaw and Sanjay Garg, *Propulsion Control Technology Development in the United States* (Washington, DC: NASA/TM—2005-213978, 2005).

743. D. Kurtz, J. W. H. Chivers, and A. A. Ned Kulite, *Sensor Requirements for Active Gas Turbine Engine Control* (Leonia, NJ: Kulite Semiconductor Products, Inc., 2001).

744. Lundin, "On the Management of National Energy R&D."

745. "Quieting the Skies," 19 September 1973, NASA Glenn History Collection, Inspection Collection, Cleveland, OH.

746. "Quieting the Skies."

747. "Refan Program Flight Tests JT8D's," *Lewis News* (24 January 1975).

748. Leonard Stitt and A. A. Medeiros, *Reduction of JT8D Powered Aircraft Noise by Engine Refanning* (Washington, DC: NASA TM X–71536, 1974).

749. "Refan Powers New Aircraft," *Lewis News* (23 December 1977).

750. Dennis Huff, "NASA Glenn's Contributions to Aircraft Engine Noise Research," *Journal of Aerospace Engineering* (April 2013).

751. Mark Bowles, *The Apollo of Aeronautics: NASA's Aircraft Energy Efficiency Program 1973–1977* (Washington, DC: NASA SP–2009–574, 2009).

752. Donald Nored, "ACEE Propulsion Overview," c mid-1970s, NASA Glenn History Collection, ACEE Collection, Cleveland, OH.

753. N. N. Anderson, "NAS 3–20636—Engine Component Improvement Program, Airline Evaluation," 19 December 1980, NASA Glenn History Collection, ACEE Collection, Cleveland, OH.

754. Carl Ciepluch, Donald Davis, and David Gray, "Results of NASA's Energy Efficient Engine Program," *Journal of Propulsion* 3, no. 6 (1987).

755. Bowles, *Apollo of Aeronautics.*

756. Bowles, *Apollo of Aeronautics.*

757. Bowles, *Apollo of Aeronautics.*

758. Bowles, *Apollo of Aeronautics.*

759. P. Genser, "Overall Atlas & Centaur Flight Record," 30 April 1980, NASA Glenn History Collection, Centaur Collection, Cleveland, OH.

760. Dawson, *Taming Liquid Hydrogen.*

761. Titan/Centaur Project Office, *Centaur Standard Shroud Cryogenic Unlatch Tests* (Washington, DC: NASA TM X–71455, 1973).

762. Lewis Research Center Staff, *Centaur Standard Shroud Heated Altitude Jettison Tests* (Washington, DC: NASA TM X–71814, 1975).

763. John Stevens, *Solar Array Experiments on the SPHINX Satellite* (Washington, DC: NASA TM X–71458, 1973).

764. Hugh Harris, NASA Lewis Research Center press release No. 74–5, 8 February 1974, NASA Glenn History Collection, Centaur Collection, Cleveland, OH.

765. R. J. Rollbuhler, *Centaur Boost Pump Turbine Icing Investigation* (Washington, DC: NASA TM X–73421, 1976).

766. "Lewis Marks Anniversary of Role in Mars Landing," *Lewis News* (22 August 1986).

767. "Titan-Centaur D-1T TC-6 and TC-7 Launch Operations and Flight Events," ELVL–1975–0000860 (1975), NASA Glenn History Collection, Centaur Collection, Cleveland, OH.

768. Jet Propulsion Laboratory, "Voyager: The Interstellar Mission" (2015), *http://voyager.jpl.nasa.gov/index.html* (accessed 15 July 2015).

769. Dawson, *Taming Liquid Hydrogen.*

770. M. J. Braun, "Technology Transfer: Lewis Research Center to Space Shuttle," 4 November 1981, NASA Glenn History Collection, Directors Collection, Cleveland, OH.

771. Blaha, "Completed Schedules of NASA Lewis Wind Tunnels."

772. J. W. Foust, *Base Pressure and Heat Transfer Tests of the 0.0225-Scale Space Shuttle Plume Simulation Model (19-OTS) in Yawed Flight Conditions in the NASA-Lewis Research Center 10x10-Foot Supersonic Wind Tunnel* (Washington, DC: NASA CR–151415, 1978).

773. B. A. Marshall and J. Marroquin, *Results of the AFRSI Detailed-Environment Test of the 0.035-Scale SSV Pressure Loads Model 84-0 in the Ames 11x11 Ft. TWT and the Lewis 8x6 Ft. and 10x10 Ft. SWT* (Washington, DC: NASA CR–167685, 1984).

774. Martin Braun, "Technology Transfer: Lewis Contributions to Space Shuttle," *Lewis News* (26 February 1982).

775. "Technology in the Service of Man."

776. Calvin Weiss, discussion with Bob Arrighi, 24 October 2008, NASA Glenn Research Center, Cleveland, OH.

777. "Technology in the Service of Man."

778. Sanford Jones, "Space Electronics Inspection Stop 5," 19 September 1973, NASA Glenn History Collection, Inspections Collection, Cleveland, OH.

779. William Rapp, Dail Ogden, and Darcy Wright, *An Overview of the Communications Technology Satellite Project—Executive Summary* (Washington, DC: NASA CR–168010, 1982).

780. Jones, "Space Electronics Inspection Stop 5."

781. "CTS Ends Tests," *Lewis News* (27 December 1974).

782. "Lewis Helps With Satellite," *Lewis News* (14 January 1976).

783. Rapp, *An Overview of the Communications Technology Satellite Project.*

784. "NASA Wins Emmy Award for CTS," *Lewis News* (2 October 1987).

785. Charles Kelsey, NASA Lewis Research Center press release 71–19, 6 May 1971.

786. Rapp, *An Overview of the Communications Technology Satellite Project.*

787. Daniel Whipple, "Proposals and Ideas for the Structuring and Design of a Large Facility Data System," November 1981, NASA Glenn History Collection, Computing Services Collection, Cleveland, OH.

788. John McCarthy, "Meet the Director," 8 November 1978, NASA Glenn History Collection, Directors Collection, Cleveland, OH.

789. "Ground Broken for One of Nation's Top Technological Computer Centers," *Lewis News* (19 January 1979).

790. "New Computer is Mental Giant," *Lewis News* (8 October 1982).

791. Daniel Whipple interview, Cleveland, OH, by Bonita Smith, 2 August 2001, NASA Glenn History Collection, Oral History Collection, Cleveland, OH.

792. Roger Schulte, *Lewis Information Network (LINK) Background and Overview* (Washington, DC: NASA TM–100162, 1987).

793. NASA Lewis Research Center, "Integrated Program Plan for Computing," February 1990, NASA Glenn History Collection, Computer Services Collection, Cleveland, OH.

794. Whipple interview by Smith.

795. "Developed by Representatives from Throughout the Center," *Lewis News* (30 October 1987).

796. Office of the Inspector General, "Audit of Lewis Information Management Systems (LIMS)," Audit Report LE–93–002, 19 March 1993, Glenn History Collection, Computer Services Collection, Cleveland, OH.

797. "LIMS Will Enhance Office Productivity," *Lewis News* (19 October 1984).

798. "State of Center Healthy, Lubarsky Tells Supervisors," *Lewis News* (11 November 1977).

799. McCarthy, "Meet the Director."

800. Dawson, *Engines and Innovation.*

801. McCarthy, "Meet the Director."

802. McCarthy, "Meet the Director."

803. Jeffrey Richelson, *The Wizards of Langley: Inside the CIA's Directorate of Science and Technology* (New York, NY: Basic Books, 2008).

804. Dawson, *Engines and Innovation.*

805. "Beggs Reaffirms—No More Staff Reductions," *Lewis News* (2 July 1982).

806. Dawson, *Engines and Innovation.*

807. James Hawker and Richard Dali, Anatomy of an Organizational Change *Effort at the Lewis Research Center* (Washington, DC: NASA CR–4146, 1988).

808. Judy Grande, "Stockman Key Figure in Future of NASA Lab," *Cleveland Plain Dealer* (8 December 1981).

809. "Budget Cuts Forcing NASA to Close Lewis Research Center," *Defense Daily* (9 December 1981).

810. Judy Grande, "There's Still Hope for Lewis," *Cleveland Plain Dealer* (12 December 1981).

811. Judy Grande, "Lewis Lab to Remain Open," *Cleveland Plain Dealer* (14 December 1981).

812. "Beggs Reaffirms—No More Staff Reductions."

813. "Week in Review," *Cleveland Plain Dealer* (7 March 1982).

Image 290: Andy Stofan speaks at the Shuttle/Centaur rollout ceremony on 23 August 1985 at General Dynamics's San Diego headquarters (GRC–1985–C–06212).

*"I want [an employee's] thoughts and his ideas, not just his arms and legs. There is an untapped talent of thoughts and ideas that is not being utilized."*

—Andy Stofan

*Image 291: The Icing Research Tunnel's drive fan as it appeared in July 1982. The IRT had been recently restored to support Lewis's reinstituted icing research program (GRC –1982–C–04302).*

# Bootstrapping the Center

"It was a damn miracle," Lewis engineer William "Red" Robbins recalled of Lewis's first official strategic planning effort, which included proposals to manage five new large projects. "Now you get one or two out of five, that's ideal. You're damned lucky. (Center Director Andy Stofan) was real lucky. He brought them all in."[814] Stofan, who replaced the outgoing John McCarthy in June 1982, expanded on McCarthy's initial efforts to reverse the reactionary mindset that had taken hold at the center. Stofan led a two-pronged attack to remake Lewis: first, conduct long-range planning with an emphasis on acquiring key Agency programs; second, transform the center's management system.

The five new projects referred to by Robbins were the next phase of the Advanced Turboprop (ATP) program, the Advanced Communications Technology Satellite (ACTS), Shuttle/Centaur, restoration of the Altitude Wind Tunnel (AWT) for icing research, and, perhaps most importantly, the electrical power system for Space Station Freedom. These programs would dominate Lewis's activities for the next decade. They not only brought Lewis back into NASA's primary space missions but led to the first new personnel additions in years and a major upgrade to the center's computing services.

Image 292: Strategic planning cartoon.

John Klineberg became Director in 1986 and bolstered the center's future by strengthening its ties with industry, universities, and the military. Lewis's growth continued with the appointment of Larry Ross as Director in 1990. The gradual reopening of Plum Brook Station during this period was a palpable sign of Lewis's comeback. President George H. Bush's Space Exploration Initiative (SEI) led to several new programs at Plum Brook. Although Lewis succeeded technologically in almost all of the new efforts, external factors dimmed the initial promise of several of the endeavors. Nonetheless the 1980s and early 1990s proved to be as productive as any in the center's history. Lewis was back.

## Recreating Lewis

As one of his final acts as Director, John McCarthy asked Robbins to initiate a long-range planning effort to prevent a recurrence of the threatened shutdown.[815] Robbins reorganized the Save the Center Committee as the Strategic Planning Committee and set to work. The 15-member group, which included representatives from Lewis's space, aeronautics, and institutional areas, began meeting in March 1982. Neither McCarthy nor incoming Director Andy Stofan participated in this process. The committee analyzed the center's history and areas of expertise while trying to predict the nation's future aerospace trends. The group then evaluated several alternatives for Lewis to pursue, their impact, and political ramifications. For the first time in nearly a decade, Lewis was becoming aggressive in regards to its future. Over the course of a week in June 1982, the committee members presented their findings to center management and then to Stofan upon his arrival shortly thereafter.[816]

Lewis hired Stofan right out of university in 1958 as a research engineer, and he was soon analyzing the sloshing effects of liquid hydrogen for the Centaur Program. Stofan was promoted through a number of positions in the program, culminating in his appointment as Head of the Titan-Centaur Project Office and Director of the Launch Vehicles Directorate.

With the end of the Titan-Centaur Project in 1978, Stofan accepted the position of Deputy Associate Administrator of Space Science at headquarters.[817] Administrator James Beggs recalled, "it didn't take us long to realize that we had a rising star in Andy."[818] After several years, however, Stofan tired of Washington, DC. When Beggs asked him to return to Cleveland as Center Director to rebuild Lewis, he looked forward to the challenge.[819] Upon his arrival in June 1982, Stofan took McCarthy's initial planning and image rehabilitation efforts and energetically applied them in a manner that complemented his attempts to restore employee morale and introduce new inclusive management techniques.

Stofan worked with the directorate heads to formalize the recommendations into a "Strategic Planning Options" report, which was released in December 1982. The planning was sometimes a difficult process because the management team was not used to being asked for their opinion on such critical decisions, but Stofan insisted on their participation.[820] For aeronautics, the plan sought to fully restore the icing research program, increase experimental work on the ATP, and continue basic propulsion research. For space, it included the pursuit of ACTS, increased electric propulsion efforts, and a new experimental microgravity research program to be conducted on the shuttle and the proposed space station. The report also proposed integrating the Centaur rocket into the shuttle system and extending the Atlas-Centaur launches.[821]

The new programs came just in time. Earlier in 1983 NASA had announced that it would phase out all of its terrestrial energy projects. In November 1982 Beggs had allowed Lewis to temporarily continue existing energy research for several years, but stated, "For the future it will be our policy to support energy programs which yield benefit to the aeronautics and space programs."[822] In addition, NASA was beginning to examine its withdrawal from the expendable launch vehicle business. This would ultimately transfer responsibility for Centaur to private industry. In an effort to realign Lewis with NASA's main space missions, the report audaciously advocated Lewis's management of the electrical power system for the newly planned space station.[823] The changes were cemented with a reorganization in February 1983. The most significant adjustment was the creation of the new Space Technology Directorate, which included the former terrestrial energy research, space power systems, and space propulsion technology areas.[824]

The strategic planning effort made it clear to the committee members that they all worked for Lewis, not for their individual divisions or fields. Stofan seized upon this and implemented what he referred to as "participative management."[825] Larry Ross later stated that the center's culture had changed from "individual entrepreneurship to one of people pulling together as a unit, as a community unto itself."[826] Under the NACA, Lewis had a great deal of autonomy and relied on the unilateral decision-making of Abe Silverstein, Bruce Lundin, and a few others to bring the center success. That autonomy eroded under NASA, and the autocratic management system did not work in the current corporate climate. Stofan's inclusion

*Image 293: Stofan talks with managers during a "Meet the Director" event at the picnic grounds in October 1986 (GRC–1982–C–06266).*

of staff from across the center in high-level decisions empowered the workforce and gave them a stake in the center's success.[827]

It was not easy to get center personnel to participate in the process, and it was not easy to change the decades-old culture. Morale was low, and many had spent their entire careers in another type of atmosphere.[828] Employees were slow to approach Stofan during informal Meet the Director sessions, so he increased the number of these types of meetings, expanded the Acquainting Wage Board, Administrative and Research Employees with New Endeavors of Special Significance (AWARENESS) program, and began converting the Utilities Building into the Employee Center.[829]

On 20 January 1981 newly elected President Ronald Reagan authorized a federal hiring freeze. The freeze was augmented by layoffs, personnel transfers, and incentives for nondefense government employees to retire.[830,831] Although this round of layoffs did not dramatically affect Lewis, many veteran employees—including top managers such as Seymour Himmel, Walter Olson, and Merwin Ault—left to take advantage of the increased retirement benefits.[832] The years of downsizing had resulted in an aging workforce that was now leaving just as new programs were being introduced.

Stofan strove to address these deficits by hiring a younger contingent of civil servants for research positions and contracting out for institutional work. In summer 1983, Lewis hired 190 new professionals, nearly all directly out of university. It was the first increase in staffing since the mid-1960s.[833] Although this increase in hiring was short-lived, it did provide a boost in morale. The utilization of contractors was controversial among the veteran personnel, however. Lewis had emerged from the NACA, which had prided itself on accomplishing all of its duties, from janitor to scientist, in-house. In the 1960s Lewis continued to handle most tasks but began contracting out for the creation of large hardware. In the 1970s NASA began contracting for additional tasks, like food service, but Lewis stubbornly resisted this trend, and tensions mounted between the center and headquarters. Administrator Beggs ordered Stofan to increase Lewis's contractor staff to bring the center more in line with the other centers.[834,835] In many

cases these contractors were retired Lewis employees who accepted positions with private firms.

## Aeronautic Achievements

The ATP program had begun in the mid-1970s with theoretical analysis and wind tunnel studies. The emerging swept-propeller design permitted fuel-efficient performance at relatively high subsonic speeds without the drag implications of normal propellers. Largely through Lewis's advocacy, the program slowly began to gain traction with the aircraft industry and military, which were feeling the effects of the energy crisis. NASA incorporated the ATP into the larger Aircraft Energy Efficiency Program (ACEE) effort in 1978.[836,837] The renewed program consisted of three phases: (1) developing the initial propeller, drive system, and airframe technologies; (2) testing larger propellers and advanced drive systems; and (3) integrating the system into an aircraft for flight-testing.

NASA contracted with Pratt & Whitney and the Allison Engine Company to develop the drive system and verify the propeller blades' ability to withstand

*Image 294: Lewis's 1983 Strategic Plan.*

the forces of a full-scale engine for long durations. In the meantime, General Electric unveiled its own turboprop design—the Unducted Fan (UDF) with gearless and counter-rotating blades. Although NASA did not foresee this development, the Agency incorporated the UDF into Lewis's ATP test program in 1984.[838] General Electric successfully ran the UDF on a test stand, and Lewis and Pratt & Whitney tested various scale models of the Pratt & Whitney engine integrated in an aircraft in the 9- by 15-Foot Low-Speed Wind Tunnel (9×15) and other facilities.

The final phase of the program, which began in October 1984, consisted of a series of flight tests for the Pratt & Whitney and General Electric engines to evaluate the propeller's structural integrity and vibrations of the propeller, as well as noise levels inside the aircraft and on the ground. Lockheed incorporated the Pratt & Whitney engine in a Gulfstream II aircraft, and Lewis pilots flight-tested the engine on the Gulfstream between 1987 and 1989.[839] Additional flight tests in 1989 measured the noise levels from the ground. General Electric flight-tested their engine on a Boeing 727 and a McDonnell Douglas MD–80 in 1986 and 1987.

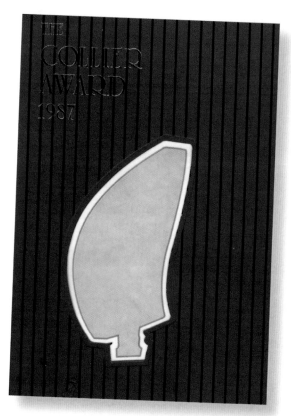

*Image 295: Collier Trophy awarded for Lewis's ATP work.*

*Image 296: The Gulfstream II and Learjet, both flown by Lewis pilots, conduct inflight noise measurements (GRC–1987–C–10756).*

*Image 297: A Propfan Test Assessment model is tested for flutter in the 9×15 wind tunnel during October 1985 (GRC–1985–C–07896).*

Lewis pilot Bill Rieke performed the difficult task of flying the center's Learjet alongside both the Gulf-stream and 727 to record the engine noise in flight. The 108 test flights during this period demonstrated that the turboprop engines consumed up to 50 percent less fuel than contemporary turbofan engines while maintaining comparable performance. Although noise levels were elevated, they were considered reasonable.[840]

NASA and the participating engine manufacturers had demonstrated a significant new propulsion technology that could save billions of dollars in fuel annually. Oil prices, however, stabilized by the mid-1980s, and the industry balked at converting its fleet of jet aircraft to turboprops. Nonetheless, NASA and its industry partners received the Collier Trophy for their efforts in 1987.[841]

❖ ❖ ❖ ❖ ❖ ❖

Icing research became an issue again in the late 1970s with the surge in the number of helicopters, small regional prop-driven airliners, and general aviation aircraft. Lewis's icing program, which had begun in the 1940s, had been suspended for nearly 20 years following the establishment of the space agency and the addition of jet engines to the nation's airline fleet. The aircraft industry occasionally used the

Icing Research Tunnel (IRT), but Lewis's own icing research was nonexistent during the 1960s and early 1970s.[842] In July 1978 Lewis hosted the International Workshop on Aircraft Icing during which participants urged Lewis to restore its icing research program to address the concerns posed by the new types of aircraft.[843] Lewis consequently brought the IRT back online and formed a small, modestly funded group of icing researchers.

Headquarters did not agree to fully restore the program until a series of icing-related commuter aircraft crashes occurred in the early 1980s.[844] The new funding allowed the center to acquire the DeHavilland Twin Otter aircraft, modernize the IRT, and initiate a study on the restoration of the AWT for helicopter icing studies. The center expended a substantial amount of time and money on the AWT effort before Congress terminated the activity in March 1985.[845] The Twin Otter and the IRT, however, continue to serve as premier icing research tools.

Lewis's new icing program was broad in scope. Researchers used the IRT, Twin Otter, and computer simulation systems to improve ice protection systems, instrumentation, testing capabilities, and computer

*Image 298: A prototype turning vane for the new AWT installed in the IRT during December 1985 (GRC–1985–C–09342).*

modeling. The tunnel, aircraft, and software complemented and validated each other's data.[846] The IRT was more active in the 1980s than at any point since 1950. Researchers investigated issues related to helicopter inlet icing, fluid deicers, pneumatic deicing boots, icing on general aviation aircraft, and a series of electromechanical deicing systems.[847]

Helicopter icing was a particular concern at the time. There were few all-weather helicopters in the United States because of the demanding Federal Aviation Administration (FAA) certification process.[848] In an effort to improve this practice, Lewis undertook an extensive effort to determine if the IRT could accelerate all-weather use. Lewis also established a Rotor Icing Consortium, consisting of industry, military, and university representatives, to consult on the effort. Sikorsky Aircraft Corporation agreed to supply a complex subscale version of its Black Hawk helicopter, and Lewis designed the test program.[849]

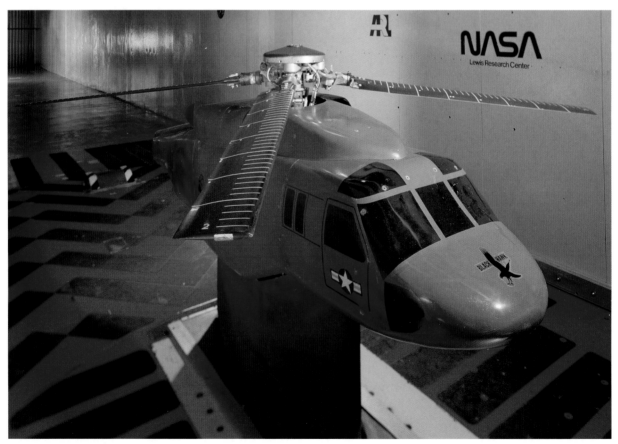

*Image 299: Rotorcraft model installed in the IRT (GRC–1993–C–03670).*

*Image 300: Lewis's Twin Otter during a wing icing study in Duluth, Minnesota (GRC–1988–C–01725).*

Because the model Sikorsky Black Hawk was the first rotorcraft to be operated in a U.S. wind tunnel, the researchers initially installed a smaller Bell OH–58 tail rotor on a special rig to serve as a trial run.[850] They found that some modifications to the tunnel and operating procedures were required, but the OH–58 tests successfully prepared Lewis for the Sikorsky model.[851,852] The researchers began testing the heavily instrumented OH–60 Black Hawk model in fall 1989 and tested again several years later with another rotor. Data from the more than 200 runs revealed that the NASA software predictions were correct for ice found in warmer temperatures but that the code needed additional work for icing at lower temperatures.[853]

❖ ❖ ❖ ❖ ❖ ❖

The first predictive icing computer codes were developed in the 1980s to offset the costs associated with wind tunnel or flight testing. The IRT and Twin Otter repeatedly verified different ice prediction software data. Lewis contributed by providing grants to universities in 1983 to develop icing codes for certain specific parameters. In 1987 Lewis combined several of these codes into a single piece of software applicable to the aircraft industry—the LEWis ICE accretion program (LEWICE). LEWICE takes into account

velocity, droplet size and trajectory, and ice layering to predict the location, shape, and rate of ice buildup over time, as well as the amount of heat required to prevent icing. The software helps aircraft designers assess performance losses for a range of icing parameters. The software predictions were validated in the IRT and with the Twin Otter. Lewis has updated the LEWICE software several times over the years, including developing a three-dimensional version. The aeronautics community continues to utilize this award-winning desktop-based version of the program.[854,855]

Like their predecessors in the 1940s and 1950s, Lewis pilots purposely flew the Twin Otter into ice-producing clouds. The Twin Otter was equipped with a heavy-duty deicing system and was resilient enough to fly through difficult weather conditions. Lewis technicians equipped the aircraft with a laser spectrometer to determine water droplet distribution, color radar, atmospheric measuring devices, and probes to measure water content in clouds.[856] In the mid-1980s Lewis employed the Twin Otter extensively to gather data on icing's effect on aircraft performance, commuter aircraft icing, stereophotography methods for measuring ice, and alternative ice-prevention systems.[857,858]

*Image 301: The Twin Otter performs an inflight icing test in Duluth, Minnesota, while a Sikorsky helicopter trails water vapor to simulate icing clouds (GRC–1988–C–01727).*

The Electro-Impulse Deicing System (EIDI) developed by Lewis in partnership with Wichita State University was one of the most successful ice-prevention systems. The EIDI used coils of copper ribbon that conducted an electrical impulse on the aircraft's leading edges to prevent ice accumulation. Test flights with the Twin Otter and a Cessna in the mid-1980s demonstrated that the EIDI system was an inexpensive and energy-efficient method of deicing wings.[859] Several companies are marketing the system for general aviation aircraft today.

Trying to locate the correct types of ice-producing clouds can be frustrating for researchers. So the Army created the Helicopter Icing Spray System (HISS) to create icing conditions in the atmosphere. HISS consisted of a Chinook helicopter equipped with an 1,800-gallon water tank and spray assembly. The helicopter would fly horizontally and spray droplets into frigid air. Then a research aircraft would follow, flying through the ensuing trail of artificial icing conditions.[860] The Lewis icing group took the Twin Otter to Minnesota during winter 1989 to help calibrate the HISS, test their own cloud sensor equipment, and generally take advantage of the region's natural icing atmosphere. Lewis pilots flew the Twin Otter through the HISS cloud, took measurements, and then photographed the ice buildup on the aircraft after exiting the cloud. Lewis was able to both improve their icing database and assist in the calibration of the Army's instrumentation.[861] The center's icing research program is flourishing today, with the IRT serving a long line of federal and commercial customers.

❖ ❖ ❖ ❖ ❖ ❖

Lewis also sought to improve the design of helicopter propulsion systems. Helicopter development had begun in the early 20th century, but the first practical machines did not appear until World War II. The arrival of gas turbine engines improved performance after the war, and the United States utilized helicopters as aerial ambulances during the Korean War. By the 1960s, the military was also successfully using helicopters as weapons.[862] It was in this atmosphere that

NASA Lewis and the Army joined forces in 1970 to design new helicopter transmissions that were lightweight and quiet.

Helicopters rely on advanced transmission systems to convert the energy from the horizontal engine to the vertically mounted rotor, but the transmissions in service during the 1960s were heavy, loud, and short-lived.[863] Erwin Zaretsky's Gearing and Transmission Section performed much of the Lewis work.[864] They used computer modeling to develop the new designs but relied on testing actual transmissions to verify the computer codes. Lewis built two full-scale transmission test stands and four component stands to support models. The Engine Research Building's 500-horsepower (hp) Helicopter Transmission Rig, which began operating in 1979, could test both traditional helicopter transmissions and the new Lewis-designed traction transmissions. In 1981 Lewis added a 3,000-hp transmission rig to measure component variables in hopes of reducing engine noise and overhauls.[865]

The team thoroughly mapped and analyzed a UH–60 Black Hawk transmission on Lewis's new 500-hp transmission rig to validate the new predictive computer codes that would be used to design advanced transmissions. Meanwhile other researchers were developing lightweight materials and lubricants and designing new types of gears that reduced the overall transmission weight and had significantly longer life-spans.[866,867] The researchers were able to demonstrate these new technologies in a Bell OH–58 helicopter transmission on the Lewis test rig.

Vibrations from the transmission gears generate a significant amount of noise within the helicopter cabin that can affect pilot performance. To reduce the noise, Lewis researchers improved the manufacturing of spiral bevel gears to reduce the geometrical errors that cause vibrations. They also constructed a dual-measurement method that significantly reduced the number of false alarms in systems that monitor the health of the engine.[868]

*Image 302: Helicopter Transmission Rig in the Engine Research Building (GRC–1978–C–4355).*

In 1987 the Army and NASA undertook a new joint initiative to develop conceptual transmissions for advanced military and cargo helicopters. These designs would reduce weight and noise while increasing reliability.[869] The effort resulted in four new transmission designs, including a modified Bell OH–58C model that utilized the Lewis bearing and gear technology and new methods of lubrication and cooling. The transmission's increased power, torque, and efficiency was the foundation for Bell's new OH–58D helicopter.[870]

### New Endeavors in Space

In the early 1980s Lewis secured two new space programs that directly involved the center with NASA's human space efforts for the first time since the early 1960s: the Shuttle/Centaur and the Space Station Power System programs. After playing only a minor role in the shuttle development, the new Lewis management team strove for inclusion in the Agency's next major space program—the space station. Lewis sought to use its 20 years of energy conversion experience to become the Agency's space power leader. The center believed that the proposed space station would be just the first of many projects.[871,872]

As the space shuttle program came to fruition in 1981, NASA resurrected its long-dormant plans to construct an orbiting space station. The station would need a large electric power system to operate electronics, conduct experiments, and maintain its orbital position. Photovoltaic power modules with solar arrays would provide the basic power requirements, and a solar dynamic power system would concentrate and store energy to maintain power for when the station was in the shade. It would be the largest space power system that had ever been attempted.[873,874]

Lewis's pursuit of the space station program was a controversial decision. Some Lewis veterans felt that such a large development program would negatively impact the center's research efforts. Other centers, especially the NASA Marshall Space Flight Center, questioned Lewis's technical capability and its ability to manage such a large program. Lewis's management, confident of the center's competence in both of these areas, opened a project office in August 1982 to determine the best way to integrate its skills with the proposed space station's needs. After nearly a year of the center's dogged advocacy, NASA assigned the power system to Lewis in late 1983.[875,876]

*Image 303: Space station solar array being set up in the PSF cleanroom in July 1990 (GRC–1989–C–00115).*

*Image 304: Space station report prepared for Andy Stofan in his new role as Associate Administrator.*

the implementation of the solar dynamic power system. Although this caused the disbanding of the Solar Dynamic Power Module Division, Lewis carried forward its work on power modules in the 1990s while continuing its efforts to meet the space station's ever-changing expectations.[881]

❖ ❖ ❖ ❖ ❖ ❖

Lewis's seminal Centaur Program began to lose steam in the 1980s. As the six Titan-Centaur launches were successfully completed in the mid-1970s, NASA was developing the space shuttle. The Agency decided to phase out its expendable launch vehicle operations—including Saturn Delta and Atlas-Centaur—and use the shuttle exclusively to launch its satellites and spacecraft.

The Agency, however, severely miscalculated the cost and frequency of the shuttle launches. As the demand for commercial and military satellites grew, the European Space Agency filled the void with its Ariane launch vehicles. NASA continued stringing the Atlas-Centaur along on a limited basis to launch satellites, but the manufacturers, General Dynamics and Lockheed, began phasing out their vehicle production. In addition Centaur was becoming somewhat outdated, and the number of resident experts on staff at Lewis, General Dynamics, and Pratt & Whitney was dwindling.[882]

The Reagan Administration's Commercial Space Launch Act of 1984 allowed private companies to begin providing launch services, but the commercial providers could not compete with the subsidized shuttle and Ariane programs. It was not until the loss of *Challenger* in 1986 that the government took serious steps to halt its planned phase-out of expendable launch vehicles for NASA missions and its plans to launch commercial satellites from the shuttle.[883]

NASA implemented a new mixed-fleet policy that utilized the shuttle and commercial expendable launch vehicles. The Agency continued to manage the launch of government payloads while General Dynamics now handled those for private industry. In addition, NASA would no longer purchase vehicles from the manufacturer, but would buy launch services. General Dynamics and Martin Marietta (formerly Lockheed) not only manufactured the Atlas-Centaurs, which were now referred to as only "Atlases," but also handled

For the first time since the Centaur Surveyor Program, Lewis had a direct role in NASA's crewed spaceflight efforts. This led to increased funding and affirmed the center's critical contributions to the Agency.[877] In 1985 Lewis created the 250-person Space Station Freedom Directorate under Ron Thomas and began constructing the Power Systems Facility (PSF). PSF served as the hub for the entire directorate's testing, provided a repository for the test data, and included a 100-foot-square, 63-foot-high class-100,000 clean room.[878] From 1984 to 1986 Lewis undertook an intensive effort to define the station's power requirements and develop an adequate system within budgetary limitations.

The overall space station design, however, changed dramatically several times in the mid-1980s, resulting in budget overruns and schedule slippage. Political arguments over the funding and technical debates regarding the design continued throughout the 1980s. The amount of requisite power changed after each of these modifications, causing numerous revisions of the power system's technical requirements.[879] In 1988 it became apparent to the Lewis staff that the latest design would stretch the 75-kilowatt system to its limits.[880]

The design changed again in March 1990 when Congress ordered NASA to reduce the station's size by 50 percent. This reduction decreased the number of externally mounted experiments and the overall power requirements. Program managers eliminated one of the four solar arrays and indefinitely delayed

Image 305: Atlas-Centaur 1 shroud jettison test at the newly reactivated Space Power Facility (SPF) for the upcoming Combined Release and Radiation Effects Satellite (CRRES) launch. It was the first major hardware test at the SPF in over 15 years (GRC–1990–C–00216).

Image 306: Shuttle/Centaur Program report.

the hardware integration and actual launches. Lewis oversaw the government missions using Atlas- or Titan-Centaurs, while the NASA Goddard Space Flight Center managed smaller vehicles like the Delta. The NASA Kennedy Space Flight Center was responsible for processing the vehicles and launching them.[884] The 25 September 1989 launch of a U.S. Navy Fleet Satellite Communications (FLTSATCOM) satellite was Lewis's final commercial launch.

Concurrently with these developments, NASA was attempting to deal with the shuttle's limited ability to launch spacecraft. The shuttle's low Earth orbit meant that some sort of booster vehicle would be needed to deploy spacecraft requiring higher altitudes, such as the upcoming Ulysses and Galileo missions to the Sun and Jupiter, respectively. In 1979 NASA began analyzing a number of options, including the use of a Centaur stage.[885,886] After much debate, in early 1981 the Agency decided to use Centaur as that booster. Again there was robust debate about which center should manage the effort. The main space centers—Johnson, Kennedy, and Marshall—argued vehemently that Shuttle/Centaur belonged at Marshall, not Lewis. Director John McCarthy argued that Lewis's existing Centaur team had experience updating vehicles, integrating payloads, and dealing with contractors.[887] Headquarters agreed and assigned responsibility for Shuttle/Centaur to Lewis in May 1981. Congress approved the decision in July 1982, providing the center with a $100 million shot in the arm just as Stofan became Director. He said at the time, "We are in a position to move promptly. We're off and running."[888]

In 1983 Lewis created a Shuttle-Centaur Program Office under Red Robbins. Lewis worked with General Dynamics to reconfigure the upper stage into shorter, stouter versions—the Centaur G and Centaur G Prime—and to develop a launching system that would fit in the shuttle's payload bay. This was the first time that Lewis had worked directly with Johnson, and tensions were strained. Lewis's expertise with Centaur was undisputed, but the center had been outside the human spaceflight program since Project Mercury. Johnson had managed the Agency's human spaceflight program since the early 1960s and was protective of its procedures and approach to safety and mission assurance in human spaceflight. Johnson also had serious concerns about transporting Centaur and its tank of liquid hydrogen alongside its astronauts. For several years Lewis and Johnson engineers wrangled over weight, redundancy, and quality assurance issues. Another issue was that Centaur was designated as both a payload and part of the shuttle system: as such it had to meet a large number of demanding requirements.[889-891]

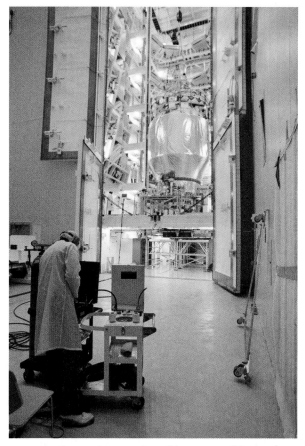

Image 307: Centaur G on a work stand being prepared for the Ulysses mission at the Vertical Processing Center on 6 February 1986 (NASA KSC–86PC–0088).

Lewis engineers worked determinedly throughout the mid-1980s to modify the Centaur and integrate it into the shuttle in time to meet the launch window for *Ulysses* and *Galileo* in late spring 1986. In January 1986 the Shuttle/Centaur underwent a successful tanking test at Cape Canaveral and appeared to be on schedule to meet the impending launch opportunities.[892]

On 28 January, however, *Challenger* exploded shortly after liftoff, and everything related to the shuttle was in question. Robbins saw the writing on the wall and retired immediately.[893] NASA grounded the shuttle fleet for nearly three years and delayed the *Galileo* and *Ulysses* missions indefinitely. The Agency canceled the Shuttle/Centaur Program five months later when officials concluded that the vehicle could not meet the shuttle's new stricter safety standards. The increase in required redundancies that would cause an untenable increase in weight.[894] The cancellation, which came after years of intensive effort, was one of Lewis's biggest disappointments. Nonetheless, Larry Ross recalled, "Our team emerged as better technical contributors, better engineers. We also ended up with some very good managers."[895]

❖ ❖ ❖ ❖ ❖ ❖

Lewis's third major space effort during this period was ACTS. As federal subsidies for U.S. satellite research decreased in the early 1980s, foreign competitors accelerated their satellite development and Japan

*Image 308: Artist's rendering of the ACTS deployment from Discovery (GRC–1987–C–01695).*

*Images 309: ACTS antennas mounted on the 8- by 6-Foot (8×6) Supersonic Wind Tunnel (GRC–1996–C–03369).*

became the premier telecommunications satellite provider. ACTS was one of the initial steps taken to reestablish the nation's communications satellite leadership.[896]

In 1978 Lewis researchers began assessing the nation's communications requirements for the 1990s. They then examined the feasibility of conceptual hardware items and took steps to get them into flight condition. In August 1984 Lewis received approval to contract with Lockheed to manufacture the satellite. Lewis created the ACTS Project Office under Richard Gedney to manage the development of the satellite's various components. The office designed the Multibeam Communications Package, analyzed the entire communications system, and created the ground station to handle all communications with the satellite. Lewis was responsible for integrating the satellite with both the ground station and the shuttle launch vehicle.[897]

*Space Shuttle Discovery* deployed ACTS on 12 September 1993. As it had for ACTS's mid-1970s predecessor, the Communications Technology Satellite, the depressed collector transmission tube generated the high-power satellite communication services. ACTS, however, used the higher-frequency Ka Band, which provided additional bandwidth. As the first high-speed digital communications satellite, ACTS served as a testbed for researchers. ACTS's ability to lift its beam from one location to another and rapidly change its relay destinations resulted in its nickname—the "Switchboard in the Sky."[898]

The satellite hosted over 65 industry and university experiments in the ensuing years. Lewis terminated its test operations in May 2000, but external users continued to use ACTS for several years. ACTS technology quickly demonstrated its application to industry, education, and telemedicine. Concurrent advances in ground-based fiber optics and cell phone technology, however, proved more effective for the current applications.[899]

## NASA's New Mission

In 1986 Stofan accepted a position at headquarters managing the space station program's tangled contractual obligations. Stofan was succeeded at Lewis by his Deputy Director, John Klineberg, who had worked for the aerospace industry in California throughout the 1960s. Klineberg had begun his NASA career in 1970 as an aeronautical research engineer at the NASA Ames Research Center. Four years later he had been named head of the Low Speed Aircraft Branch at headquarters, and in 1978 NASA Administrator Robert Frosch had asked Klineberg to serve as John McCarthy's Deputy Director at Lewis.[900,901]

As Director, Klineberg saw several of the center's new programs come to fruition, including the ATP and ACTS. He also managed the space station power effort as the design continually transformed. Klineberg continued Stofan's ongoing strategic planning efforts and provided the staff the flexibility and freedom to carry out their research. In 1986 he oversaw the creation of an Employee Center which consolidated personnel services such as the cafeteria, the credit union, the exchange, and medical services.[902] Two of Klineberg's most notable achievements were the restoration of Plum Brook Station and the establishment of the Ohio Aerospace Institute (OAI).

OAI is a nonprofit partnership between Lewis, the Air Force Research Laboratory, the local business community, researchers from the various institutions,

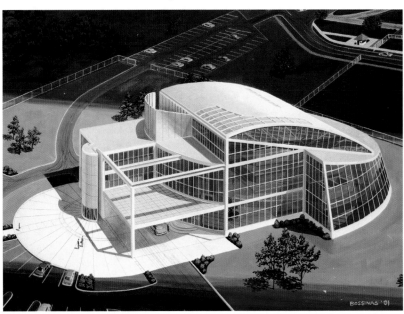

*Image 310: OAI (GRC–1991–C–01824).*

*Image 311: John Klineberg (GRC–1987–C–75023).*

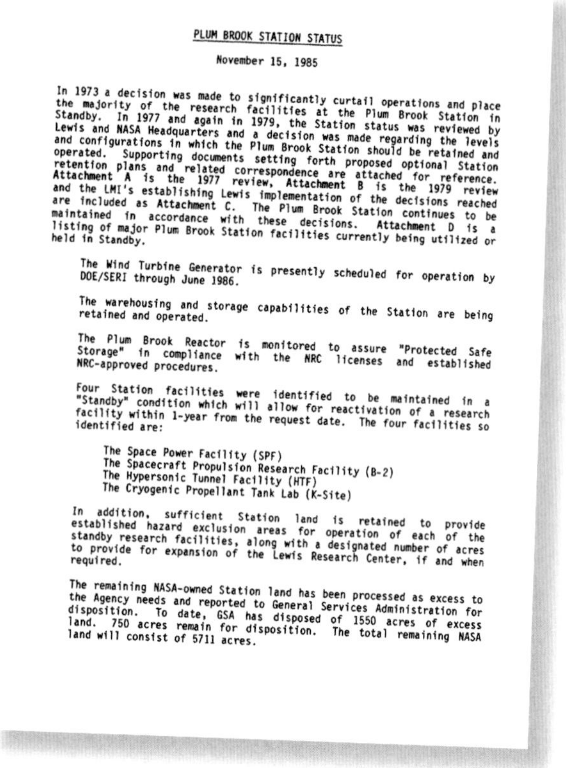

PLUM BROOK STATION STATUS
November 15, 1985

In 1973 a decision was made to significantly curtail operations and place the majority of the research facilities at the Plum Brook Station in Standby. In 1977 and again in 1979, the Station status was reviewed by Lewis and NASA Headquarters and a decision was made regarding the levels and configurations in which the Plum Brook Station should be retained and operated. Supporting documents setting forth proposed optional Station retention plans and related correspondence are attached for reference. Attachment A is the 1977 review, Attachment B is the 1979 review and the LMI's establishing Lewis implementation of the decisions reached are included as Attachment C. The Plum Brook Station continues to be maintained in accordance with these decisions. Attachment D is a listing of major Plum Brook Station facilities currently being utilized or held in Standby.

The Wind Turbine Generator is presently scheduled for operation by DOE/SERI through June 1986.

The warehousing and storage capabilities of the Station are being retained and operated.

The Plum Brook Reactor is monitored to assure "Protected Safe Storage" in compliance with the NRC licenses and established NRC-approved procedures.

Four Station facilities were identified to be maintained in a "Standby" condition which will allow for reactivation of a research facility within 1-year from the request date. The four facilities so identified are:

    The Space Power Facility (SPF)
    The Spacecraft Propulsion Research Facility (B-2)
    The Hypersonic Tunnel Facility (HTF)
    The Cryogenic Propellant Tank Lab (K-Site)

In addition, sufficient Station land is retained to provide established hazard exclusion areas for operation of each of the standby research facilities, along with a designated number of acres to provide for expansion of the Lewis Research Center, if and when required.

The remaining NASA-owned Station land has been processed as excess to the Agency needs and reported to General Services Administration for disposition. To date, GSA has disposed of 1550 acres of excess land. 750 acres remain for disposition. The total remaining NASA land will consist of 5711 acres.

*Image 312: Plum Brook status report.*

and nine Ohio universities. Established in 1989 with Michael Salkind as president, the consortium facilitates collaboration on aerospace research, continuing education, and technology transfer. Lewis provided eight acres at the edge of its West Area for construction of a 70,000-square-foot home for the group. Ohio funded the $10.7 million construction, and the federal government supplied three-fourths of the annual operating budgets.[903] Ohio Governor Richard Celeste and other officials participated in the groundbreaking ceremony in October 1990, and the eye-catching steel-and-glass facility opened in October 1992.[904]

OAI offers free classroom and modern communication facilities for students, researchers, and faculty.[905] It also awards fellowships and scholarships for graduate students pursuing aerospace and science degrees at Ohio universities. The students are afforded the opportunity to work with experts in the field while working on their degrees.[906,907] OAI includes distance learning facilities that allow Lewis engineers to earn graduate degrees from remote universities.

❖ ❖ ❖ ❖ ❖ ❖

The desolation at Plum Brook Station was interrupted in 1980 when NASA granted Garrett Corporation permission to modify the Space Power Facility (SPF) for a five-year uranium centrifuge demonstration. As that project was winding down in 1985, the Office

of Management and Budget (OMB) reviewed Plum Brook's situation and recommended that NASA remove the test facilities and excess the land. Robert Kozar, a former Plum Brook engineer, led a multiagency commission that reviewed the OMB's findings. The commission concluded that NASA should fully reactivate Plum Brook Station, not excess it. NASA accepted the commission's recommendations and assigned Kozar the responsibility for restarting the four largest facilities—the SPF, the Space Propulsion Research Facility (B–2), the Cryogenic Propellant Tank Facility (K Site), and the Hypersonic Tunnel Facility (HTF).[908] The facilities would be operated on a strict pay-as-you-go policy. The Rocket Systems Area and the reactor facility would not be resurrected.

Plum Brook engineers had to reverse the serious modifications that Garrett had made to the SPF before it could be reactivated. The work proceeded rather quickly, and the vacuum chamber was reactivated in 1988.[909] In the 1990s engineers used the SPF to conduct shroud jettison tests for Atlas, Titan, and Ariane rockets. NASA restored K Site by 1989 and used it to investigate the use of slush hydrogen as a propellant in the 1990s. NASA brought the B–2 vacuum chamber back online in 1987 for the military's

Image 313: Interior of the B–2 test chamber with the lid removed in April 1987 (GRC–1987–C–02665).

Space Power Experiments Aboard Rockets program, and restored the facility's liquid-hydrogen system in 1996.[910] The restoration of the HTF, which began in 1990, took six years to complete.[911] Although it was not as active as it had been in the 1960s, Plum Brook Station and its unique test facilities were once again contributing to the national space program.

## Space Exploration Initiatives

In April 1990 Larry Ross replaced John Klineberg as Lewis Center Director when the NASA Administrator asked the latter to serve as Director of the Goddard Space Flight Center. Like Stofan, Ross had begun his career at Lewis in the early 1960s as a test engineer right out of university and had worked his way up through the Centaur Program. Ross was active in the Centaur's Surveyor and orbiting space telescope missions. In 1976, he was named Director of the Titan-Centaur Program and, in 1978, Chief of the Launch Vehicles Directorate. Klineberg selected Ross to serve as Deputy Director in 1987.[912,913]

Image 315: Larry Ross (right) meets with Bob Angus in the Administration Building during August 1993 (GRC–1993–C–06422).

January 1, 1993

# Update
### with LARRY ROSS

Communicating on a real and personal level with the Center is a much more difficult proposition than one might realize. I've discovered that it takes a variety of approaches to reach the diverse audiences at Lewis, and this is just one attempt to reach more people. This column will enable me to pass on information and say what's on my mind without the usual obstacles getting in the way.

Lately, I've found that my greatest challenge is bringing the Strategic Plan alive for everyone at the Center and successfully implementing the Total Quality initiative. I truly feel that our plans for implementing these changes are strong and that the people who have signed-on up to this point are dedicated to a successful outcome. The most important tests are still to come, however, as we move to improve the entire Center.

The important thing to realize about strategic plans and management initiatives is that they only work if everyone in the organization really wants them to. Nothing much is accomplished if only a few champions are attempting to carry the entire weight of change. However, changing the way people think is an extremely difficult thing to do—as anyone who has ever encountered the "but we've always done it this way" attitude can attest. The business of bringing people around to these changes in culture and process is going to require a lot of work and a great deal of patience. Everyone's ideas, talents, and determination will be needed if this is to be a successful effort. That's why it's so important that we share our thoughts and feelings about our Center's future.

Although this column offers an opportunity for me to speak my mind, it's still only a one-way dialogue. Through various means, many of you have expressed a desire to see me and discuss your concerns and ideas in an informal manner. I've been holding "dutch lunches" (we each pay our own way) on a fairly regular basis for the past several months with a cross-section of the Center. People sign up to join me for lunch, say what's on their mind, and ask any questions. The rules are simple: No topic is off limits; your comments are not for attribution, but mine are unless otherwise requested. These one-on-ones have been wonderful and I've learned more about the pulse of the Center in this manner than any other single format. Recently, much to my delight, two enterprising groups of secretaries got on my calendar and scheduled their own dutch lunches. It seems to be catching on, so anyone wanting to help me get "ground truth" should feel free to call my office or simply drop me a note. ◆

Image 314: Ross instituted a regular column in the Lewis News.

Image 316: Ross as a Centaur test engineer in the early 1960s (GRC–2008–C–01424).

Lewis research thrived during Ross's four-year tenure. The center's annual budget exceeded $1 billion for the only time in its history, and the center published more technical reports in 1990 and 1991 than at any point since 1974. In addition, it added 200 new civil servant positions and helped to establish a new business park outside its west gate to house contractors.[914] The reactivation of Plum Brook Station continued, the ACTS and microgravity programs thrived, and the Atlas I-Centaur launched the CRRES satellite— the first NASA satellite to be launched by General

*Image 317: Conceptual drawing of a lunar-oxygen-augmented nuclear thermal rocket (NTR) (GRC–1997–C–00827).*

Dynamics under the new commercial launch business model. Ross began holding private one-on-one lunches with employees to increase dialogue at the center, and he instituted a regular column in the *Lewis News*.

The world was undergoing a dramatic transformation during this period. In 1985 Mikhail Gorbachev became leader of the Soviet Union and introduced his perestroika reforms. In rapid succession Gorbachev reorganized the Soviet ruling structure, brokered an arms reduction agreement with the United States, and loosened control of Russia's satellite states. The dismantling of the Berlin Wall in 1990 proved to be the most enduring symbol of the new atmosphere. In December 1991 the Soviet Union dissolved, bringing the nearly 50 years of the Cold War to an end.[915]

President George H. Bush took advantage of the ease in tensions to introduce his new Space Exploration Initiative (SEI). The initiative, announced in July 1989 on the 20th anniversary of the Apollo 11 landing, called for completion of the space station, a return to the Moon, and an eventual human voyage to Mars.

The president, however, did not obtain a reliable cost appraisal for the endeavor until after the announcement. Ohio Congresswoman Mary Rose Oakar rued the fact that the president "did not add some muscle to his plan."[916] The eventual $500-billion-dollar estimate and poor planning effectively killed the initiative after about 18 months, although the official cancellation did not occur until early 1992.[917]

Although the SEI was short-lived, it did revive interest in the long-duration human missions that had been conceived in the early 1960s. These complex endeavors required two technologies that Lewis had investigated extensively in the 1960s—nuclear propulsion and the long-term storage of cryogenic fuels. Lewis researchers took up these subjects once again in the 1990s.

❖ ❖ ❖ ❖ ❖ ❖

Researchers consider nuclear propulsion to be the most viable method for transporting humans to Mars. With twice the thrust of traditional chemical rockets, NTRs would significantly reduce travel time and extend launch windows. In 1990,

NASA, the Department of Defense, and the Department of Energy jointly analyzed 17 nuclear rocket designs. They concluded that solid core designs based on Nuclear Engine for Rocket Vehicle Application (NERVA) technologies from the 1960s were the safest, most feasible, and quickest to develop. They were also aware of the public's hesitancy regarding nuclear energy and the need for new ground testing centers.[918]

Lewis, the NASA center with the most nuclear propulsion experience, established a 13-person Nuclear Propulsion Office in May 1991 to oversee NASA's new nuclear efforts.[919] Researcher Stanley Borowski and his colleagues published a number of papers examining different aspects of NTR missions—including the use of liquid oxygen extracted from the martian atmosphere to power an afterburner-type configuration.[920] In addition, Frank Rom and other Lewis veterans resumed the nuclear work that had been suspended almost 20 years before.

As the studies progressed, researchers argued that a bimodal nuclear rocket would be the most cost-effective design. Theoretically, NASA's proposed nuclear stage for heavy lift work would have plenty of fuel remaining after it completed its initial full-power boost. Borowski and others argued that the rocket system could be modified to utilize the remaining fissionable material to generate lower levels of propulsion for the remaining journey to Mars and possibly a return trip to Earth for reuse.[921] The SEI effort and its proposed mission to Mars did not come to fruition, but the center continues studying the technology for the inevitable resurgence of the concept.

❖ ❖ ❖ ❖ ❖ ❖

In the 1960s Lewis had extensively studied methods of insulating storage tanks for cryogenic propellants to support future human missions to Mars, but NASA canceled the research as plans for those types of missions faded with NASA's post-Apollo budget cuts. The issues with the long-term storage of cryogenic propellants resurfaced with the SEI's proposed mission to Mars. It was critical that propellants be stored and transferred in space so that the spacecraft would not have to carry all the fuel needed for a long-duration flight during Earth-to-orbit launch.

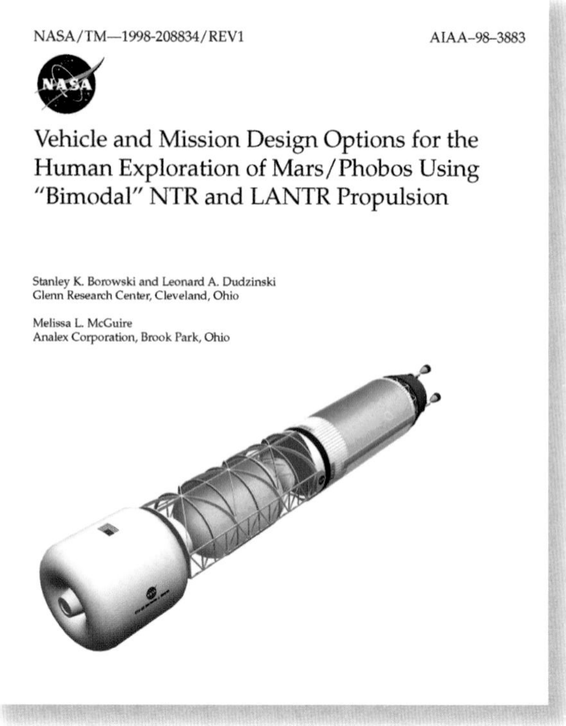

NASA/TM—1998-208834/REV1        AIAA-98-3883

Vehicle and Mission Design Options for the Human Exploration of Mars/Phobos Using "Bimodal" NTR and LANTR Propulsion

Stanley K. Borowski and Leonard A. Dudzinski
Glenn Research Center, Cleveland, Ohio

Melissa L. McGuire
Analex Corporation, Brook Park, Ohio

*Image 318: NASA Technical Memorandum about NTR.*

Lewis had conducted a great deal of research in simulated space conditions, but there had not yet been a test in space. Lewis's Cryogenic Fluids Technology Office had begun work on a shuttle-based cryogenic fluid management experiment, but NASA canceled this effort after the loss of *Challenger*. When the SEI emerged, Lewis campaigned for a free-flying cryogenic fluid management demonstration on an expendable rocket. The result was the Cryogenic On-Orbit Liquid Depot Storage, Acquisition, and Transfer Satellite (COLD–SAT).[922]

Lewis devised COLD–SAT to be launched on an Atlas in 1997. It included 13 experiments to study the pressurization, venting, fuel transfer, insulation, and filling of cryogenic propellant tanks.[923] Lewis reactivated K Site at Plum Brook Station in 1988 to support COLD–SAT, but NASA then canceled this proposed $200-million effort following an extensive feasibility study in 1990. As a part of this effort, two Tank Pressure Control Experiments using Freon were flown aboard the space shuttles in the early 1990s.

Nonetheless, by August 1989 plans were under way for a new research program at K Site to study slush hydrogen. Researchers felt that they could reduce the volume of liquid hydrogen by 15-percent by lowering its temperature and forming slush. This would decrease the size of the tank and overall vehicle. The new interest in slush hydrogen was largely driven by the experimental National Aerospace Plane (NASP) that was born out of the X-plane and Dynasoar efforts in the 1960s. NASA and the military had secretly initiated the technology demonstration effort, also known as the X–30, in the early 1980s. President Reagan formally announced the program in his 1986 State of the Union Message, calling it the Orient Express. NASP would be an extremely lightweight, runway-based vehicle that could fly into orbit at hypersonic speeds.

A number of new technologies were required to meet those specifications, and Lewis was heavily involved

with three of these fields: high-temperature materials and seals, regenerative cooling, and cryogenic propellant management that included slush hydrogen.[924] Bruce Steinetz developed a unique high-temperature flexible fiber seal to prevent the hot engine gases from escaping through the vehicle's moving panels. In 1996 NASA named the seal the Government Invention of the Year.[925]

Lewis had briefly explored slush hydrogen in Plum Brook's Boiling Fluids Rig during the 1960s, but it was the National Bureau of Standards (now the National Institute of Standards and Technology) that had generated much of the existing data.[926] The recently renovated K Site provided Lewis an optimal facility for performing basic slush-hydrogen research. Lewis installed a hydrogen densification system at K Site

*Image 319: William Klein examines settings for the slush hydrogen test rig in the K Site vacuum chamber during June 1991 (GRC–1991–C–07458).*

*Image 320: A 13-foot-diameter propellant tank installed in K Site's 25-foot-diameter vacuum chamber (GRC–1967–C–03315).*

that could generate over 700 gallons of slush hydrogen each day. It was the largest slush hydrogen system in the world.[927]

In 1987 Lewis coordinated a new national slush-hydrogen development team to acquire basic slush-hydrogen data, design handling systems, and generate computer models for pumping, transferring, and storing the hydrogen. The Space Vehicle Propulsion Branch used K Site to analyze the tank pressurization and flow of the slush from the tank, and Lewis created computer models to predict this activity.[928]

The utilization of slush hydrogen to cool the NASP vehicle was a major breakthrough in the program. The vehicle needed to reach Mach 25 to achieve orbit, but early studies indicated that the high speeds would produce intense aerodynamic heating that would overheat the exterior surfaces. This hurdle was overcome with the introduction of a new flow system that released slush hydrogen to the leading edges. The heat converted the slush to liquid, which was then used as the vehicle's propellant. Early calculations revealed that the amount of liquid hydrogen required for cooling was greater than that required for propellant.[929]

Lewis researchers also pursued the use of a heat-absorbing chemical reaction that used two unique forms of hydrogen. This process reduced the amount of hydrogen required for cooling. Although much of the research was conducted in the Cryogenics Components Laboratory,[930] engine concept tests were run in the 10- by 10-Foot Supersonic Wind Tunnel and the PSL, external burning studies were conducted in the 8×6, and significant work on high-temperature seals and materials was performed.[931]

Military funding for the program ceased in 1993 as the Cold War ended. Lewis had been involved in many aspects of the program, and the researchers had developed a number of new technologies. Nonetheless, NASP was far from being ready for flight.

### Lewis's First 50 Years

Lewis's year-long celebration of its semicentennial commenced on a frozen 23 January 1991—50 years after the original groundbreaking. A local high-school marching band roused the 300 attendees as officials unveiled a time capsule in front of the Administration Building. The capsule, which was incorporated into

*Image 321: Historian Virginia Dawson conducting research in 1984 for her history of the center,* Engines and Innovation *(GRC–1984–C–03885).*

a 9-foot-tall sculpture, contained 58 items, including reports, videos, photographs, and articles.[932,933] At this time, Lewis management decided to name the Employee Center in honor of former Director Ray Sharp.[934] To further commemorate the anniversary, NASA commissioned historian Virginia Dawson to write the history of the center. Dawson interviewed many of the center's iconic leaders and researchers and retrieved invaluable center documents from the National Archives. Her *Engines and Innovation: Lewis Laboratory and American Propulsion Technology* remains the definitive description of Lewis's history.[935]

The extraordinary accomplishments of the center's first 25 years were in some ways matched in the second 25 years in ways that the NACA could not have foreseen in 1941. The successes were tempered by the unrelenting budgetary realities of the Agency. Lewis, nonetheless, had turned a corner. The center was more successful in the late 1980s and early 1990s than at any point since the 1960s, managing the space station power system, the ACTS satellite, and the launch vehicle operations for NASA's missions. In addition, the center was heavily involved in the development of new technologies for future high-speed and subsonic aircraft.

*Image 322: Time capsule installed in front of the Administration Building in 1991 (GRC–1991–C–08985).*

## Endnotes for Chapter 8

814. William Robbins interview, Cleveland, OH, by Virginia Dawson, 15 May 1986, NASA Glenn History Collection, Oral History Collection, Cleveland, OH.

815. Robbins interview by Dawson.

816. William Robbins interview, Cleveland, OH, by Michael McMahon and Virginia Dawson, 15 May 1986, NASA Glenn History Collection, Oral History Collection, Cleveland, OH.

817. Lewis Public Information Office, "Biographical Sketch of Andrew J. Stofan," January 1984, NASA Glenn History Collection, Directors Collection, Cleveland, OH.

818. "Budget Cuts Forcing NASA to Close Lewis Research Center," *Defense Daily* (9 December 1981).

819. Andrew Stofan interview, Cleveland, OH, by Tom Farmer, "This Way Up: Voices Climbing the Wind, WVIZ Documentary," 1991, NASA Glenn History Collection, Oral History Collection, Cleveland, OH.

820. Stofan interview by Farmer.

821. "Strategic Planning Options for Lewis Research Center," June 1982, NASA Glenn History Collection, Directors Collection, Cleveland, OH.

822. "Lewis' Energy Work Gets a Qualified Go-Ahead," *Lewis News* (5 November 1982).

823. "Strategic Planning Options."

824. "Reorganization Focuses on National Goals," *Lewis News* (4 February 1983).

825. W. Warner Burke, Edward Richley, and Louis DeAngelis, "Changing Leadership and Planning Processes at Lewis Research Center," *Human Resource Management* (Spring 1985).

826. "The Present: Lewis People and Their Missions, A 50-Year Voyage: Cleveland's Saga of Air and Space," 1991, Glenn History Collection, Directors Collection, Cleveland, OH.

827. James Hawker and Richard Dali, *Anatomy of an Organizational Change Effort at the Lewis Research Center* (Washington, DC: NASA CR–4146, 1988).

828. Stofan interview by Farmer.

829. Hawker, *Anatomy of an Organizational Change Effort.*

830. John David Lees and Michael Turner, *Reagan's First Four Years: A New Beginning* (Manchester, UK: Manchester University Press, 1988).

831. Ronald Reagan, "Memorandum Directing a Federal Employee Hiring Freeze" (20 January 1981), Online by Gerhard Peters and John T. Woolley, The American Presidency Project, *http://www.presidency.ucsb.edu/ws/?pid=43601* (accessed 16 July 2015).

832. John Klineberg talk, c1981, NASA Glenn History Collection, Directors Collection, Cleveland, OH.

833. "190 New Hires Biggest Boost to Lewis Workforce in 20 Years," *Lewis News* (29 July 1983).

834. Stofan interview by Farmer.

835. Hawker, *Anatomy of an Organizational Change Effort.*

836. D. A. Sagerser and S. G. Ludemann, *Large-Scale Advanced Propfan Program Progress Report* (Washington, DC: NASA TM–87067, 1985).

837. Herman Mark interview, Cleveland, OH, by Virginia Dawson, 12 March 1985, NASA Glenn History Collection, Oral History Collection, Cleveland, OH.

838. Roy Hager and Deborah Vrabel, *Advanced Turboprop Project* (Washington, DC: NASA SP–495, 1988).

839. Edwin Graber, "Overview of NASA PTA Propfan Flight Test Program," *Aeropropulsion 1987* (Washington, DC: NASA, 1992).

840. Hager, *Advanced Turboprop Project.*

841. Mark Bowles, *The Apollo of Aeronautics: NASA's Aircraft Energy Efficiency Program 1973–1977* (Washington, DC: NASA SP–2009–574, 2009).

842. "Icing Research Tunnel History: 50 Years of Icing, 1944–1994," 9 September 1994, NASA Glenn History Office, Test Facilities Collection, Cleveland, OH.

843. *Aircraft Icing* (Washington, DC: NASA CP–2086, 1979).

844. William Leary, *We Freeze to Please: A History of NASA's Icing Research Tunnel and the Quest for Flight Safety* (Washington, DC: NASA SP–2002–4226, 2002).

845. Congressional Advisory Committee on Aeronautics Assessment, *NASA Aeronautics Budget for FY86*, Org 2600, Box 55, "Response to Congressional Advisory Committee on Aeronautics" folder,

846. J. J. Reinmann, R. J. Shaw, and W. A. Olsen, Jr., *Aircraft Icing Research at NASA* (Washington, DC: NASA TM–82919, 1982).

847. Reinmann, *Aircraft Icing Research.*

848. A. A. Peterson, L. Dedona, and D. Bevan, *Rotorcraft Aviation Icing Research Requirements, Research Review, and Recommendations* (Washington, DC: NASA CR–165344, 1981).

849. Thomas Miller and Thomas Bond, *Icing Research Tunnel Test of a Model Helicopter Rotor* (Washington, DC: NASA TM–101978, 1989).

850. Miller, *Icing Research Tunnel Test.*

851. "Lewis Team Uses Model Helicopter Rotor To Develop Techniques for First Test of Helicopter Model in Icing Research Tunnel," *Lewis News* (30 March 1990).

852. Reinmann, *Aircraft Icing Research.*

853. Randall Britton and Thomas Bond, *An Overview of a Model Rotor Icing Test in the NASA Lewis Icing Research Tunnel* (Washington, DC: NASA TM–106471, 1994).

854. John Reinmann, Robert J. Shaw, and Richard Ranaudo, *NASA's Program on Icing Research and Technology* (Washington, DC: NASA TM–101989, 1989).

855. NASA Glenn Research Center, "LEWICE" (2014), *http://icebox.grc.nasa.gov/design/lewice.html* (accessed 16 July 2015).

856. "Otter Is Research Workhorse," *Lewis News* (23 April 1982).

857. Reinmann, *NASA's Program.*

858. NASA Glenn Research Center, "LEWICE."

859. Mark G. Potapczuk, "Aircraft Icing Research at NASA Glenn Research Center," *Journal of Aerospace Engineering* 26 (2013): 260–276.

860. Peterson, *Rotorcraft Aviation Icing.*

861. "Lewis and Army Teams Join Forces To Improve Icing Flight Research," *Lewis News* (30 March 1990).

862. Stanley McCowen, Helicopters: *An Illustrated History of Their Impact (Weapons and Warfare)* (Santa Barbara, CA: ABC–CLIO, 24 May 2005).

863. John Acurio, "Propulsion Directorate," *Lewis News* (9 January 1987).

864. "Helicopter Group Saluted," *Lewis News* (27 July 1984).

865. "Helicopter Test Rig is Operational," *Lewis News* (16 July 1982).

866. "Novel Face Gears Proved Feasible for Advanced Rotorcraft Transmissions," *Research & Technology 1992* (Washington, DC: NASA TM–105924, 1992).

867. Timothy Krantz, *NASA/Army Rotorcraft Transmission Research, A Review of Recent Significant Accomplishments* (Washington, DC: NASA TM–106508, 1994).

868. Ajay Misra and Leslie Greenbauer-Seng, "Aerospace Propulsion and Power Materials and Structures Research at NASA Glenn Research Center" *Journal of Aerospace Engineering* 26, (2013): 459–490.

869. John Coy, Dennis Townsend, and Harold Coe, *Results of NASA/Army Transmission Research* (Washington, DC: NASA TR–87–C–3, 1987).

870. "Advanced Technology Helicopter Transmission," *Lewis Research and Technology Report 1985* (Washington, DC: NASA, 1985).

871. Stofan interview by Farmer.

872. "New Lewis Office Seeks Space Station Involvement," *Lewis News* (27 August 1982).

873. Donald Nored and George Halinan, *Electrical Power System for the U.S. Space Station* (Washington, DC: NASA TM–88856, 1986).

874. "Why the Power Systems Facility is Needed," *Lewis News* (2 October 1987).

875. Stofan interview by Farmer.

876. "New Lewis Office."

877. Robbins interview by Dawson.

878. "Why the Power Systems Facility is Needed."

879. Andy Stofan interview, Cleveland, OH, by Adam Gruen, 7 March 1988, NASA Glenn History Collection, Oral History Collection, Cleveland, OH.

880. William Broad, "Space Station Plan Found Unworkable," *Cleveland Plain Dealer* (19 March 1990).

881. Thomas Gerdel, "Lewis Space Station Project Changing," *Cleveland Plain Dealer* (6 March 1991).

882. Joseph Nieberding and Francis Spurlock, "U.S. ELVs: A Perspective on Their Past and Future," *Space Commerce* 1 (1990): 29–34.

883. Virginia Dawson and Mark Bowles, *Taming Liquid Hydrogen: The Centaur Upper Stage Rocket 1958–2002* (Washington, DC: NASA SP 2004–4230, 2004).

884. "Field Center Expendable Launch Vehicle (ELV) Roles, Responsibilities, and Organizational Structure," 17 December 1987, NASA Glenn History Collection, Centaur Collection, Cleveland, OH.

885. Centaur G Subagreement to the NASA/DOD Memorandum of Understanding on Management and Operation of the Space Transportation System," November 1982, NASA Glenn History Collection, Centaur Collection, Cleveland, OH.

886. "Shuttle/Centaur Overview," Lewis Research Center press release, July 1985, NASA Glenn History Collection, Centaur Collection, Cleveland, OH.

887. John McCarthy to Alan Lovelace, 25 March 1981, NASA Glenn History Collection, Centaur Collection, Cleveland, OH.

888. Karen Long, "Lewis Gets 50% Boost in Budget," *Cleveland Plain Dealer* (28 July 1982).

889. Lutha Shaw interview, Cleveland, OH, by Virginia Dawson and Joe Nieberding, 10 November 1999, NASA Glenn History Collection, Centaur Collection, Cleveland, OH.

890. Larry Ross interview, Cleveland, OH, by Mark Bowles, 29 February 2000, NASA Glenn History Collection, Oral History Collection, Cleveland, OH.

891. Vernon Weyers interview, Cleveland, OH, by Mark Bowles, 8 April 2000, NASA Glenn History Collection, Centaur Collection, Cleveland, OH.

892. Dawson, *Taming Liquid Hydrogen* (NASA SP 2004–4230).

893. Robbins interview by Dawson.

894. James Fletcher to Slade Gorton, 24 June 1986, NASA Glenn History Collection, Centaur Collection, Cleveland, OH.

895. John Mangels, "Long-Forgotten Shuttle/Centaur Boosted Cleveland's NASA Center Into Manned Space Program and Controversy," *Cleveland Plain Dealer* (11 December 2011).

896. "The ACTS Initiative: Pioneering the Next Generation of Space Communications," *NASA Tech Briefs* (May 1988).

897. "Working to Meet the Communication Needs of the 1990s," *Lewis News* (1 November 1985).

898. "Switchboard in the Sky: The Advanced Communications Technology Satellite (ACTS)," *NASA Facts* (Washington, DC: FS–2002–060013–GRC, June 2002).

899. "Switchboard in the Sky."

900. Kenneth Atchison, "Truly Appoints Klineberg as Director of Goddard Space Flight Center," NASA Release 90–51, 9 April 1990.

901. "Profile: John Klineberg." *Space News* (11 October 1999).

902. "Employee Center Dedicated," *Lewis News* (8 August 1986).

903. Frank Montegani, "OAI Questions," *Lewis News* (24 April 1992).

904. John Klineberg to Thomas J. Coyne, 31 January 1990, NASA Glenn History Collection, Directors Collection, Cleveland, OH.

905. "Hopes Soar with OAI," *Lewis News* (4 December 1992).

906. "The Present: Lewis People and Their Missions."

907. Ohio Aerospace Institute (Cleveland, OH: The Encyclopedia of Cleveland History, 2004), *http://ech.case.edu/cgi/article.pl?id=OAI* (accessed 22 July 2015).

908. Robert Kozar interview, Cleveland, OH, by Virginia Dawson, 30 March 2000, NASA Glenn History Collection, Oral History Collection, Cleveland, OH.

909. Roger Smith, *Space Power Facility Readiness for Space Station Power System Testing* (Washington, DC: NASA TM–106829, 1995).

910. "Well Preserved Plum Brook Facilities Could Be Used for Space Station and Aerospace Plane Testing," *Lewis News* (11 December 1987).

911. Scott Thomas and Jinho Lee, *The Mothball, Sustainment, and Proposed Reactivation of the Hypersonic Tunnel Facility (HTF) at NASA Glenn Research Center Plum Brook Station* (Washington, DC: NASA/TM—2010-216936, 2010).

912. "Ross Is Named to Head Launch Vehicles Group," *Lewis News* (21 March 1978).

913. "Ross Appointed Deputy Director," *Lewis News* (2 October 1987).

914. "State of the Center Address Highlights Past Achievements," *Lewis News* (10 May 1991).

915. Jack Matlock, Jr., *Reagan and Gorbachev: How the Cold War Ended* (New York, NY: Random House Publishing, 2005).

916. "NASA Lewis Spirits Soar Over Space Plans," *Cleveland Plain Dealer* (21 July 1989).

917. Thor Hogan, *Mars Wars: The Rise and Fall of the Space Exploration Initiative* (Washington, DC: NASA SP–2007–4410, 2007).

918. Stanley Borowski, et al., "Nuclear Thermal Rockets: Key to the Moon-Mars Exploration," *Aerospace America* 30, no. 7 (July 1992): 34–37.

919. Vernon Weyers to Lewis Supervisors, "Establishment of the Nuclear Propulsion Office," 1 May 1991, NASA Glenn History Collection, Directors Collection, Cleveland, OH.

920. Stanley Borowski, Leonard Dudzinski, and Melissa McGuire, *Vehicle and Mission Design Options for the Human Exploration of Mars/Phobos Using "Bimodal" NTR and LANTR Propulsion* (Washington, DC: NASA/TM—1998-208834, 1998).

921. Borowski, *Vehicle and Mission Design Options.*

922. David DeFelice to Robert Arrighi, email, 23 February 2015, NASA Glenn History Collection, Cleveland, OH.

923. John Schuster, Edwin Russ, and Joseph Wachter, *Cryogenic On-Orbit Liquid Depot Storage, Acquisition, and Transfer Satellite* (COLD–SAT) (Washington, DC: NASA CR–185249, 1990).

924. "The National Aero-Space Plane," *Round Trip to Orbit: Human Space Flight Alternatives* (Washington, DC: U.S. Government Printing Office OTA–ISC–419, August 1989).

925. "Lewis-Developed Seal Wins Acclaim," *Lewis News* (June 1997).

926. John Klineberg to J. R. Thompson, 2 April 1990, NASA Glenn History Collection, Directors Collection, Cleveland, OH.

927. "Plum Brook's K-Site Facility Reactivated for Slush Hydrogen Tests," *Lewis News* (11 May 1990).

928. "Fueling and Cooling the National Aero-Space Plane," *Lewis News* (11 May 1990).

929. "The National Aero-Space Plane."

930. "Converting Parahydrogen to Orthohydrogen May Cool the National Aero-Space Plane," *Lewis News* (11 May 1990).

931. "Airbreathing Propulsion for Hypersonic Flight: The Lewis Perspective," 2 April 1996, NASA Glenn History Collection, Directors Collection, Cleveland, OH.

932. "Time Capsule Contributions," *Lewis News* (22 November 1991).

933. Karen Long, "Celebrating a Half-Century of Innovation," *Cleveland Plain Dealer* (24 January 1991).

934. "Sharp Center Dedication Remembers the Past," *Lewis News* (11 October 1991).

935. Virginia Dawson, *Engines and Innovation: Lewis Laboratory and American Propulsion Technology* (Washington, DC: NASA SP–4306, 1991).

Image 323: Fluids and Combustion Facility during testing at the
Structural Dynamics Laboratory (GRC–2004–C–01827).

## 9. Reformation

*"My challenge is to convince you that you could do more, do it a little better, do it for less if we use more innovative management techniques and utilize the individual capabilities of each and every NASA employee."*

—Dan Goldin

*Image 324: Technician analyzes a NASA Solar Electric Power Technology Application Readiness (NSTAR) thruster in the Electric Propulsion Laboratory (EPL) during December 1993 (GRC–1993–C–08855).*

$\mathcal{B}$y the time that President Bill Clinton took office on 20 January 1993, the recent Cold War victory had reduced international tensions but had left the federal government strapped with a massive federal deficit. Clinton's inauguration coincided with NASA's latest review of the space station effort. The review determined that the program was over budget and behind schedule once again. The space station served as a palpable example of what many considered was wrong with the government: a sprawling federal program that had been sold to the nation without its true costs and schedule revealed. Clinton ordered NASA Administrator Dan Goldin to conduct a massive restructuring of the program to expedite the station's construction while significantly reducing its funding. This effort led to the dissolution of NASA Lewis Research Center's Space Station Freedom Directorate.

The primary theme throughout Goldin's nearly 10-year term, the longest of any NASA Administrator, was "Faster, Better, Cheaper." This meant not only doing more with less, but being smarter and willing to replace single expensive missions with a greater number of smaller, less expensive endeavors. In 1995 the White House charged Goldin with reforming the entire Agency to meet imminent federal budget cuts. Goldin subjected the NASA centers to dramatic reductions, reorganizations, and consolidations in an attempt to create a more efficient agency. He instructed each center to analyze methods for streamlining activities and reducing overhead and staff. Despite the self-imposed downsizing, Congress reduced NASA's funding even further.[936]

In January 1994 Goldin appointed Donald Campbell as Lewis's Center Director. The Agency's turmoil made Campbell's decade-long tenure, which roughly paralleled Goldin's, among the most difficult periods in the center's history. A large portion of its space station work and several of its traditional roles had been transferred elsewhere; other areas, like electric propulsion, microgravity, and aeropropulsion, flourished. Nonetheless, Lewis, which had survived the layoffs in the 1970s and the near closure in 1981, persevered once again during the Agency's latest reductions. The center marked its transformation in 1999 with its redesignation as the John H. Glenn Research Center.

## Zero Base Review

Upon becoming NASA Administrator in April 1992, Dan Goldin quickly introduced a new "Faster, Better, Cheaper" philosophy in hopes of reforming the Agency's culture of large programs, which did not mesh with the new atmosphere of federal downsizing. The massive Space Exploration Initiative had failed almost immediately, and the Agency was floundering under the costly space station, shuttle, and Hubble Space Telescope programs. Under Goldin, NASA would focus on a series of smaller, less expensive advanced technology missions, reduce management levels at headquarters, and improve planning efforts for future missions. Goldin began taking steps to restructure the shuttle and space station programs and canceled expensive, lower priority programs like the National Aerospace Plane.[937]

With Congress poised to cut the estimated $30-billion space station, NASA had to reduce the program's cost, finalize the design, and adhere to a schedule. In February 1993 newly elected President Bill Clinton gave Goldin three months to reinvent the space station with a significantly reduced budget and development schedule.[938] This monumental effort was complicated

Image 325: *Dan Goldin began his professional career in 1962 working on electric propulsion systems at Lewis. He left NASA for TRW in 1967 and worked his way up to be the firm's vice president (GRC–1962–C–60944).*

by the announcement in April that the Russians would be playing a role in the program. The Soviet Union had broken apart just over a year before, and the Clinton Administration was seeking to forge new ties. In addition, management hoped that Russian participation would reduce U.S. expenditures, accelerate construction, and provide alternative access to the station.[939]

NASA's intensive redesign resulted in three possible configurations. Each exceeded the mandated budget restrictions, but all three were significantly less expensive than the original plan. In June 1993 Clinton approved a combination of two of the designs and included the solar power units from the third. In one of its most significant decisions regarding NASA's human space program, Congress approved continuation of the pared-down program by a single vote.[940-942]

In fall 1993 the Clinton Administration announced that Russia would not be just contributing to the space station but would be a full partner. NASA released the final space station redesign several days later, and Space Station Freedom became the International Space Station (ISS).[943,944] After nearly a decade and $11.2 billion spent, very little space station hardware had been built, let alone launched into orbit. Now concrete steps were finally in place to begin the actual construction in 1998.[945,946]

As part of the new direction, NASA transferred all management functions to the ISS project office at Johnson Space Center and all systems engineering to the Boeing Company. This restructuring resulted in the closure of the program's headquarters in Reston, Virginia, and large cutbacks at the other centers, including Lewis.[947,948] The center disbanded its four-division Space Station Freedom Directorate in January 1994.

❖ ❖ ❖ ❖ ❖ ❖

As the space station saga played out, Washington, DC, was undergoing its own transformation. In spring 1993, the House of Representatives introduced balanced-budget and deficit-reduction bills, and Vice President

*Image 326: Artist's rendering of Space Station Freedom's Alpha design (GRC–1994–C–00566).*

Al Gore announced the Administration's strategy to reinvent government. One of the plan's many goals was a dramatic reduction in the federal workforce. By 1995, NASA was able to reduce its civil servant staff by nearly 4,000 through a hard hiring freeze, buyouts, and attrition. The downsizing and other consolidation efforts generated billions in budget reductions over Goldin's first three years at NASA.[949,950]

As part of this process, Goldin followed through on an effort initiated by his predecessor to identify institutional changes that would improve the Agency. The recommendations of the internal report, which sparked harsh criticism when it was leaked to the media in fall 1993, included the transfer of Lewis's space propulsion research to Marshall Space Flight Center, the closure of Plum Brook Station, and the relocation of the center's aircraft to the Dryden Flight Research Center (now Armstrong Flight Research Center). Although these recommendations were not immediately implemented, they presaged actions in the near future.[951]

On 6 January 1994 NASA announced the appointment of new directors at five NASA centers—including Lewis. Goldin asked Larry Ross to manage a feasibility study for a new national wind tunnel complex and named Donald Campbell as the center's eighth Center Director.[952,953] Campbell had spent his career at the Wright-Patterson Air Force base. Over the course of 30 years he had worked his way up from aircraft engine test engineer to program manager to the Director of Aeropropulsion and Power.[954]

The Campbell appointment came at the same time that the center dissolved its large Space Station Directorate and the Agency learned that its budget would be substantially reduced in the coming years.[955] In November 1994 the Republican Party won control of both houses of Congress and promised to balance the federal budget. With further cutbacks eminent, Goldin had taken steps to reduce the Agency's budget by $35 billion over the next five years. Despite this effort, the Clinton Administration ordered the Agency to cut another $5 billion in December 1994.[956]

*Image 327: Goldin and Campbell converse in the Lewis hangar in August 1998 (GRC–1998–C–01638).*

Goldin instructed the centers to perform a zero-based self-assessment of all their functions to identify methods for streamlining activities and creating efficiencies in preparation for the impending fiscal year 1996 budget submission. This type of review did not base current expenditures on those from previous years but required a new justification for each line item. NASA planned to reduce its civil servant staff to its lowest levels since the early 1960s, remove unneeded facilities, and minimize management overhead.[957,958] The reanalysis led to NASA's reassignment of civil servants in operations work to research activities, introduced full-cost accounting and standardization, and transferred the management of certain science programs to university partners. The effort also included the establishment of "centers of excellence" and the consolidation of communications, information technology, and certain administrative services.[959]

NASA issued the "A Budget Reduction Strategy" report in February 1995 to spur discussions regarding the realignment of center roles.[960,961] The document recommended that Lewis's responsibility for expendable launch vehicles, communications, hypersonics, and flight research should be relinquished, and that Plum Brook Station, the Propulsion Systems Laboratory (PSL), and the Rocket Engine Test Facility (RETF) should be closed. Campbell and Lewis managers countered most of these suggestions, but Lewis's internal review found that the center could decrease its spending by almost $30 million through consolidation and personnel reductions.[962]

Lewis and the other centers reported their Zero Base Review findings to headquarters in mid-March 1995,[963] and Goldin announced the results on 19 May 1995. Even with NASA's self-imposed reductions, Congress ordered the Agency to cut more. There would have to be some fundamental changes in the Agency's way of doing business.[964,965] The Agency identified Lewis's primary missions as aeropropulsion and commercial communications and named Lewis NASA's center of excellence for turbomachinery—although the designation was later extended to microgravity research in fluid physics, combustion, and to a lesser extent, materials and fundamental physics microgravity research.[966]

Lewis would maintain Plum Brook Station and the PSL but would close the RETF. In a huge hit to its

**Space Experiments Division**

**Zero-Based Review**

*December 19-21, 1994*

Image 328: *Space Experiments Division's internal review for NASA's Zero Base Review.*

morale, the center also would have to transfer its launch vehicles responsibilities to Kennedy Space Center, flight operations to Dryden, and space propulsion to Marshall. Campbell reorganized the center in October 1996 to focus on the responsibilities set out by the Zero Base Review. Campbell stated, "We're moving into a new era within the Agency where we will be accountable for all of our resources."[967]

## Space Station Transformation

Under the new U.S.-Russian space agreement, NASA would assist with the Russian space station Mir, which had been in orbit since 1986. The Mir collaboration would provide NASA with hands-on space station experience and the Russians with superior U.S. technologies such as the space power system. In February 1994, NASA assigned Lewis the management of the joint U.S.-Russian Mir Cooperative Solar Array program. The program sought to extend the life of Mir's ailing power system and provide energy for U.S. experiments on the station. The United States provided the photovoltaic power modules and a solar collector to concentrate the energy, both of which Lewis had designed for the canceled Space Station Freedom program. The accordionlike power module was 9- by 52-feet and contained over 6,000 flexible solar cells. Russia designed the array's support structure.[968]

Lewis researchers traveled to Kennedy in early 1996 to verify the panel's electrical performance after its shipment to the launch site.[969] The system was launched

*Image 329: ISS solar array testing in the Space Power Facility (SPF) (GRC–2000–C–00279).*

*Image 330: Testing for the Mir Cooperative Solar Array (GRC–1995–C–00994).*

on *Space Shuttle Atlantis* in May 1996 and successfully installed on Mir. The Mir power system provided data that helped engineers predict the lifespan of the ISS solar panels.

Meanwhile researcher Richard Shaltens led the development of the solar concentrator, which uses a large mirror to reflect solar radiation into a Brayton cycle energy conversion system that stores the energy.[970] Shaltens tested the system in an EPL vacuum chamber in early 1995. Researchers operated the system for over 365 hours under various simulated solar conditions. It was the nation's first operation of a solar-dynamic power system in realistic space conditions.[971] NASA and the Russian Space Agency intended to verify the system aboard Mir in 1997, but NASA's restructuring of its Mir activities led to the cancellation of the mission.[972] Lewis continued working on the system and testing the hardware, much of which had already been constructed, for possible future U.S. deployment.[973]

The ISS employed a larger version of the photovoltaic power system that Lewis had developed for Space Station Freedom and Mir. This 110-kilowatt system was the largest and most advanced in existence. Construction of the ISS began in 1998, and in December 2000, astronauts installed the station's eight 35-foot-long solar arrays, which included 250,000 solar cells.[974]

Although Lewis did not lead the ISS power program, the center's Power Systems Project Office supported the ongoing effort by life-cycle testing of the nickel-hydrogen batteries and testing the gimbals that rotated the solar arrays. Lewis researchers also designed a plasma contactor, which eradicates the electrical charges that build up on the ISS exterior and can injure astronauts working on the solar arrays. The power system requires a large radiator to dissipate heat from the energy conversion process and the on-board equipment. From 1996 to 1999 Lewis engineers tested the radiator in

*Image 331: Titan-Centaur launch of the Cassini/Huygens spacecraft on 15 October, 1997 (NASA KSC–97PC–1543).*

simulated space conditions inside the SPF at Plum Brook Station. The center continues to be responsible for the testing and operations of the power system flight hardware and for the computer modeling used to predict the power system's performance.[975,976]

## Launch Vehicles Travails

One of the more painful results of the Zero Base Review was the transfer of Lewis's responsibility for the Atlas-Centaur launch vehicles to Kennedy. NASA had turned over its launch services to private industry in the 1980s, but Lewis had continued to manage launches carrying NASA payloads. In this role, Lewis had purchased launch services from the manufacturer, integrated the payload into the launch vehicle, and identified the proper trajectory and launch window.[977] During the Zero Base Review, representatives from Goddard Space Flight Center, Kennedy, and Lewis met several times to discuss potential methods of streamlining NASA's expendable launch vehicle activities. In November 1996 the Agency decided to consolidate all launch services work at Kennedy.[978]

Lewis's final two Centaur launches were joint missions with NASA and the European Space Agency (ESA). The first, the Solar Heliospheric Observatory (SOHO), was designed to conduct in-depth studies of the Sun. An Atlas-Centaur launched SOHO in December 1995. The second, Cassini/Huygens, was even more ambitious.[979] The mission would use the Jet Propulsion Laboratory- (JPL) designed Cassini

orbiter and ESA's Huygens lander to explore Saturn's largest moon, Titan. The $2 billion spacecraft weighed 3 tons and was the largest and most complex interplanetary spacecraft ever assembled. NASA selected the powerful Titan IV-Centaur launch vehicle for the effort. The size of the spacecraft, its power source, and the long distance to Saturn complicated the launch. Lewis's involvement included the 1990 testing of the Titan IV shroud in the SPF, integrating the spacecraft into the rocket, and determining the launch window and trajectory.[980]

The Titan IV-Centaur lifted off in the early morning hours of 15 October 1997 with Lewis personnel assisting with the launch operations and controlling the Centaur until it separated from the spacecraft, sending Cassini/Huygens on a seven-year journey to Titan.[981] In 2005 Huygens became the first spacecraft to land on an object in the outer solar system. Cassini brought an end to Lewis's supervision of NASA's launches. During its 35 years in the business, Lewis had managed the launches of 17 interplanetary missions, 21 lunar vehicles, several telescopes, and dozens of satellites.[982]

❖ ❖ ❖ ❖ ❖ ❖

NASA did not perform any planetary missions throughout most of the 1980s, but the situation changed rapidly in the 1990s with Cassini/Huygens and a series of nine smaller, less costly technology-based missions. Early satellites and spacecraft were small because launch vehicle capability was limited. As

*Image 332: Twin peaks on the martian surface as seen by Sojourner (NASA).*

*Image 333: Pathfinder airbag drop in the SPF (GRC–1995–C–01614).*

Atlas-Centaur and other rockets, including the shuttle, became commonplace, the payloads increased. However, the number of these larger, more expensive missions diminished as the Agency's budget shrank in the late 1970s. In the early 1990s NASA began launching a series of small spacecraft, and in 1992, Dan Goldin used this philosophy as the basis for his Faster, Better, Cheaper plan for the Agency.

NASA would increase its space science efforts by concentrating on simple, rapidly developed, and reasonably priced missions that could be flown more frequently. The failure of such a mission would not deal a crushing blow to the Agency.[983]

The *Mars Pathfinder* was one of NASA's first efforts in this new realm. JPL designed both the *Pathfinder* lander and its Sojourner rover, the first wheeled vehicle on Mars. *Mars Pathfinder* sought to demonstrate that a low-cost mission could be sent to the Martian surface at a reasonable cost to evaluate the performance of a rover. *Pathfinder* would be NASA's first return to the Martian surface since *Viking* in the mid-1970s.[984] *Viking* had used retrorockets to slow its descent and soft-land on Mars. JPL engineers, however, did not want to contaminate the *Pathfinder* landing site with rocket exhaust. Instead they designed a unique landing system that used a parachute and rocket braking to slow the descent and a collection of airbags to cushion the impact. The airbags allowed *Pathfinder* to safely bounce multiple times before coming to a stop.[985]

Because JPL engineers were concerned that the cloth airbags might rip open on Mars's rocky surface, they asked Lewis to develop a series of tests in the SPF's large vacuum chamber to verify the airbags' integrity. In a simulated martian atmospheric environment, the test engineers slammed the bags and lander model down from the top of the

❖ ❖ ❖ ❖ ❖ ❖

November 1996

## Lewis experiments will land on Mars in 1997

(continued from page 5)
NASA Lewis also developed a method to control the potentially damaging electrostatic charge buildup on the rover resulting from the combination of its movement in the very dry Martian environment and the atmospheric conditions and pressure. A series of very fine Tungsten wire points were machined and placed at strategic locations on the rover to dissipate this charge.

The in-house effort to develop the experiments involved a cross-divisional team made up of employees from the Advanced Space Analysis Office, Power Technology Division, Instrumentation and Controls Branch, Structures Division, Office of Safety and Mission Assurance, Space Experiments Division, Resources Analysis and Management Office, and Procurement Division. The project operated under minimal supervision from management and focused on maximum civil servant usage with very little outside contracting.

"This resulted in about half the cost and time compared to the old way of doing business," Stevenson said.

Information gathered through the NASA Lewis experiments and JPL rover operations will be relayed back to Earth throughout the one-month mission on Mars. The NASA Lewis team will work with JPL researchers and others to evaluate the findings.

While the Viking landers that touched down on the planet in 1976 didn't detect life, orbital photographs showed evidence that water once flowed on the Martian surface, leaving deep channels, river deltas, and lake beds. Stevenson said that Mars Pathfinder's destination is an excellent choice in terms of looking for evidence of ancient water flows and a variety of geological features of interest, including those possibly harboring evidence of early life. The spacecraft is planned to land at the mouth of an ancient water channel called Ares Vallis, just north of the Mars equator, where rocks have presumably tumbled down from the highlands.

"If life did arise on an Earth-like early Mars, fossil remains should be preserved in the surface rocks. Mars Pathfinder will lay the groundwork for future missions that would bring back samples to Earth," he said.

Eight additional low-cost, unmanned missions are scheduled to land on the planet over the next decade. ◆

**Editor's Note:** *Additional information about the NASA Lewis experiments that will be part of the Mars Pathfinder mission can be found on the Internet:*

http://powerweb.lerc.nasa.gov/pv/SolarMars.html

(Left) NASA Lewis researchers applied recently developed technology in new ways to create three experiments ideally suited to the Pathfinder microrover Sojourner.

Photos courtesy of the Jet Propulsion Laboratory

(Right) This uniquely designed wheel is the heart and soul of the NASA Lewis experiment that will determine the abrasive effects of Mars surface materials. Knowledge gained from the Wheel Abrasion Experiment will help researchers design future Martian rovers.

6

*Image 334: Lewis News article about Lewis's experiments on Pathfinder.*

122-foot-tall chamber onto an angled board that had rocklike materials affixed to it. They simulated different types of terrain by adjusting the board's attitude from horizontal to steep angles.

The need for the test program became evident when several of the single-layer bags tore open. JPL engineers repeatedly modified the bags and systems without success. Finally the design team created a 17-square-foot collection of 24 bags composed of multiple lightweight layers instead of a single thick layer of fabric.[986,987] The new design was then successfully inflated in simulated space conditions inside the Spacecraft Propulsion Research Facility (B–2) test chamber.[988] *Pathfinder* was launched on 2 December 1996, and the airbag system worked flawlessly as *Pathfinder* descended onto a rocky ancient floodplain on 4 July 1997.[989]

In the early 1990s Lewis's Photovoltaic Research Branch created a computer model to predict the different types of solar radiation present in the martian atmosphere. JPL engineers used this model to determine that there was enough solar energy to operate the lander and rover. The team also used the model in designing the solar panels to power the vehicles. The *Pathfinder* mission was the first use of solar-powered technology on Mars.[990]

JPL asked Lewis to contribute three experiments to the *Pathfinder* mission. Geoffrey Landis and other Lewis researchers were interested in using Sojourner to verify the amount of solar energy on Mars and determine the effect of dust on solar array performance. The rover included an instrumented solar cell encased by a retractable glass window. Once each day, the cover was drawn back to expose the cell directly to the sunlight. Another sensor with two vibrating crystals— one covered and one exposed daily—was used to measure the amount of dust that settled on the unit. A comparison of these data to the readings with the dust-covered glass door closed revealed moderate power losses.[991,992]

Lewis also worked with the rover's wheels. There was concern that the wheels might accumulate static electricity that could overload the vehicle's batteries. Lewis developed and tested small tungsten discharge plates to attract any electrical charges and discharge them into the atmosphere. JPL added the plates to Sojourner, and they protected the rover as intended. In addition, engineers covered one of the rover's wheels in a metallic coating to measure the abrasiveness of the surface. A small sensor relayed data on the erosion of the coating over time.[993]

Although the martian surface mission had been designed for 30 days, *Pathfinder* and Sojourner operated for nearly three months. The mission provided a wealth of information and led NASA researchers to conclude that the planet once had liquid water and a thicker atmosphere.[994] *Pathfinder* spurred a series of increasingly large rover missions to Mars that continues today.

## Electric Propulsion in Space

NASA introduced its New Millennium Program in 1994 as part of its broader effort to launch more frequent and cost-effective science missions. The program's primary objective was the demonstration of various advanced technologies that could be utilized by future spacecraft. JPL partnered with Orbital to design the *Deep Space 1* spacecraft to validate a variety of these new technologies—including the Lewis-designed thruster and power-processing unit for the ion propulsion system and its complementary solar concentrator arrays.[995]

*Deep Space 1* was NASA's first space mission using ion thrusters as its primary mode of propulsion. The spacecraft's ion propulsion system was based on an electron bombardment thruster whose origin can be traced back to the original thruster that Harold Kaufman invented at Lewis in 1958. The center had been pursuing electric propulsion since the late 1950s, and each successive generation of thruster improved upon its predecessor. In the late 1970s Lewis researchers began using xenon, instead of mercury and cesium, as the preferred working gas because it is nontoxic, it can be stored at high pressure, and its high atomic mass provided favorable performance.[996]

Lewis began developing 30-centimeter- (cm) diameter xenon thrusters and a solar array and concentrator in the mid-1980s. The solar power system, which included a unique magnifying lens that concentrated solar radiation onto two high-power solar arrays, could generate up to 20 percent more power than contemporary systems.[997] These developments led to Lewis's November 1992 partnership with JPL on the NASA Solar Electric Power Technology Application Readiness (NSTAR), a solar-powered xenon-gas-based electric propulsion system. Lewis created the initial prototype, and JPL tested it for over 8,000 hours in simulated space conditions. The actual NSTAR flight engine was industrially manufactured on the basis of the Lewis design.[998]

*Deep Space 1* was launched into space on a Delta II rocket on 24 October 1998. The solar arrays deployed within hours of the launch, and the ion thruster began operation as scheduled 30 days later.[999] *Deep Space 1* successfully demonstrated its new technologies during its first 90 days, including operation of the NSTAR thruster. After reaching the asteroid Braille in April 1999, NASA extended the mission. NASA issued its final commands on 18 December 2001, three months after the spacecraft encountered the Borrelly comet.

The NSTAR system operated for over 16,000 hours, far surpassing *Space Electric Rocket Test II's (SERT–II's)* record-setting performance.[1000,1001]

NSTAR propulsion system was also utilized on JPL's 2007 *Dawn* mission to explore the two largest objects in the asteroid belt, the protoplanet Vesta and dwarf planet Ceres. *Dawn* was the first scientific mission to use solar electric power. The propulsion system included three solar-powered NSTAR ion thrusters that were operated sequentially. Glenn manufactured several of the components and oversaw the hardware review process. *Dawn* successfully visited Vesta in 2011 and Ceres in 2015.[1002]

*Image 335: NSTAR thruster test (GRC–2015–C–06537).*

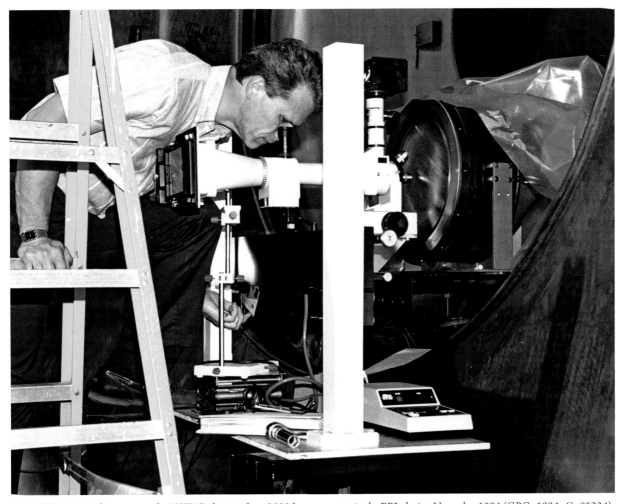

*Image 336: A researcher prepares the NSTAR thruster for a 2000-hour wear test in the EPL during November 1994 (GRC–1994–C–05234).*

## Aeronautics

One of the Zero Base Review's more controversial recommendations was the transfer of the center's flight operations work to Dryden. If fully implemented, the proposal would be the death knell for the center's 50-plus-year flight research program. The hangar, which was the center's first and most recognizable structure, would be utilized only to host visiting aircraft, and the 14-person Flight Operations staff would be reduced to 2.

The Zero Base Review instructed the centers to consolidate all required research aircraft at Dryden while excessing the rest. At the time, Lewis possessed six aircraft—a Learjet, a Twin Otter, a DC–9, a Gulfstream, a North American Rockwell OV–10A, a Beechcraft T–34, and the NASA 5. The latter four were quickly transferred, but there was resistance to transferring the others because of their importance to Lewis's research efforts.

*Image 337: Flight Operations veterans Kurt Blankenship and Bill Rieke flying the center's Learjet (GRC–2001–C–003108).*

Sheet1

| LEWIS RESEARCH CENTER -- AIRCRAFT CONSOLIDATION PLAN | | | |
|---|---|---|---|
| **Aircraft Identification** | **Program Plan** | **Status** | **Consolidation Plan Action** |
| DHC-6 | Transfer to Dryden | Pending | Transfer |
| T-34B | Transfer to WPAF - tentative in September | Pending | Decommission |
| T-34C | Transfer to NAVY - tentative in September | Pending | Transfer |
| T-34C | Transfer to NAVY - tentative in September | Pending | Transfer |
| Lear 25 | Transfer to Dryden | Pending | Transfer |
| OV-10A | Transfer to Langley | Complete | Decommision |
| OV-10D (615) | Transfer to Other Gov't Agency or Davis Monthan | Pending | Decommision |
| OV-10D (617) | Transfer to Other Gov't Agency or Davis Monthan | Pending | Decommision |
| DC-9 | Transfer to Microgravity Institute or Dryden | Pending | Transfer |
| As depicted above, there are only three (3) Lewis Research Center aircraft that are scheduled to be transferred to Dryden -- the DHC-6, Lear, and DC-9. The latter is contingent upon whether or not the proposed Microgravity Research Institute desires to perform this function/site to be determined. | | | |

*Image 338: Aircraft consolidation plan.*

The Twin Otter was the workhorse for the center's icing research program. So that the program could still use the Twin Otter's services, Lewis began negotiations to transfer the Twin Otter to the Canadian National Research Council. The council would then lease the aircraft's services back to NASA's icing program. The photovoltaic program relied on the Learjet to calibrate their advanced solar cells. It would have to make arrangements to continue this work at Dryden. The microgravity program utilized the modified DC–9 before it was transferred in September 1997. After that, the researchers conducted all microgravity flights on Johnson's KC–135. Operations of the Learjet and Twin Otter were set to be terminated in early 2000.

This time, however, there was strong pushback from several members of Congress, particularly regarding the projected savings and the negative impact on the research programs. An Inspector General investigation in 2000 found that the NASA Zero Base Review incorrectly assumed that Dryden could handle all of the Agency's aircraft without increases in personnel. More importantly, the review did not address the

*Image 339: Icing researcher Judy Van Zandt with pilots Rich Ranaudo and Tom Ratvasky beside the Twin Otter (GRC–1997–C–03962).*

effect of the transfers on the individual research programs.[1003] The Inspector General determined that the Twin Otter and Learjet should be maintained in Cleveland,[1004] and NASA canceled its plans to eliminate Lewis's Flight Operations efforts.

❖ ❖ ❖ ❖ ❖ ❖

The Zero Base Review also impacted the center's aeronautical research, including the closure or threatened-closure of some aeronautical facilities, in particular wind tunnels that were thought to be duplicated elsewhere. Nonetheless, Lewis participated in three large multicenter aeronautics programs— High Speed Research (HSR), Advanced Subsonic Technology (AST), and Ultra-Efficient Engine Technology (UEET). Congressman Conrad Burns stressed the national importance of these efforts during a 1995 NASA appropriations hearing. "It is estimated that the first country to market … an [HSR] aircraft stands to gain $200 billion in sales and 140,000 new jobs….

[The AST] market … generates 1 million jobs and contributes over $25 billion annually to the U.S. trade balance. These programs are moneymakers, and it is in the national interest to give them the support they need."[1005]

By the 1990s, fear of another energy crisis faded and a strong energy market emerged. This led to a renewed interest in supersonic passenger aircraft. The nation's first attempt at supersonic transport in the 1960s had failed as opposition arose over its potential noise and pollution. A second effort in the 1970s produced some new technology, but the declining market for supersonic transports and budget cuts led to its cancellation in 1981.[1006] With stable energy prices and the impending retirement of the European Concorde, analysts predicted that the market for larger, more economical supersonic transports would open up in the 2000s.[1007]

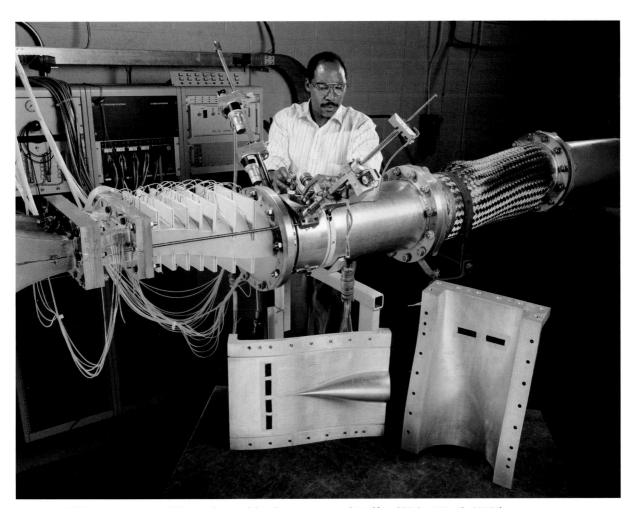

*Image 340: Bill Darby prepares an HSR inlet duct model in the Engine Research Building (GRC–1995–C–02109).*

In 1990 NASA undertook the HSR program to develop technology for the high-speed civil transport vehicles expected to emerge in the coming years. The HSR sought to develop basic technologies for a theoretical aircraft that could transport 300 passengers across the ocean at 1,500 mph. It would be up to industry to utilize the technology to design and manufacture the vehicle.[1008]

The HSR program involved all of NASA's aeronautics centers, with Langley Research Center managing the overall effort, and Boeing, Pratt & Whitney, and General Electric as the key partners. Lewis was responsible for the propulsion system.[1009] The first phase of the program confirmed that the HSR concept was feasible. Then in 1996 NASA began developing component technology to make the system environmentally and economically viable. The propulsion system had to significantly reduce emissions, meet airport noise requirements, and be cost effective. To meet these goals, Lewis concentrated on combustors, exhaust nozzles, engine inlets, and fans.[1010]

The combustor, or combustion chamber, was the key to lowering the nitrogen oxide ($NO_x$) emissions that occurred when the engines raced during takeoff. Reducing ozone-depleting $NO_x$ was essential to the overall HSR design. To address this need, the design team modified the combustor concept from the 1970s that used premixed and prevaporized fuel. Testing of this combustor in the Engine Research Building demonstrated that it could meet the program's reduced emissions goal, but Lewis continued its efforts because of concerns about combustion instability and loss of engine power.[1011]

In addition to the emissions reduction, Lewis lowered noise levels using a mixer-nozzle, developed a combustor liner with ceramic matrix composite materials, and designed an advanced supersonic inlet. These efforts involved testing the advanced inlet on an engine in the 10- by 10-Foot Supersonic Wind Tunnel, small-scale combustors and high-temperature ceramic matrix composite materials in the Engine Research Building, engine emissions in the PSL, and nozzles in the Aero-Acoustic Propulsion Laboratory.[1012]

The HSR was terminated prematurely in 1999 when new economic studies indicated that the market for supersonic transport was not materializing. There was also concern that technologies meeting certain short-term environmental levels might not be applicable to unknown future standards. Finally, the supersonic transport was geared toward international flights, and NASA management was increasingly seeking to concentrate on making improvements to domestic air travel.[1013] NASA shelved the premixed, prevaporized combustor concept as the supersonic transport application disappeared. Nonetheless the program showed that a cost effective, environmentally friendly supersonic transport engine could eventually be produced.[1014]

❖ ❖ ❖ ❖ ❖ ❖

The United States dominated the airliner manufacturing market for decades, but that share began decreasing with the emergence of Europe's Airbus in the 1970s and 1980s. The U.S. airline industry was hampered by overcrowded airports, the lack of life-prediction technologies for older aircraft, and new environmental regulations. Several NASA centers and manufacturing corporations initiated the AST program in 1996 to enhance the technological base for large U.S. civil transport aircraft, focusing on the technologies with the highest payoffs. AST sought to increase airline profitability through increased productivity, better efficiency, and reduced cost. The ambitious effort integrated aircraft, airline operators, airspace systems, safety, and environmental aspects into a single program.[1015]

Lewis worked closely with engine manufacturers on AST's propulsion aspects, including noise reduction, emissions control, and engine systems components. Engine designers found that increased internal pressure improved engine performance but also elevated emissions. The Advanced Subsonic Combustion Rig, which was added to the Engine Research Building in the mid-1990s, was essential to the Lewis effort. It was the nation's only facility that could test full-scale combustors under operating conditions similar to those experienced in new high-pressure-ratio engines.[1016,1017]

In the early 1990s Lewis undertook a long-term effort to design a new lean fuel combustor that did not premix the fuel and air. This lean direct injection (LDI) system employed numerous injectors that added fuel directly to the flame. Researchers have demonstrated that the LDI system performs well at mid and high power levels. The center continues to develop the LDI concept, improving component fabrication, fuel spray, and active controls.[1018,1019]

*Image 341: Advanced Subsonic Combustion Rig in the Engine Research Building (GRC–1999–C–00249).*

Under the AST program, General Electric developed its successful twin-annular pre-mixing swirler (TAPS) combustor, which was based on its lean dual annular design from the 1970s. TAPS resolved previous problems with uneven temperatures and high carbon dioxide levels, resulting in a more stable combustor that maintained low emissions during both high- and low-power operation. During low-power cruising, the TAPS combustor produced roughly half of the $NO_x$ that fuel-rich fuel combustors produced.[1020]

The Department of Energy developed a new combustor type in the 1980s that was diametrically different from the Lewis lean-burn concept. The Rich-burn/Quick-mix/Lean-burn (RQL) combustor quickly transitions a rich fuel input into a lean burn area. This generates additional soot but enables the combustor to meet desired low-$NO_x$ levels and simplifies operation. In the AST program, Pratt & Whitney decided to pursue the RQL technology with its Technology for Advanced LOw $NO_x$ (TALON) combustors.

The AST program also examined different sawtooth chevron nozzle configurations to quiet engine exhaust. The jagged nozzle edge facilitates the mixing of the hot engine exhaust and the cool atmosphere. This interaction has been a significant source of engine noise. Researchers tested 54 small-scale nozzle designs in Glenn's Aero-Acoustic Propulsion Laboratory. Ensuing flight tests in 2001 on the center's Learjet demonstrated that the chevron nozzles reduced noise significantly with only negligible impact on the engine's thrust.[1021]

❖ ❖ ❖ ❖ ❖ ❖

Congress sharply reduced funding for the AST effort in 1999, but NASA centers sustained separate portions of the program.[1022] Lewis established the new UEET program in October 1999 to carry on its pollution-reduction work. UEET addressed pollution by decreasing fuel consumption through improved efficiency and continuing the development of low-emission combustors. To improve efficiency, Lewis studied highly loaded turbomachinery, new lightweight composite materials for compressors and turbines, the aerodynamics of engine-airframe integration, and intelligent propulsion controls.[1023]

NASA had also continued some AST work under the new Vehicle Systems Program (VSP). VSP was a multicenter and industry partnership that sought to develop an array of new technologies for future civilian aircraft of all sizes and applications. Its broad goals were distilled into four research areas—lower emissions, new energy sources, quiet engines, and improved aerodynamics for fuel efficiency. In 2003 the VSP officially incorporated Lewis's UEET efforts.[1024]

The five-year UEET project was geared toward developing specific near-term technological advances that aircraft manufacturers could employ to minimize pollution while maintaining high performance. The overall VSP program sought to reduce $NO_x$ emissions by 70 percent during takeoff and landing and by up to 90 percent during normal cruising. Again the combustors were the key to achieving these goals.[1025]

**UEET**

Ultra-Efficient Engine Technology Program

*Image 342: The UEET logo.*

*Image 343: Burner rig heating a titanium-aluminide alloy sample (GRC–1999–C–01995).*

Under Glenn's UEET management, Pratt & Whitney and General Electric continued development of their respective combustors, the RQL-based TALON and the lean mix TAPS. The manufacturers overcame a new wave of issues resulting from the significant increase of operating pressures required by these new engines. These engines operated at much higher pressures than previous generations of engines, but they produced only half of the emissions. Pratt & Whitney utilized the TALON in its PW1000G engine series, which powers Airbus's A320 airliners and other aircraft. General Electric incorporated the TAPS combustor into their next-generation (GEnx) engine design.

The GEnx engine, which powers Boeing's 787 Dreamliner, was one of the most significant results of the AST and UEET programs. Glenn's contributions to the GEnx include the combustor, the noise-reducing chevron nozzles, and several new materials technologies.

*Image 344: Pratt & Whitney combustor test setup in an Engine Research Building test cell during September 1998 (GRC–1998–C–01995).*

The GEnx was the first jet engine to include a fan case and blades made entirely from composite materials. Glenn researchers had investigated fiber-and-matrix composite materials in the 1990s and established a method for designing an all-composite fan. Glenn researchers had also demonstrated that the lightweight titanium-aluminide alloy was strong enough to withstand impacts. Consequently, GEnx employs titanium-aluminide turbine blades. Glenn also worked with General Electric on the implementation of a nickel-aluminide alloy as a coating for the high-pressure turbine blades on the GEnx. In the 1990s, Glenn, General Electric, and Pratt & Whitney developed the ME3 alloy, which could perform in temperatures up to 1300°F. General Electric incorporated ME3 turbine disks into the GEnx engine.[1026]

NASA restructured the VSP program in early 2005 resulting in the cancellation of the UEET effort. NASA's aeronautics budget in the 1990s was the highest that it had ever been as the nation sought to poise its aviation industry for increased competition in the coming years. The cancellation of HSR and AST in 1999 returned the aeronautics budget to the level that it had been at since the early 1970s. The program

terminations did not cause any Agency layoffs or facility closures, but they signaled the beginning of a decline in aeronautics research. NASA began transferring personnel to space programs that combined aeronautics and space technologies.[1027] Lewis's aeronautics budget declined from $250 million in 1998 to $158 million in 2000, and the center considered shutting down or mothballing nearly all of its aeronautics facilities except the Icing Research Tunnel.[1028]

❖ ❖ ❖ ❖ ❖ ❖

The center was also involved in less advanced types of aircraft. The popularity of privately piloted general aviation spiked in the late 1970s before increases in fuel prices, aircraft costs, and complexity of operation caused the market to plummet. Analysts predicted that the field was ripe for resurgence in the mid-1990s, but the industry had performed little research and development in the interim, particularly in regards to engines. The existing engines were reliable, but they were complicated, noisy, and expensive.[1029]

In 1996 Lewis and the Federal Aviation Administration (FAA) created the General Aviation Propulsion (GAP) program to quickly develop new technologies that

*Image 345: GEnx engine with a chevron nozzle on an Air India Boeing 787 Dreamliner. By Oliver Cleynen (own work) [cc BY-SA 3.0 (http://creativecommons.org/licenses/by-sa/3.0)], via Wikimedia Commons.*

*Image 346: Cessna 206 general aviation aircraft (GRC–1980–C–05641).*

would result in inexpensive general aviation engines with low emissions and noise. The GAP program addressed both piston engines for private aircraft and turbojets for small business aircraft. The researchers sought to simplify the engine design and manufacturing process to reduce costs while improving performance and decreasing noise.[1030]

NASA worked with Teledyne Continental Motors to develop an inexpensive 200-horsepower piston engine that was quiet, lightweight, and easy to operate. The cost savings were achieved primarily by incorporating many components into a single aluminum casting. In addition, the diesel engine could run on jet fuel, which was less expensive than aviation gas, and pilots could operate the engine with a single lever that controlled the fuel flow.[1031,1032] In the end, however, the engine was not certified. Aircraft manufacturers were wary of its low power and advanced design. Despite the diesel's popularity overseas, the U.S. aircraft industry remains lukewarm to the general concept.[1033,1034]

The jet engine phase of the program sought to reduce engine costs by a factor of 10.[1035] NASA worked with Williams International to develop the FJX–2 turbofan engine. This 700-pound-thrust engine weighed less than 100 pounds. Lewis researchers analyzed the engine on test rigs in 1997. Then the complete engine was operated for over 500 hours during the next three years, including under simulated altitude conditions in the PSL. An independent analysis showed that the cost of the FJX–2 was on par with contemporary piston engines of the same size. Williams incorporated the FJX–2 into its V-Jet II concept aircraft. After performing successful test flights, Williams presented the V-Jet II at the 1997 Oshkosh Airshow.[1036]

The following year, Eclipse Aviation sought to modify the V-Jet II and make it available commercially. The result—the Eclipse 500—was flight-tested in 2002 with the NASA engine, which had been renamed the EJ22. The test revealed that Eclipse's modifications to the aircraft design had increased weight to the point that the lightweight engines were not powerful enough to sustain peak performance. The company decided to use traditional jet engines instead. The aircraft went on to be a successful inexpensive business jet, but the projected cost-savings and industry revolution were not realized. In the end, neither of the GAP engines made it to the FAA certification process, but the program demonstrated that quiet, reliable engines could be produced at a low cost.[1037,1038]

## Microgravity Research Blossoms

Lewis's microgravity program grew rapidly in the late 1980s and reached its apex in the 1990s. Lewis's two drop towers were beehives of experimental activity, new research aircraft flew low-gravity parabolas, and the center constructed specialized facilities to prepare experiments for shuttle missions. The new space shuttle program did not attract the number of research customers that NASA officials had predicted. As a result, Glenn was frequently able to utilize the vehicle to expand its microgravity studies. The shuttle would carry over 200 Glenn microgravity experiments before being retired in 2011.[1039]

Building on its early 1960s research that revealed how liquid hydrogen would behave in space and determined the cause of the *Apollo 1* fire, Lewis now dove deep into microgravity research involving combustion, materials, and physics. The work in the Zero Gravity Research Facility (Zero-G) in the late 1960s made it apparent that the microgravity environment provided a unique setting for studying the basic elements of combustion and fluid physics, which had as many applications on Earth as in space.[1040] The center established the Aerospace Environment Branch in the 1970s to further pursue the research.

In 1980 Lewis acquired a Learjet and modified it to serve as a multipurpose testbed that could perform microgravity missions. As with the AJ-2 in the 1960s, the pilot flew a series of parabolas, each of which produced up to 20 seconds of microgravity. Lewis opened the Microgravity Materials Science Laboratory in

September 1985 to assist researchers in modeling and planning materials experiments for the shuttle.[1041,1042] In 1986, as work in the Agency and center expanded significantly, the center created the Space Experiments Division to develop shuttle experiments.[1043]

In the early 1990s Lewis ramped up its microgravity efforts even further by significantly upgrading its two drop towers and adding the $7.1 million Space Experiments Laboratory to the Zero-G in 1993. The new facility, which contained a high bay and several clean rooms, allowed the Space Experiments Division to consolidate its shuttle preparation work, which had previously been performed at 12 different facilities.[1044] In October 1993 Lewis acquired a McDonnell Douglas DC-9 aircraft to expand its low-gravity flight

*Image 347: Lewis's DC-9 on the downward slope of a microgravity-inducing parabola (GRC-2001-C-00615).*

*Image 348: View down the 2.2-Second Drop Tower (GRC–1994–C–05830).*

capabilities even further. The Lewis Flight Operations staff removed the passenger seats, installed padding, and added new electrical and data systems. The DC–9 flew the same parabolas as the Learjet (yielding 20 seconds of microgravity), but the DC–9 could accommodate a great deal more experimental equipment and personnel than the Learjet could.[1045,1046] Between 1995 and 1997 the DC–9 flew over 430 hours while hosting more than 70 experiments.[1047] Lewis also utilized sounding rockets to provide up to 6 minutes of microgravity for combustion and fluid experiments.

The center had the most comprehensive set of microgravity facilities anywhere, advanced new diagnostics techniques, and an engineering and technical staff to support that research. In-house researchers, industrial partners, and academic communities from all over the world utilized Lewis's unique tools to either validate

textbook doctrine or to develop new insights into fundamental combustion, fluid, and materials phenomena. The number of pending grants and proposals increased from 50 to several hundred over the course of a few years. The 2.2-Second Drop Tower often supported 12 tests per day, the Zero-G conducted larger, more precise drop experiments daily, and Lewis aircraft carried experiments from across all disciplines most of the year.[1048]

❖ ❖ ❖ ❖ ❖ ❖

In the 1960s Lewis researchers were occasionally able to conduct longer-duration microgravity studies using NASA spacecraft. These experiments, however, were governed by the mission's available physical space, power, and duration. The new space shuttle program provided researchers with abundant resources for experiments lasting up to two weeks. It was the ESA, however, that initially led the way in utilizing the shuttle for microgravity research. In March 1982 NASA

*Image 349: Simon Ostrach floats in the KC–135 as Renato Colantonio monitors a microgravity experiment (GRC–1996–C–02847).*

included a series of ESA microgravity experiments on the third shuttle mission. The following year the ESA created the Spacelab module—a reusable platform that could house multiple microgravity experiments that would be operated by shuttle astronauts during their flight. In April 1985 U.S. researchers began conceiving their own experiments for Spacelab, and in January 1992 STS–45 carried the International Microgravity Laboratory, which included experiments from numerous international investigators, including Lewis.[1049]

The U.S. Microgravity Laboratory 1 (USML–1), which flew in a Spacelab module on STS–50 for 14 days in June–July 1992, was the nation's first major collection of shuttle experiments and remains the pinnacle of Lewis's shuttle-based microgravity. USML–1 contained 31 different government, university, and industry experiments, including 7 from Lewis. As Lewis's first significant presence on a shuttle mission, the center gave it top priority.

Simon Ostrach, the former Lewis researcher who became internationally renowned for his work on buoyancy-driven flows, devised the Surface-Tension-Driven Convection Experiment (STDCE) to study fluid flows in the absence of such phenomena. STDCE was Lewis's largest payload to date and the center's number one concern at the time. Lewis was faced with a critical prelaunch decision when the STDCE hardware accidentally fell from a hoist at the Kennedy Space Center. Preliminary tests suggested that it had not been damaged, so the team decided to fly the hardware "as is." In flight, STDCE and the six other experiments performed flawlessly. Five of these were small, hand-operated experiments that were performed in a glovebox in Spacelab. They were conceived, developed, and managed by a group of inexperienced, early career people at the center. These same experimenters, owing in large part to this early, hands-on experience with spaceflight hardware, are managing major projects at the center today.[1050,1051]

One of the USML–1 experiments, which its own investigators labeled as "perhaps the most trivial experiment to ever fly on the shuttle," sought to determine if a candle flame would burn in zero gravity. At the time, there was a widespread belief that the candle would not burn in a low-gravity atmosphere because there would be no buoyant-driven convection to bring fresh oxygen to the flame. Some people believed, however, that a process known as molecular diffusion would deliver sufficient oxygen to the flame. Three researchers—Daniel Dietrich, Howard Ross, and a Case Western Reserve University professor, James T'ien—set out to answer the question once and for all by attempting the experiment in space. Candle experiments conducted on the shuttle and later on Mir proved that not only would the candle flame burn, albeit weakly, but it would survive up to four times longer than if it burned on Earth. The flame had a round shape because of the lack of buoyant convection in the weightless environment.[1052]

This simple experiment eventually led to new important technology in a completely different field. In order to fly on the Mir, the Russians required the Glenn researchers to incorporate oxygen and carbon dioxide sensors into the experiment. They created new smaller and more accurate sensors. Ten years later one of those sensors was incorporated into the mask of pilots flying the F–22 fighter aircraft. At the time, F–22 pilots were occasionally passing out in flight for unknown reasons. Glenn's new sensors confirmed suspicions that the problem stemmed from an oxygen deficiency. So, from a simple curious question about candles, came a sensor that helped solve a problem in state-of-the-art military aircraft.[1053,1054]

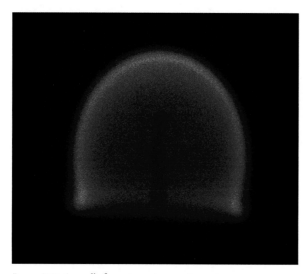

Image 350: A candle flame in a microgravity environment (GRC–1998–C–00486).

USML–1 paved the way for a new generation of U.S. microgravity research.[1055,1056] Virtually every space shuttle mission of the 1990s included at least one Lewis microgravity experiment, and several flew multiple Lewis investigations. Lewis worked with the Marshall payload operations personnel to integrate the flight experiments with Spacelab prior to each launch. The researchers and engineers were also highly involved with the development of crew procedures and the frequent, face-to-face training of the astronauts who would perform the experiments.[1057]

Some of the hardware was rapidly redesigned to study fluid oscillations and was reflown in October 1995 as USML–2. The three U.S. Microgravity Payload missions in the early 1990s were notable because the experiments were controlled remotely from the new Telescience Support Center at Lewis.[1058,1059] The Microgravity Science Laboratory (MSL), which was flown in July 1997, contained experiments from several universities that helped professors rewrite textbooks. MSL also included 11 Lewis tests, including Combustion Module-1. The module was the largest package of Lewis experiments ever flown and the shuttle's most complex set of experimental payloads yet.[1060]

Significant Lewis shuttle experiments during this period included the creation of what today are still the weakest flames (1 watt) ever observed in nature, the identification of the universal relationship between soot creation and slow-burning diffusion flames (which has medical, fire prevention, engine manufacturing, and industrial applications),[1061] critical property measurements during dendrite crystal growth (which helps improve industrial metal casting processes),[1062] the first observation of dendritic growth in crystals with small, evenly dispersed particles (which revealed the value of microgravity for studying the behavior of gel-like materials),[1063] and the determination that microgravity significantly affected the process of using heat to compact metal without liquefying it (which allowed cost reductions in the manufacturing of metal cutting tools).[1064]

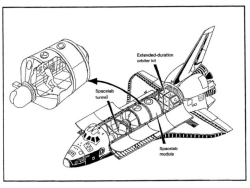

**Gravitational Role in Liquid-Phase Sintering**

National Aeronautics and
Space Administration

**Lewis Research Center**
Cleveland, Ohio 44135

**NASA Headquarters**
Office of Life and Microgravity Sciences
and Applications
Microgravity Science and
Applications Division

**NASA Lewis Research Center**
Space Experiments Division
Materials Division

*Image 351: Liquid-Phase Sintering brochure.*

Among the most notable achievements was the shattering of long-held misconceptions about how flames would behave in the absence of gravity. The now-outdated ideas that flames would be weak and not survive without a buoyant flow were proven wrong, and fires were discovered to be potentially more hazardous in spacecraft than on Earth. Fundamental experiments showed that a steadily propagating flame can exist in microgravity; Sandra Olson's experiments in this area revealed that materials would burn at lower oxygen concentrations and, when imposed with a forced-airflow typical of spacecraft ventilation systems, actually burn faster than their normal gravity counterparts. The experiments and associated theoretical and numerical studies showed the importance of radiation heat transfer in combustion systems and its criticality in determining flame burning rates and flammability limits.[1065,1066]

One of the key factors of flying experiments in space was knowing the residual gravity levels. The center

designed and flew the Space Acceleration Measurement System (SAMS) to determine variations in the microgravity levels during flight. SAMS flew on every mission, even those without other center hardware. The SAMS unit that completed the most flight hours of any experimental hardware during the shuttle era is now on display in the Smithsonian Institution. SAMS units continue to fly today on the International Space Station.[1067,1068]

❖ ❖ ❖ ❖ ❖ ❖

The center's Microgravity Program was at its peak in the mid-1990s and involved over 700 civil servant, contractor, and university personnel. It was at this point that NASA initiated the Zero Base Review. Despite protests by NASA's director of Life and Microgravity Sciences and others at headquarters (including independent assessment teams), Agency leadership assigned Marshall as the lead center for

April 1997  *Lewis NEWS*

**Director's Corner**
with DONALD CAMPBELL

**National Center for Microgravity Research**

I'm pleased to announce that NASA Lewis has entered into an agreement with the Universities Space Research Association (USRA) and Case Western Reserve University (CWRU) to form a new National Center for Microgravity Research on Fluids and Combustion. The new partnership was formally created on March 13 when NASA Administrator Daniel Goldin signed an agreement during a ceremony at CWRU attended by Dr. Paul Coleman, president of URSA; Dean Thomas Kicher, CWRU School of Engineering; and me.

The strategy for continued microgravity research was identified in the Zero Base Review proposal by Lewis. The intent was to establish a microgravity institute for combustion and fluids research. The proposal ran into difficulty because of civil servant issues with an institute. The strategy for continued, important research with Lewis involvement was not abandoned, and the concept of a center was envisioned as an alternate approach.

A proposal for this new entity surfaced last fall, and it took a great deal of creativity, perseverance, and tact on the part of our legal and procurement divisions to get the final agreement ready and approved by NASA Headquarters. Equally important to the success of this proposal was the commitment of the people in the Microgravity Science Division. They did an outstanding job in assessing the proposal and focusing on the needs of a new National Center that will be responsible for key microgravity science functions. Also, there was great cooperation and assistance from key people in Code UG at Headquarters and at Marshall Space Flight Center.

Based on the terms of the agreement, the administrative offices of the National Center will be located at CWRU, and the staff scientists and technicians will be housed at Lewis. CWRU has agreed to provide office space to the program, as well as two tenured faculty positions. Our goal is to create an atmosphere where the university research community and NASA employees jointly pursue new avenues of research and technology. It is also our hope that the new National Center will attract industrial research organizations into the Agency's microgravity research program. If all goes well, the National Center should be up and running at its full complement of 30 people by December 1997.

I think our new partners will help to enhance the image of Lewis and solidify our role as the nation's center of excellence in microgravity research in fluids and combustion. I want to thank the support service contractors at Lewis who are currently performing this function. You have all done an outstanding job, and I hope to see many of you at the new National Center.

.

*Image 352: Lewis News article by Don Campbell about the National Center for Microgravity Research.*

overall management of microgravity science. The new designation remained in place despite vocal opposition by Lewis management and members of Congress.[1069]

At the urging of some at headquarters, NASA maintained Lewis's responsibility for fluids and combustion research, but Marshall gained oversight of the overall microgravity science program. Eventually, as part of the Agency's downsizing, Lewis transformed its Space Experiments Division into the Microgravity Science Division and excessed the DC–9. In its place, arrangements were made for Johnson's KC–135 aircraft to periodically travel to Cleveland to carry out microgravity flight tests.[1070] Although staffing in this field dropped dramatically, microgravity payloads in fluids and combustion science, as well as SAMS, fly today on the International Space Station, and Glenn remains the renowned leader in these fields.

As part of Administrator Goldin's desire to create a new close university tie for each NASA center, the National Center for Microgravity Research was instituted at Case Western Reserve University in March 1997 under the leadership of Simon Ostrach. This new scientific community utilized Lewis's microgravity facilities and experts.[1071] The institute thrived for nearly 10 years before NASA cut its funding.[1072]

❖ ❖ ❖ ❖ ❖ ❖

Researchers in Lewis's Electro-Physics Branch developed one of the more unique spinoffs of space technology. Bruce Banks, Kim de Groh, and others found that in space single atoms of oxygen, referred to as atomic oxygen, caused some degradation and embrittlement to external components like solar arrays. Glenn researchers undertook an extensive effort in the 1990s to understand and prevent this damage. Since oxygen does not remain as a single atom in the Earth's atmosphere, they created a vacuum facility that enabled atomic oxygen to be applied to a large sample surface area in a simulated space environment. The researchers also conducted numerous experiments on both the shuttle and the ISS.[1073,1074] They found that atomic oxygen causes most organic materials to become gaseous carbon monoxide or carbon dioxide. To mitigate the corrosion, the group developed a method of coating solar arrays with a thin film that blocks atomic oxygen without impeding solar cell performance.

Image 353: *Demonstration of the restoration of a fire-damaged painting from Saint Alban's church (GRC–2011–C–00516).*

Banks and Sharon Miller began pursuing ways to utilize the destructive nature of atomic oxygen for terrestrial applications, such as cleaning delicate materials. The numerous uses identified included the sterilization of medical implants, the decontamination of aircraft components, the creation of better seals, and most notably, the restoration of damaged artwork.[1075] Paintings with surfaces damaged by soot, ink, or other markings are very difficult to repair without harming the pigment. Miller and Banks found that atomic oxygen slowly gasifies the damaging hydrocarbons on the surface without impacting the underlying paint.

They experimented with several purposely damaged paintings in their vacuum test facility. Over the course of several days, they were able to successfully remove the contaminants. This restoration could be performed on what were thought to be permanently damaged works. Then an art conservator could apply a binder to protect the actual paint.[1076] The researchers created a portable atomic oxygen device that could be used remotely at museums and churches. The device has been used successfully to restore works by Jackson Pollack, Andy Warhol, Roy Lichtenstein, and other artists, with works dating back to the Renaissance.[1077]

## Physical Downsizing Begins

NASA's downsizing in the late 1990s led to a push to remove duplicative or unused infrastructure. As missions and staffing decreased, the Office of Inspector General felt that the need for facilities should diminish correspondingly,[1078] but this assessment did not take into account that maintaining usable facilities was less expensive than rebuilding them when the need resurfaced. Lewis had a long history of repurposing or upgrading its facilities when missions changed. Nonetheless the Office of Inspector General began an unremitting campaign to convince NASA to reduce its physical assets. The center began taking steps to remove two historical facilities—the RETF and the Plum Brook Reactor Facility (PBRF). The former was still in use, but the reactor had been in a safe-protected mode since early 1973.

*Image 354: Rocket engine firing at the RETF in July 1995 (GRC–1995–C–02448).*

Lewis constructed the RETF engine test stand in the mid-1950s to study engines that used high-energy liquid propellants. In 1984 the National Park Service placed the RETF on its National Register of Historic Places for its contributions to the 1960s space program.[1079] Researchers continued to use the RETF in the 1990s, including for demonstrations of a new fuel injector designed to reduce the cost of launching payloads into space.[1080]

The RETF was set back in a ravine at the far western end of the campus. The neighboring Cleveland Hopkins International Airport had long sought to extend its runways through the area, but Lewis had steadfastly refused to cede the property. As part of the Agency's streamlining in the mid-1990s, NASA began consolidating all of its space propulsion activities at Marshall and Stennis Space Center. The resulting cancellation of Lewis's chemical rocket program led to a 1995 agreement with the City of Cleveland to remove the RETF. NASA demolished the structure in 2003, and the new runway opened in 2004.[1081]

In 1998 the center undertook an ambitious plan to finally remove the PBRF, the Agency's only nuclear reactor. Efforts to repurpose the facility after its 1973 shut down had been fruitless. NASA had commissioned multiple studies in the late 1970s and 1980s to identify the necessary costs and procedures to remove the reactor, but in each case the Agency considered the effort too expensive. The estimates only increased over the years, however, and at the urging of the Nuclear Regulatory Commission (NRC), NASA finally decided to proceed with the decommissioning in 1998. The cost was exponentially higher than the proposals had been in the 1970s.[1082,1083]

A joint NASA-U.S. Army Corps of Engineering team spent three years developing an extensive decommissioning plan, which the NRC approved in 2002. Crews began stripping the facility of all of its internal components, piping, and equipment and removed the radioactive material from the site. Once this first phase of the decommissioning work was completed, NASA suspended the project until additional funding was provided. The final demolition proceeded quickly once the effort resumed, and by October 2012, the 27-acre site had been remediated back to its original condition.

*Image 355: The gutted remains of the reactor's Hot Lab, which was used to remotely examine irradiated test specimens (NASA SPF 1697).*

### New Name

The center's near closure in 1981 was fresh in the minds of local Congressmen as they responded to NASA's Zero Base Review recommendations in the mid-1990s. Although there was no current danger of shutting down the center, the continuing cutbacks and Agency redirection were threatening to render the center impotent. In September 1998 Senator Mike DeWine initiated an effort to bolster the center's wounded standing by renaming it after his colleague John Glenn.

John Glenn retired from Congress in 1997 after serving as a Senator from Ohio for 24 years. In addition, Glenn, who was famously the first American to orbit the Earth, would soon return to space as a crew member of the STS–95 shuttle mission. The renaming not only honored Glenn's contributions to NASA, but ostensibly elevated the center's public visibility. John Glenn was universally recognized, while George Lewis was relatively unknown outside the aerospace community.

DeWine attached the action to the Veterans Administration—Housing and Urban Development Appropriation for 1999. President Clinton approved the proposal on 21 October 1998, a week before Glenn's shuttle mission. The center officially became the John H. Glenn Research Center on 1 March 1999, and the Cleveland campus was named Lewis Field.[1084]

Glenn later recalled, "Quite apart from whether my name was on it or associated with it, I was proud of the fact that we were calling attention to some of these advances in research and engineering that had come from the center, and which make a big difference for our country. And the fact that my name was going to be connected with that, I was particularly proud."[1085] It was the fourth name since the center's establishment in 1941. Although several NASA centers have changed names, this was the first time that one honorary designation was replaced by another.

Image 356: John and Annie Glenn were feted during the center's renaming ceremony on 7 May 1999. The activities included an F-16 flyby, a parade, a picnic, and a renaming ceremony in the hangar (GRC_1999_C_01153).

The world changed dramatically during Don Campbell's final years as Director. Upon hearing news that an airliner had crashed into the World Trade Center on 11 September 2001, Campbell and Bill Wessel, Director of Safety and Assurance, hurried into a conference room to watch the developing news. As soon as the second tower was struck, they initiated plans to evacuate the center. The situation in Cleveland was tense because of concern about a Delta airliner with a possible bomb heading toward Hopkins Airport, Cleveland Mayor Michael White ordered the closure of all federal and city buildings and the schools, while Hopkins officials emptied the airport as the aircraft approached. The aircraft landed safely and was ordered to park away from the terminal but near the NASA hangar. As a precaution, Glenn decided to evacuate all of its staff out the rear gate at the opposite end of the center, and the ensuing traffic jam took 90 minutes to clear. Later, an FBI inspection of the aircraft and passengers revealed no weapons.[1086]

Glenn reopened two days later under eerily empty skies as the endless march of airliners into and out of the adjacent airport was suspended. The center added increased security measures, random vehicle searches, and new communications plans and closed the Visitor Center indefinitely.

The 1990s were a difficult period for both the Agency and the center. The Faster, Better, Cheaper and Zero Base Review initiatives produced mixed results. NASA's successes with *Mars Pathfinder* and *Deep Space 1* were tempered by embarrassing losses of other Mars missions. The Agency overstated some predicted reductions, made poor decisions regarding facility closures, and performed inaccurate cost/benefit analyses of its actions. An Office of Inspector General report specifically castigated the Agency for the attempt to consolidate its aircraft and close Plum Brook Station.[1087] Nonetheless, an in-house study interviewed hundreds of NASA and NASA-related personnel and found wide support for what the effort accomplished.[1088] Although, the harsh budget reductions and program transfers would continue into the next millennium, the center achieved a number of major accomplishments in aeronautics and space in the 1990s.

*Image 357: Don Campbell addresses the audience at NASA Day in Dayton, Ohio (GRC–2003–C–02406).*

## Endnotes for Chapter 9

936. Nancy Kassebaum, *National Aeronautics and Space Administration Authorization Act, Fiscal Year, 1996*, Senate Congressional Record, vol. 141, no. 1662, 19 October 1995.

937. Jon Boyle, "Working 'Faster, Better, Cheaper': A Federal Research Agency in Transition" (Dissertation submitted to the Faculty of the Virginia Polytechnic Institute and State University, 2002).

938. William Broad, "Space Station Trimming Back, Survives the Ax," *New York Times* (19 February 1993).

939. William Broad, "Large Role for Russia Expected on Station," *New York Times* (13 April 1993).

940. White House Press Release, *Space Station Redesign Decision Reduces Costs, Preserves Research, Ensures International Cooperation*, 17 June 1993.

941. Marcia Smith, *NASA's Space Station Program: Evolution and Current Status, Testimony Before the House Science Committee*, 4 April 2001.

942. "President Approves Space Station Redesign," *Lewis News* (2 July 1993).

943. Warren Leary, "With Russian Aid, Better Space Lab," *New York Times* (5 November 1993).

944. Smith, *NASA's Space Station Program*.

945. Smith, *NASA's Space Station Program*.

946. "President Approves Space Station Redesign."

947. "Questions Related to Closing of Reston Space Station Program Office," 17 August 1993, NASA Glenn History Collection, Space Station Collection, Cleveland, OH.

948. William Broad, "Space Station Work Gets Tight Controls," *New York Times* (18 August 1993).

949. Kassebaum, *National Aeronautics and Space Administration Authorization Act*.

950. Laurie Boeder and Dwayne Brown, "Review Team Proposes Sweeping Management, Organizational Changes at NASA," NASA press release 95–73, 19 May 1995, NASA Glenn History Collection, Directors Collection, Cleveland, OH.

951. Larry Ross, "Roles and Missions," *Lewis News* (19 November 1993).

952. Larry Ross interoffice memorandum, "Evolution," 6 January 1994, NASA Glenn History Collection, Directors Collection, Cleveland, OH.

953. Jeff Vincent, "NASA Administrator Announces Management Changes," NASA Headquarters press release 94-3, 6 January 1994, NASA Glenn History Collection, Directors Collection, Cleveland, OH.

954. "Ross to Head Up Wind Tunnel Program Office," *Lewis News* (28 January 1994).

955. Kristin Wilson, "NASA Takes Steps To Get More Out of Future Budgets," *Lewis News* (25 February 1994).

956. Kassebaum, *National Aeronautics and Space Administration Authorization Act*.

957. General Accounting Office, *NASA Infrastructure: Challenges to Achieving Reductions and Efficiencies* (Washington, DC: GAO/NSIAD–96–187, September 1996).

958. Boeder, "Review Team Proposes Sweeping Management."

959. Boeder, "Review Team Proposes Sweeping Management."

960. General Accounting Office, "NASA Infrastructure."

961. Boeder, "Review Team Proposes Sweeping Management."

962. "Center Can Save $29.5 Million Through Innovative Activities," *Lewis News* (14 April 1995).

963. "Center Can Save $29.5 Million."

964. "Administer Goldin Explains Streamlining," *Lewis News* (17 February 1995).

965. Boeder, "Review Team Proposes Sweeping Management."

966. Robert Fails, "ZBR Implementation Plan Briefing to Office of Aeronautics," 6 February 1996, NASA Glenn History Collection, Directors Collection, Cleveland, OH.

967. Jenise Veris, "Realignment Will Optimize NASA Lewis Capabilities," *Lewis News* (November 1996).

968. Mike Skor and Dave J. Hoffman, "Mir Cooperative Solar Array," *Research & Technology 1996* (Washington, DC: NASA TM–107350, 1997).

969. T. W. Kerslake, D. A. Scheiman, and D. J. Hoffman, "Dark Forward Electrical Testing on the Mir Cooperative Solar Array," *Research & Technology 1996* (Washington, DC: NASA TM–107350, 1997).

970. Richard R. Secunde, Thomas L. Labus, and Ronal Lovely, *Solar Dynamic Power Module Design* (Washington, DC: NASA TM–102055, 1989).

971. R. K. Shaltens and R. V. Boyle, *Update of the 2 kW Solar Dynamic Ground Test Demonstration Program* (Washington, DC: NASA TM–106730, 1994).

972. John Mansfield to Donald Campbell, "Solar Dynamics," 15 February 1996, NASA Glenn History Collection, Directors Collection, Cleveland, OH.

973. Thomas Kerslake, Lee Mason, and Hal Strumpf, *High-Flux, High-Temperature Thermal Vacuum Qualification Testing of a Solar Receiver Aperture Shield* (Washington, DC: NASA TM–107505, 1997).

974. Valerie Lyons, "Power and Propulsion at NASA Glenn Research Center: Historic Perspective of Major Accomplishments," *Journal of Aerospace Engineering* 26 (April 2013): 288–299.

975. "Powering the Future," *NASA Facts* (Washington, DC: NASA PB–00537–0611, 2011).

976. Lyons, "Power and Propulsion."

977. Larry Ross, Lew Allen, and Forrest McCartney, "Memorandum of Agreement Among the Lewis Research Center and the Jet Propulsion Laboratory and the John F. Kennedy Space Center Concerning Planetary Missions on a Titan IV/Centaur Launch Vehicle," 5 September 1990, NASA Glenn History Collection, Centaur Collection, Cleveland, OH.

978. "Transition Plan for Transferring Program Management of Intermediate Expendable Launch Vehicle (IELV) Launch Services From Lewis Research Center to Kennedy Space Center," 14 May 1997, NASA Glenn History Collection, Directors Collection, Cleveland, OH.

979. Michael Meltzer, *The Cassini-Huygens Visit to Saturn: An Historic Mission to the Ringed Planet* (New York, NY: Spring Praxis Books, 2014).

980. "A Big Boost for Cassini," *NASA Facts* (Washington, DC: NASA FS–1999–06–004–GRC, June 1996).

981. "A Big Boost."

982. Joe Nieberding, "History of Launch Vehicles at Lewis Research Center," 3 March 2010, NASA Glenn History Collection, Centaur Collection, Cleveland, OH.

983. Howard E. McCurdy, *Faster, Better, Cheaper: Low-Cost Innovation in the U.S. Space Program* (Baltimore, MD: Johns Hopkins University Press, 2001).

984. Steven Stevenson, *Mars Pathfinder Rover—Lewis Research Center Technology Experiments Program* (Washington, DC: NASA TM–107449, 1997).

985. "Mars Pathfinder Air Bag Landing Tests" (21 May 2013), *http://www.nasa.gov/centers/glenn/about/history/marspbag.html* (accessed 1 August 2014).

986. Stevenson, *Mars Pathfinder Rover.*

987. McCurdy, *Faster, Better, Cheaper.*

988. Kristin Wilson, "Plum Brook Tests Put Bounce in Pathfinder Landing System," *Lewis News* (November 1996).

989. "Mars Pathfinder, Exploring Mars With the Sojourner Rover" (10 August 2012), *http://www.nasa.gov/mission_pages/mars-pathfinder/* (accessed 28 March 2015).

990. McCurdy, *Faster, Better, Cheaper.*

991. Geoffrey Landis, "Pathfinder—A Retrospective," *Spaceflight* 44 (August 2002).

992. McCurdy, *Faster, Better, Cheaper.*

993. McCurdy, *Faster, Better, Cheaper.*

994. "Mars Pathfinder," *http://www.nasa.gov/mission_pages/mars-pathfinder/*

995. "Deep Space 1 Launch" NASA Press Kit, October 1998.

996. Michael Patterson and James Sovey, "History of Electric Propulsion at NASA Glenn Research Center: 1956 to Present," *Journal of Aerospace Engineering* 26 (2013): 300–316.

997. Doreen Zudell, "Glenn's Ion Engine Earns Its Wings," *Aerospace Frontiers* (February 2002).

998. Patterson, "History of Electric Propulsion."

999. "Deep Space 1 Ion Propulsion System Operation Sequence and Status" (21 May 2008), *http://www.nasa.gov/centers/glenn/about/history/ds1opseq.html* (accessed 22 July 2015).

1000. Patterson, "History of Electric Propulsion."

1001. "Glenn Contributions to Deep Space 1" (21 May 2008), *http://www.nasa.gov/centers/glenn/about/history/ds1.html* (accessed 22 July 2015).

1002. Patterson, "History of Electric Propulsion."

1003. Carroll Little to Distribution, "Consolidation of Aircraft at the Dryden Flight Research Center," 5 October 1995, NASA Glenn History Collection, Directors Collection, Cleveland, OH.

1004. Edward Heffernman to Mike DeWine, 14 August 2000, NASA Glenn History Collection, Directors Collection, Cleveland, OH.

1005. Kassebaum, *National Aeronautics and Space Administration Authorization Act.*

1006. Edward McLean, *Supersonic Cruise Technology* (Washington, DC: NASA SP–472, 1985).

1007. "The High-Speed Research Program: A Conversation With Project Manager Joe Shaw," *Lewis News* (25 May 1990).

1008. "The High-Speed Research Program."

1009. "NASA's High-Speed Research Program," NASA Facts Online, 2004, NASA Glenn History Collection, High-Speed Research Collection, Cleveland, OH.

1010. Dhanireddy R. Reddy, "Seventy Years of Aeropropulsion Research at NASA Glenn Research Center," *Journal of Aerospace Engineering* (April 2013).

1011. C. T. Chang, "NASA Glenn Combustion Research for Aeronautical Propulsion," *Journal of Aerospace Engineering* (April 2013).

1012. *Research & Technology 1997* (Washington, DC: NASA/TM—1998-206312, 1998).

1013. "Impact of the Termination of NASA's High Speed Research Program and the Redirection of NASA's Advanced Subsonic Technology Program," Report to Congress, 4 December 2000.

1014. Reddy, "Seventy Years of Aeropropulsion Research."

1015. "Advanced Subsonic Technology," *Spinoff* (Washington, DC: NASA NP–217, 1995).

1016. Pete Pachlhofer and Peg Whalen, "Lewis' Advanced Subsonic Combustion Rig Unique Facility Will Develop Advanced Subsonic Technology for Commercial Aircraft," *Lewis News* (29 July 1994).

1017. Kenneth Zaremba, "Improving Aviation Through Cutting Edge Research," *Lewis News* (April 1997).

1018. C. T. Chang, "NASA Glenn Combustion Research for Aeronautical Propulsion," *Journal of Aerospace Engineering* (April 2013).

1019. "Pure Power PW1000G Engine" (2015), *http://www.pw.utc.com/PurePowerPW1000G_Engine* (accessed 11 October 2015).

1020. Chang, "NASA Glenn Combustion Research."

1021. K. B. M. Q. Zaman, J. E. Bridges, and D. L. Huff, "Evolution from 'Tabs' to 'Chevron Technology'— A Review," *International Journal of Aeroacoustics* 10, no. 5 (October 2011): 685–710.

1022. "Impact of the Termination."

1023. Shaw, R. J., "Ultra-efficient Engine Technology (UEET) Project," *Research & Technology 2003* (Washington, DC: NASA/TM—2004-212729, 2004).

1024. Carol Ginty, "Overview of the Ultra-Efficient Engine Technology and Quiet Aircraft Technology Projects" 9 November 2004, NASA Glenn History Collection, Directors Collection, Cleveland, OH.

1025. Reddy, "Seventy Years of Aeropropulsion Research," 26.

1026. Michael Nathal and Jenise Veris, "Glenn Takes a Bow for Impact on GEnx Engine," *Aerospace Frontiers* (July 2008).

1027. "Impact of the Termination."

1028. "Impact on Glenn Research Center of HSR/AST Termination," 4 December 2000, NASA Glenn History Collection, Directors Collection, Cleveland, OH.

1029. "Making Future Light Aircraft Safer, Smoother, Quieter, and More Affordable," *NASA Facts* (Washington, DC: FS-1996-07-001-LeRC. 1996).

1030. "Revolutionary GA Engines on the Way?" *Flying Magazine* (March 1997).

1031. "Small Aircraft Propulsion: The Future Is Here," *NASA Facts* (Washington, DC: FS–2000–04–001–GRC, 2000).

1032. Mike Busch, "GAP Engine Update," AVweb (27 July 2000), *http://www.avweb.com/news/reviews/182838-1.html?zkPrintable=true* (accessed 28 March 2015).

1033. Bill Brogdon, "Aircraft Diesel Engines," 10 May 2012, NASA Glenn History Collection, General Aviation Program Collection, Cleveland, OH.

1034. Stephen Pope, "Diesel Aircraft Engines Revolution," Flying Magazine (28 October 2013), *http://www.flyingmag.com/aircraft/diesel-aircraft-engines-revolution* (accessed 24 March 2015).

1035. Lori Rachul, "NASA Initiates New General Aviation Propulsion Program," *NASA News* 96–46 (3 August 1996).

1036. Williams International, *The General Aviation Propulsion (GAP) Program* (Washington, DC: NASA/CR—2008-215266, 2008).

1037. David Nolan, "The Little Engine That Couldn't," *Air & Space Magazine* (November 2005).

1038. "Air Taxi at Your Service," *NASA Spinoff 2002* (Washington, DC: NASA/NP–2002–09–290–HQ, 2002): 68–69.

1039. "NASA, "GRC Microgravity and Technology Flight Experiments: Accomplished," ISS Research Project (1 April 2013), *https://issresearchproject.grc.nasa.gov/accomplished.php* (accessed 3 December 2015).

1040. Lauren M. Sharp, Daniel L. Dietrich, and Brian J. Motil, "Microgravity Fluids and Combustion Research at NASA Glenn Research Center," *Journal of Aerospace Engineering* 26 (April 2013): 439–450.

1041. Thomas Glasgow, Microgravity Materials Science Laboratory pamphlet B–0115, February 1988, NASA Glenn History Collection, Microgravity Collection, Cleveland, OH.

1042. "Center Opens New Microgravity Materials Lab," *Lewis News* (6 September 1985).

1043. Larry Ross, "Space Flight Systems Directorate," *Lewis News* (9 January 1987).

1044. Dallas Lauderdale and Jack Lekan, "Space Experiments Lab/Zero G Rehab," *Lewis News* (7 June 1991).

1045. John Yaniec, *Users Guide for NASA Lewis Research Center DC-9 Reduced Gravity Aircraft Program* (Washington, DC: NASA TM–106755, 1995).

1046. "Microgravity in the Sky," *Lewis News* (31 March 1995).

1047. Sharp, "Microgravity Fluids and Combustion Research."

1048. Howard Ross email to Robert Arrighi, "New Improved Microgravity Section," 19 November 2015, NASA Glenn History Collection, Oral History Collection, Cleveland, OH.

1049. T. Y. Miller, *First International Microgravity Laboratory Experiment Descriptions* (Washington, DC: NASA TM–4353, 1992).

1050. Ross email to Arrighi.

1051. Simon Ostrach and Y. Kamotani, *Surface Tension Driven Convection Experiment (STDCE)* (Washington, DC: 1996 (NASA CR–198476, 1996).

1052. H.D. Ross, D.L. Dietrich, and J.S. Tien. "Candle Flames in Microgravity: USML–1 results– 1 Year Later," *Joint Launch + One Year Science Review of USML–1 and USMP–1 With the Microgravity Measurement Group* (Washington, DC: NASA CP-3272-Vol-2, 1994).

1053. Ross email to Arrighi.

1054. "NASA Technology Transfer Program: "Portable Unit for Metabolic Analysis (PUMA)" *http://technology.nasa.gov/patent/LEW-TOPS-16* (accessed 3 December 2015).

1055. Simon Ostrach interview, Cleveland, OH, 2013, NASA Glenn History Collection, Oral History Collection, Cleveland, OH.

1056. N. Ramachandran, D. O. Frazier, and S. L. Lehoczky, *Joint Launch + One Year Science Review of USML-1 and USMP-1 With the Microgravity Measurement Group* (Washington, DC: NASA CP–3272, 1994).

1057. Ross email to Arrighi.

1058. Ostrach interview.

1059. Ramachandran, "Joint Launch."

1060. "STS–83 Columbia Microgravity Science Laboratory–1 (MSL–1)," Kennedy Space Center Release No. 42–97, March 1997. *http://www-pao.ksc.nasa.gov/release/1997/42-97.htm* (accessed 3 December 2015).

1061. P. B. Sunderland, D. L. Urban, and Z. G. Yuan, *Laminar Soot Processes Experiment: Findings from Space Flight Measurements* (Washington, DC: NASA/CP—2003-212376).

1062. Diane Malarik and Martin Glicksman, "Isothermal Dendritic Growth Experiment (IDGE) Is the First United States Microgravity Experiment Controlled From the Principal Investigator's University" 1997, *http://hdl.handle.net/2060/20050177156*

1063. Jerri Ling and Michael Doherty, "Physics of Hard Spheres Experiment (PhaSE) or "Making Jello in Space" *http://hdl.handle.net/2060/20050177227*

1064. Anish Upadhyaya, Ronald Iacocca, and Randall German, *Gravitational Role in Liquid Phase Sintering* (Washington, DC: NASA/CP–1998-208868, 1998).

1065. Ross email to Arrighi.

1066. Sandra L. Olson and Gary A. Ruff, *Microgravity Flame Spread in Exploration Atmospheres: Pressure, Oxygen, and Velocity Effects on Opposed and Concurrent Flame Spread* (Washington, DC: NASA TM—2008-215260, 2008).

1067. Ross email to Arrighi.

1068. Richard DeLombard and S. Ryaboukha, "Space Acceleration Measurement System" Investigation 8.8.1 Summary Report, *http://history.nasa.gov/SP-4225/science/sams.pdf*

1069. Mike DeWine, et al., "FY2001 Action Plan for NASA Glenn Research Center," 30 May 2001, NASA Glenn History Collection, Directors Collection, Cleveland, OH.

1070. Kristin Wilson, "Research Aircraft Flies Over Center for Last Time," *Lewis News* (September 1997).

1071. "The New USRA Center for Microgravity Research," USRA Researcher (Fall 1997).

1072. Ostrach interview.

1073. "Large-Area Atomic Oxygen Facility Used to Clean Fire-Damaged Artwork," *Research & Technology 1999* (Washington, DC: NASA/TM—2000-209639, 2000).

1074. Tori Woods, "Out of Thin Air" (17 February 2011), *http://www.nasa.gov/topics/technology/features/atomic_oxygen.html* (accessed 30 July 2015).

1075. "Large-Area Atomic Oxygen Facility."

1076. "Large-Area Atomic Oxygen Facility."

1077. Bruce Banks, "Destructive Power of Atomic Oxygen Used to Restore Artwork" (4 January 2005), *http://www.nasa.gov/centers/glenn/business/AtomicOxRestoration.html* (accessed 23 July 2015).

1078. Thomas Schulz, "NASA Facilities: Challenges to Achieving Reductions and Efficiencies," Testimony Before the Subcommittee on National Security, International Affairs, and Criminal Justice, Committee on Government Reform and Oversight, House of Representatives, 11 September 1996.

1079. National Parks Service, National Historic Landmarks Program, "Rocket Engine Test Facility" (5 September 2014), *http://www.nps.gov/nhl/find/withdrawn/rocket.htm* (accessed 23 March 2015).

1080. Gordon Dressler and Martin Bauer, *TRW Pintle Engine Heritage and Performance Characteristics* (Reston, VA: AIAA 2000–3871, 2000).

1081. Virginia Dawson, *Ideas Into Hardware: A History of the Rocket Engine Test Facility at the NASA Glenn Research Center* (Washington, DC: NASA, 2004).

1082. Mark Bowles, *Science in Flux: NASA's Nuclear Program at Plum Brook Station 1955–2005* (Washington, DC: NASA SP–2006–4317, 2006).

1083. *Of Ashes and Atoms: The Story of NASA's Plum Brook Reactor Facility* (Washington, DC: NASA SP–2005–4605, 2005).

1084. Donald Campbell to Lewis Employees and On-Site Contractors, "Center Name Change," 1999, NASA Glenn History Collection, Directors Collection, Cleveland, OH.

1085. John Glenn interview, Cleveland, OH, by David DeFelice, 6 March 2015, NASA Glenn History Collection, Oral History Collection, Cleveland, OH.

1086. James Renner, "Plan 9/11 From Cyberspace the Body Snatchers of United 93 and Other Tales of Terror From Cleveland," *Cleveland Free Times* 14, no. 20 (6 September 2006).

1087. General Accounting Office, "NASA Infrastructure."

1088. Tony Spear, *NASA Faster, Better, Cheaper Task Final Report* (Washington, DC: NASA, 2000).

# 10. Changing Missions

*"There's a difference between accomplishment and achievement. An accomplishment is doing your taxes. A true achievement is a one-of-kind experience for the center, you as an indvidual."*

—Jim Free

Image 359: The Ares I–X rocket lifts off from launch Pad 39B at Kennedy Space Center in Florida on 28 October 2009 (KSC–2009–5963).

$\mathcal{A}$t 9 a.m. on 1 February 2003 the Space Shuttle Columbia broke apart while reentering the clear skies over Texas, killing seven astronauts. In the aftermath, the Columbia Accident Investigation Board (CAIB) issued a highly critical report that not only determined the cause of the accident, but challenged NASA's goals, management approach, and culture. The board demanded that NASA institute a number of changes before launching another shuttle mission. The CAIB also strongly recommended that NASA issue an unambiguous statement of the long-range goals of its human spaceflight endeavors, and noted the critical need for the budget to align with that objective.[1089] The NASA Glenn Research Center made several contributions to expedite the shuttle's return to flight and played an important role in the Agency's new exploration plans.

After months of congressional hearings regarding human spaceflight and behind-the-scenes planning by NASA and the White House, President George W. Bush unveiled the Vision for Space Exploration (VSE) on 14 January 2004. The VSE was a wide-ranging space exploration initiative that included robotic missions, a base on the Moon, and eventual human journeys to Mars. It was the first attempt to send humans beyond low Earth orbit since Apollo. The plan included phasing out the space shuttle to free up funds to develop new space vehicles and technologies for these extended missions. NASA would rely on foreign nations and commercial rockets for transportation to the International Space Station (ISS).[1090]

The VSE dramatically affected activities at Glenn from 2005 to 2010. NASA assigned the center several roles, including management of the service module that provided critical life support and communications systems for the new crew vehicle, and responsibility for manufacturing an upper-stage mass model for the first developmental rocket launch—Ares I–X.[1091] Unrelated to VSE, NASA also tasked Glenn with the development of nuclear-based ion thrusters to send a space probe to Jupiter's moons. The government's modest funding for the new exploration effort, however, forced NASA to move hundreds of millions of dollars away from life and microgravity science—and to a lesser extent, from aeronautics—to help pay for the new VSE flight hardware.[1092]

Despite these steps, the lack of additional needed funding prevented NASA from maintaining the VSE's intended schedule. After an independent assessment in 2009, President Barak Obama decided to restructure the effort, resulting in changes throughout the Agency, including a diminishment of work at most NASA centers, including Glenn. President Obama, however, also recommended an increase in advanced spaceflight technology development. Glenn secured leadership roles in several of these new technology programs, including solar electric propulsion. The center reorganized in 2014 to address the post-VSE environment and adjust to new budgetary and staffing realities.[1093]

**Return to Flight**

In December 2001, President Bush selected Sean O'Keefe to replace Dan Goldin as NASA Administrator. O'Keefe, a Washington, DC, financial administrator with no aerospace experience, sought to improve NASA's fiscal accountability and to unify the centers and directorates under the "One NASA" theme. In an April 2002 address, he stressed that "NASA's mission…must be driven by the science, not by destination. And while policy and politics and

*Image 360: Administrator O'Keefe visits Glenn in March 2004 (GRC–2004–C–00453).*

economics are inevitable factors, science must be the preeminent factor."[1094] To facilitate the completion of the ISS, O'Keefe sought to improve the shuttle, develop a new space plane, and create a replacement for the shuttle. The loss of *Columbia* on 1 February 2003 scuttled those initial plans.[1095]

Within hours of the disaster, the independent CAIB was established to determine both the root and systematic causes of the accident. The grounding of the shuttle fleet prevented the launching of satellites and spacecraft, disrupted space science experiment programs, and delayed construction of the ISS.[1096] The ability to focus its efforts during times of crisis is one of NASA's best characteristics. The *Columbia* investigation was NASA's most pressing issue throughout 2003, and everything became secondary to the goal of returning the shuttle to service.[1097]

❖ ❖ ❖ ❖ ❖ ❖

The loss of *Columbia* had ramifications on Glenn's microgravity program. The shuttle had been carrying six of Glenn's microgravity experiments and the Combustion Module-2 (CM–2) test rack. The CM–2 was the largest and most complex pressurized system ever

flown on the shuttle.[1098,1099] Because the three Glenn studies on CM–2 were a significant component of the mission, the entire *Columbia* crew had visited the center in late January 2001 to learn how to operate the experiments. Glenn researchers had trained the astronauts how to operate the hardware and had demonstrated physical adjustments that could be made to improve the experiment data. As preparations proceeded during the months leading up to the launch, the crew and center personnel formed friendships.[1100]

The bond between Glenn researchers and the astronauts was demonstrated while the astronauts were conducting the Mist experiment, which sought to investigate ways to extinguish fires using a water mist.[1101] The experiment developed a small leak that could not be stopped. In a scene reminiscent of *Apollo 13*, the Glenn ground team developed a solution using parts they knew were onboard *Columbia*. Astronauts Mike Anderson and Kalpana (KC) Chawla, who sacrificed her own precious Earth observation time, pulled the hardware from the CM–2 and successfully made the repair. Despite a significant loss of time, the astronauts and Glenn team were able to complete over 90 percent of the planned Mist experiments. The repair

*Image 361: Greg Fedor (in white) briefs members of the STS–107 crew on the Space Acceleration Measurement System (SAMS) in the Power Systems Facility (PSF). Left to right: Laurel Clark (obscured), Ilan Ramon, David Brown, Kalpana Chawla, William McCool, and Mike Anderson (GRC–2001–C–00235).*

was one of the highlights of the mission.[1102] Although the loss of the orbiter impacted the Glenn researchers both professionally and personally, most of the test data had been transmitted to the center prior to the orbiter's destruction.[1103]

❖ ❖ ❖ ❖ ❖ ❖

Glenn immediately began supporting the investigation into the cause of the accident. The center's Materials and Processes Failure Analysis team analyzed high-temperature reactions with wing leading-edge materials in simulated reentry and breakup conditions.[1104] Glenn's Structural Mechanics and Dynamics Branch studied the effect of shuttle insulation impacting the reinforced carbon-carbon (RCC) material in the center's new Ballistics Impact Laboratory. The facility was designed to study projectile aerodynamics and test their effect on different materials. It includes three guns that can fire objects at more than 2,000 mph in a simulated space environment.

Crash investigators were particularly interested in the effect of a sizeable piece of the insulating foam that had separated from the external tank during launch and struck the RCC panels on the shuttle's left wing.[1105] The Glenn team found that the firing of very small pieces of the foam bent and cracked RCC samples.[1106-1108]

In June the Southwest Research Institute in San Antonio tested the foam's impact on full-scale samples of an actual shuttle wing. They were surprised to find that a lightweight piece of foam (under 2 pounds) produced a 16-inch hole in the wing's RCC panels.[1109] Glenn's ballistics studies supported the San Antonio work and helped to fine tune the investigation's computer simulations. Glenn also developed extensive computer simulations to help predict damage caused by debris and developed models to validate the large-scale tests.[1110-1112]

In addition, Glenn supported the development of new technologies for future flights. NASA, which was reviewing all aspects of the shuttle external tank insulation, asked the center to study the two large foam ramps that were used to improve aerodynamics along the tank's cable trays. There was concern that

*Image 362: The Glenn Ballistics Laboratory in the Materials and Structures Laboratory (GRC–2000–C–00447).*

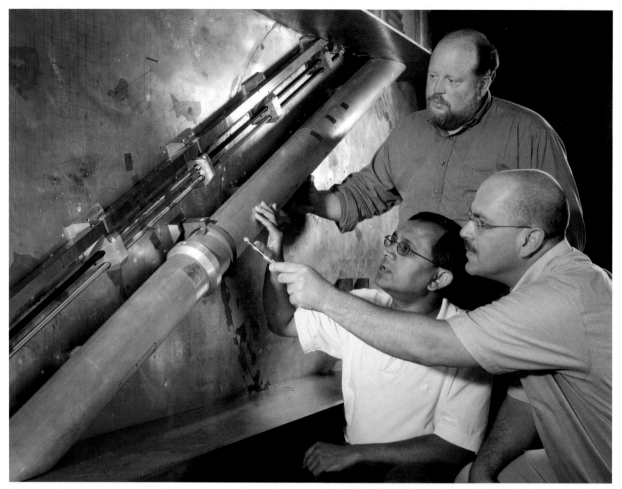

*Image 363: Jayanta Panda, Scott Williamson, and Daniel Sutliff examine the shuttle ramp protuberance test setup in the 8- by 6-Foot Supersonic Wind Tunnel (8×6) (GRC–2007–C–01848).*

this foam might break off under certain conditions. In late summer 2003 Scott Williamson led aerodynamic testing of the ramps in the 8×6.[1113] Although the tests demonstrated that the ramps were safe, NASA engineers recommended that other alternatives be considered.[1114] NASA decided to eliminate the ramps and reengineer the tank design after a large piece broke free during the next shuttle launch.[1115,1116]

By early July the CAIB officially announced that the piece of foam had damaged the wing's protective RCC panels. During reentry, the resulting breach had admitted the heat that caused the vehicle to disintegrate. The board's recommendations included the development of new methods of inspecting the RCC panels prior to launch, an inflight repair kit, and new requirements for analyzing the vehicle on orbit.[1117] The center's Ceramics Branch created and tested the Glenn Refractory Adhesive for Bonding and Exterior Repair

*Image 364: The CAIB report.*

to fix cracks in the RCC panels on future missions, and the Mechanical Components Branch analyzed the lubrication and wear of the actuator mechanisms that were part of the shuttle's landing gear system.[1118-1120]
At 10:39 a.m., 26 July 2006, the shuttle returned to space with the launch of *Discovery*. The crew and vehicle safely returned to Earth nearly 14 days later. The three-year Return to Flight effort was complete.

Although NASA was able to identify and remedy the physical causes of the *Columbia* accident, it would be much more difficult to address the CAIB's cultural and managerial findings. In the weeks leading up to the report's release, Administrator O'Keefe debated with other NASA officials regarding the Agency's response to what were likely to be harsh determinations. In the end, O'Keefe was able to convince his colleagues that it would be best for NASA to not only accept but embrace the findings.[1121]

## Vision Takes Shape

After the loss of *Columbia*, O'Keefe and NASA officials examined alternatives for the U.S. space program, including a new space plane and new launch vehicles. There was no specific mission identified for these vehicles. Meanwhile some low-level White House staffers began informally discussing the future of human space missions. As the CAIB was finalizing its report in the early summer of 2003, the informal White House meetings became more official. The expanding group, which now included higher-level officials and NASA representatives, began parsing out a long-term post-*Columbia* space mission. In an effort to respond to the CAIB report, NASA needed to justify continued human spaceflight and develop a plan that had clear objectives and consistent funding. The inter-agency committee privately reviewed a full spectrum of options for future space exploration. By late summer, the group began estimating costs for the various proposals and transforming ideas into policy. The

*Image 365: Human space exploration concepts created by Glenn contractor artist Les Bossinas in 1989 (GRC–1989–C–07306).*

Image 366: Fabrication of a lifesize cutaway model of the Crew Exploration Vehicle's (CEV) service module in December 2005 (GRC–2005–C–01868).

committee eventually recommended that NASA send humans to the Moon once again.[1122]

Throughout this period congressional committees held a spate of hearings on the *Columbia* accident, NASA's response to the CAIB report, and restructuring of the Agency. For the first time in decades, Congress questioned whether the nation should continue to engage in human spaceflight. Contemporary human spaceflight did not have the political implications of the Apollo Program, predicted financial returns from activities such as lunar mining never materialized, and some argued that robotic missions produced more scientific data than crewed missions. The costs and dangers associated with human spaceflight needed new justification.[1123]

Others, including aerospace and technology consultant Michael Griffin, argued that human spaceflight was justified by man's innate drive to explore. Griffin predicted that hundreds of years from now our descendants would remember the current era for the Apollo Moon landings much in the same way that we remember the Early Modern Period for Columbus's voyages.[1124]

Yet Apollo's legacy efforts, the shuttle and space station, were expensive endeavors that did not inspire the public's thirst for exploration. Dr. Wesley Huntress of the Carnegie Institution for Science contended that if astronauts were going to risk their lives "it should be for extraordinarily challenging reasons…not to endlessly circle the block." He stressed the need for a unified national vision consisting of both robotic and human spaceflight to systematically achieve specific new exploration goals beyond low Earth orbit.[1125]

❖ ❖ ❖ ❖ ❖ ❖

NASA was asking for a significant budget increase to fund the new space proposals, but President Bush had ordered the restriction of all nonmilitary funding in the coming years.[1126] O'Keefe and many within NASA felt that it was unrealistic to introduce a new Moon exploration plan without increased funding. In October 2003 Administrator O'Keefe began an intense lobbying campaign. Eventually the White House promised a $1 billion increase for the Agency over the next five years. By late November NASA devised a complex strategy to come up with the remainder of the required funding. The plan included

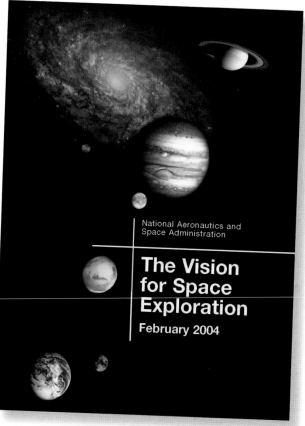

National Aeronautics and Space Administration

# The Vision for Space Exploration
### February 2004

*Image 367: NASA's VSE.*

the cancellation of the new space plane and launch vehicles, and termination of the shuttle program.[1127]

On 19 December 2003 O'Keefe and other NASA officials met with President Bush, Dick Cheney, and other high-level advisors to go over the final proposal. The president approved the controversial budget increase but urged the group to devise a more significant exploration effort. NASA quickly added a crewed mission to Mars—which some had previously advocated—to the plan.[1128] On 14 January 2004 President Bush announced the VSE.[1129] Exploration—with its potential scientific, security, and economical dividends—became NASA's new focus. In what may have been an omen of VSE's fate, the president did not mention the initiative less than a week later in his 2004 State of the Union Address.

The VSE included robotic explorer missions to the Moon and Mars that would develop technologies for more complex efforts in the future, such as Project Constellation—the most ambitious aspect of the plan. Constellation would establish a multipurpose lunar base that could serve as a stepping stone to eventual human missions to Mars. NASA would restore the shuttle for a limited number of flights to complete work on the ISS; then it would phase it out. Russia and private U.S. companies would supply the required vehicles to fill the gap between the shuttle retirement and the implementation of a new NASA human-rated launch vehicle and crew capsule.[1130] O'Keefe created a new Exploration Enterprise and began working with a host of entities, including scientists, the media, politicians, and the public, for input on the VSE. To expedite the multidecade design process, NASA decided to base the new launch vehicles and CEV on Apollo-era technology.

In 2003 O'Keefe asked Donald Campbell to head NASA's Special Projects Office for Nuclear Power Systems. O'Keefe appointed Glenn veteran Julian Earls to the Center Director position that October.

Earls had joined the center in 1965 as a physicist in the Health Physics Office. Three years later he became the Head of the Health Physics and Licensing Section of the Nuclear Systems Division. In the 1980s he was named Chief of the Health, Safety, and Security Division, Acting Director of the Administration and Computer Services Directorate, and Director for the Office of Health Services. Earls progressed through a string of upper management positions throughout the 1990s that included his appointment as Deputy Director in 2002.[1131]

## Propelling NASA Across Space

Administrator O'Keefe was an ardent supporter of alternative propulsion systems that could accelerate the exploration schedule. More than a year before the release of the VSE plan, he requested proposals for a nuclear-based deep space mission. In early 2003 the Jet Propulsion Laboratory (JPL) announced plans for a new spacecraft that would use nuclear energy to power ion thrusters. NASA would demonstrate the technology by sending *Prometheus*, the largest and most powerful electric propulsion vehicle to date, on a mission to three of Jupiter's moons.[1132]

The project required the services of several centers, the Department of Energy, and industry partners. Glenn was responsible for the reactor, the electric propulsion, Brayton power conversion, cooling, and the communications systems. The center also would support the program through testing and integration of the spacecraft into the launch vehicle. Glenn managed the development of four electric propulsion options, including its own High Power Electric Propulsion (HiPEP) thruster, which surpassed contemporary ion engines in efficiency, power, and longevity.[1133]

Image 368: *One of the first activities of Earls's two-year tenure as Glenn Center Director was hosting the Realizing the Dream of Flight Symposium on 5 November 2003 at the Great Lakes Science Center. A panel of preeminent aerospace historians discussed an array of individuals who shaped the development of the nation's aerospace technology. The symposium was one of a series of activities held across the nation in 2003 to commemorate the 100th anniversary of the Wright Brothers' initial flight in 1903 (GRC–2003–C–02005).*

HiPEP's rectangular shape, which was different from the circular grids utilized on previous thrusters, was

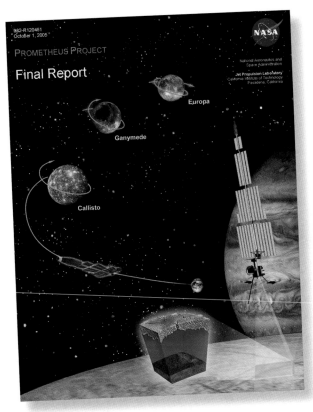

Image 369: NASA's Prometheus Project Final Report.

selected to address the requirements of high power and long life for this mission.[1134] Glenn conducted preliminary design work and began performance testing in 2003.[1135] The thruster underwent a 2,000-hour wear test in the Electric Propulsion Laboratory (EPL) vacuum facilities during 2004 and 2005.[1136]

NASA engineers combined elements from HiPEP and a JPL-designed thruster to create the final *Prometheus* propulsion system, referred to as "Herakles." *Prometheus* also included a gas-cooled reactor with a Brayton power conversion system. From 2003 through 2005 Glenn conducted full-system Brayton demonstrations, vibration analysis, and a variety of other tests on the system. In addition, Glenn completed a performance test program of a commercially manufactured Herakles prototype ion thruster.[1137]

Glenn researchers continued to pursue both solar-electric and nuclear-electric propulsion options throughout the 2000s for applications ranging from primary propulsion for science missions to station-keeping for communications spacecraft. Two of the

Image 370: A rectangular HiPEP thruster being removed from the EPL vacuum tank in June 2005 after the duration test (GRC–2005–C–01061).

*Image 371: Testing of a prototype NEXT thruster in the EPL during May 2006 (GRC–2006–C–01260).*

most prominent efforts were the development of an ion thruster system referred to as "NASA's Evolutionary Xenon Thruster (NEXT)" and multiple Hall thruster systems.

In August 2002 NASA initiated the NEXT program to build upon the NASA Solar Electric Power Technology Application Readiness (NSTAR) thrusters developed in the 1990s. Although the NSTAR thrusters performed well on the *Deep Space 1* and *Dawn* missions, improvements were necessary to perform longer duration flights to the outer solar system. Glenn researchers produced a 40-centimeter-diameter solar-powered thruster design that operates on the same principles as NSTAR but has a larger power throttling range, higher input power capability, higher efficiency, higher specific impulse, and significantly longer lifetime. These capabilities enable more ambitious missions to the outer reaches of our solar system, while potentially playing significant propulsion roles for military and communications satellites near Earth.[1138]

Glenn designed and manufactured six engineering model NEXT thrusters and a prototype power processor. For over a decade Glenn tested the technology extensively in vacuum facilities at the center, JPL, and the Aerospace Corporation.

As a result of the Glenn design and testing, Aerojet Rocketdyne developed a prototype thruster and flight-like propellant management system, JPL created a breadboard gimbal, and L–3 Communications developed a prototype power processor. Glenn performed acceptance testing of these various subsystems both

individually and in integrated assemblies. In 2013 Glenn completed a long-duration life test of an engineering model NEXT thruster. During the more than 50,000 hours of operation, the thruster processed over 900 kilograms of xenon propellant.[1139] Glenn recently awarded a contract to Aerojet Rocketdyne to fabricate two NEXT flight systems (thrusters and power processors) for use on a future NASA science mission.[1140]

❖ ❖ ❖ ❖ ❖ ❖

An alternative thruster design, referred to as a "Hall thruster," is more applicable to time-critical missions than gridded ion thrusters are. Like gridded ion thrusters, Hall thrusters ionize inert gases such as xenon or krypton and then accelerate the ions to produce thrust. The primary difference between these two devices is that Hall thrusters do not utilize grids to accelerate the ions. Instead, the ions are accelerated using an axial electric field that is created in the presence of a radial magnetic field. Hall thrusters are generally less complex and physically smaller than ion thrusters, which often lowers the cost of the system.[1141]

Lewis researchers investigated Hall thrusters in the early 1960s but eventually suspended their efforts

*Image 372: Ivanovich Anatoli Vassine poses with a Russian T–160E Hall thruster being tested at the center in November 1997 (GRC–1997–C–04095).*

when they were unable to achieve acceptable efficiency. Soviet researchers, however, successfully developed the technology and began utilizing Hall thrusters in the 1970s. In 1991 a team of U.S. researchers, including Lewis personnel, traveled to Russia to learn more about the technology. Lewis acquired a Russian SPT–100 and performed extensive testing in 1992 and 1993.[1142] Lewis continued efforts to transfer Russian Hall thruster technology to the U.S. user community throughout the 1990s.[1143] These efforts culminated in the flight system integration of the Electric Propulsion Demonstration Module (EPDM) that was composed of a Russian-produced D–55 Hall thruster and U.S. power-processing technology. In 1998, the Space Technology Experiments (STEX) spacecraft utilized the EPDM system. It was the first flight of a Hall thruster on a Western spacecraft.[1144,1145]

Glenn researchers then pursued the development of several other Hall thrusters with input power levels ranging from 1 to 50 kilowatts to support a variety of future NASA missions. One of these devices was the High Voltage Hall Accelerator (HiVHAc) thruster

that Glenn and Aerojet developed in the early 2000s. The HiVHAc thruster was developed to satisfy the performance and lifetime requirements of cost-capped NASA missions such as Discovery-class missions at a lower cost than with ion thruster systems.[1146] Engineers performed mission analyses which revealed that the HiVHAc system could perform a variety of space exploration missions more efficiently than the NSTAR or NEXT thrusters could.[1147] In 2005 Glenn researchers tested the HiVHAc thruster over a wide range of power input levels and then operated it for nearly 5,000 hours in an EPL vacuum facility.[1148]

Glenn also pursued the development of 50-kW-class Hall thrusters including the NASA–457M, which was operated at a power level of over 70-kW during tests in 2002. The following year, the researchers investigated the use of alternative propellants such as krypton. In 2003 they applied this technology to a higher fidelity high-power Hall thruster design, the NASA–400M, and began incorporating components from both models into a new 300M design.[1149-1151]

*Image 373: HiVAC thruster installed in tank 5 of the EPL during May 2013 (GRC–2013–C–03742).*

## Institutional Impacts

O'Keefe, who was appointed Administrator to apply his fiscal management background to NASA, closely examined NASA's infrastructure. He considered all options, including the closure of some centers and the consolidation of all aeronautics work. Although O'Keefe did not carry out those ideas, he did expand the full-cost accounting effort initiated by Dan Goldin in the late 1990s. This allowed the Agency to track, manage, and forecast the number of full-time equivalent personnel on each project. NASA had previously funded its staff separately from its projects, enabling flexibility for assigning people as needed. Now projects would have to predict their needs more closely and take into account and budget for their own staffing. The inability to predict precisely and the loss of flexibility proved to be problematic. In reality employees were frequently assigned to multiple projects. So if funding for one was canceled, a percentage of that employee's salary was "uncovered." To compensate for this gap, NASA charged the projects higher fees for staff. Project managers began transferring work outside of NASA to reduce costs. This resulted in more uncovered staff and even higher fees.[1152]

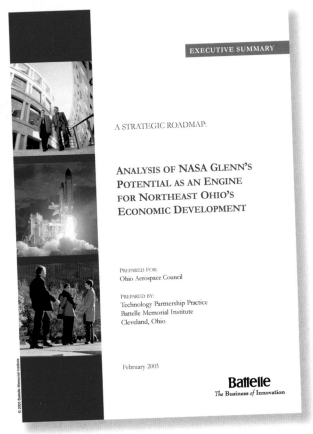

*Image 374: Glenn economic impact document.*

This payroll disconnect was aggravated by NASA's intentional transfer of more of its exploration work to industry. This raised project staffing fees in other fields such as aeronautics, resulting in the cancellation of nonspace projects despite otherwise normal budgets. The Marshall Space Flight Center and Langley, Ames, and Glenn research centers were particularly affected as the number of uncovered personnel quickly escalated in 2004. Productivity and morale plummeted as the technical staff scrambled to find enough work under the full-cost system. With layoffs looming, many young employees left NASA for industry or academia.[1153]

Meanwhile, the presidentially appointed Aldridge Commission sought to determine the best strategy for implementing the VSE. Their primary concern was the sustainability of support and funding for the long-term effort over multiple administrations and congresses. The commission solicited input from scientists, NASA managers, and academics on specific activities to incorporate into the VSE. Not surprisingly, conflicting agendas quickly muddled the VSE objectives, and the human exploration of Mars began to overshadow the Moon goal. After more than a year, little concrete progress had been made.[1154]

It also was becoming clear that funding would be the critical issue for the VSE. The Bush Administration only contributed an extra $1 billion for the VSE's initial five-year period. NASA was responsible for providing the additional $11 billion required for that term through restructuring and physical downsizing—the same process that the Agency had painfully pursued less than 10 years before during the Zero Base Review.[1155-1157] The situation was exacerbated by the erosion of NASA's budget as the Return to Flight and ISS construction efforts dragged on longer than expected.

After just three years, Administrator O'Keefe tendered his resignation in December 2004. The unforeseen Return to Flight struggle and the widespread criticism of his decision not to service the Hubble Space Telescope for safety reasons had drained him.[1158] O'Keefe's financial management and workforce planning efforts did not produce the desired results for the Agency. His greatest legacy was establishing the VSE—which reinvigorated human spaceflight—and his indefatigable leadership in the days and weeks immediately after the loss of *Columbia*.

## Refocusing the Vision

In April 2005 President Bush appointed former NASA engineer and aerospace consultant Mike Griffin as Administrator. Griffin began reshaping the VSE initiatives to deal with budgetary realities. This included canceling *Prometheus* and a number of nonspace programs in an effort to pay for the VSE. Griffin terminated the Aldridge Commission's ill-defined activities and initiated a new study to rapidly identify VSE's human missions and the vehicles necessary to carry them out. NASA released its findings on 12 July 2005, just two weeks prior to the shuttle's successful return to flight.[1159]

Constellation would include two Ares launch vehicles—Ares I for humans and the larger Ares V for cargo—as well as the multipurpose *CEV*, now renamed *Orion*. The *Orion* vehicle—Constellation's marquee element—was designed to replace and supersede the shuttle's crew-carrying capability. *Orion* was

Image 375: *Michael Griffin visits Glenn on 16 May 2005, one month after being confirmed as NASA Administrator (GRC–2005–C–00704).*

Image 376: *While NASA was still defining the Constellation mission and parameters, Glenn sought ways to participate in the program. In 2005 Glenn teamed with Marshall to design the Orion service module mock-up. The Glenn Fabrication Shop then created an 18-foot-diameter full-scale model with a 180° cutaway highlighting the module's internal components.*[1160,1161] *Ever since, it has been on display next to the Administration Building (GRC–2007–C–00566).*

*Image 377: Model of the Ares V vehicle built behind the Administration Building (GRC–2009–C–01301).*

intended to work with either Ares booster and to ferry up to six astronauts to the ISS or four to the Moon. It consists of two main units: (1) the command module carrying the crew and (2) the service module containing the propulsion, power, and communications systems. *Orion* also includes solar arrays to generate electrical power and radiators to dissipate heat.[1162] A number of vehicle elements would be based on upgraded Apollo and shuttle hardware rather than new technology, in order to minimize costs and programmatic risk.[1163]

Just as NASA finally established its Constellation plans in 2005, Congress reduced funding for the VSE even further as the Iraq War and Hurricane Katrina drained federal resources. Not only were the promised annual increases eliminated from the fiscal year 2006 budget, but NASA's overall appropriation was

reduced.[1164] Rather than delaying or downsizing Constellation, Griffin subtracted 75 percent of that deficit from its primary microgravity, life science, and aeronautics programs. He also stemmed the growth of NASA's expanding Earth Science's initiative.[1165]

❖ ❖ ❖ ❖ ❖ ❖

This had significant implications for Glenn. As nearly all NASA funds were shifted to exploration, the center was faced with the elimination of positions, program cancellations, and the threatened closure of large test facilities.[1166] Paradoxically, Glenn's shrinking budget and lack of a near-term test program also threatened Plum Brook Station, much of which was built specifically for advanced post-Apollo space missions like the VSE. The center eliminated hundreds of positions through buyouts and layoffs.[1167-1169]

The nearly 25 percent decrease in NASA's aeronautics budget between 2005 and 2007 forced NASA to terminate its work on the Ultra-Efficient Engine Technology (UEET) program.[1170,1171] In 2005 NASA reorganized its aeronautics work into four large programs—the Fundamental Aeronautics Program to handle basic research, the Airspace Systems Program for air traffic, the Aeronautics Test Program to maintain fundamental capabilities, and the Aviation Safety Program to reduce fatal aircraft accidents.[1172] The VSE restructuring also led to the termination of Glenn's solar cell research effort, the transformation of microgravity science programs into engineering endeavors, and a 20-percent staff reduction at the National Center for Microgravity Research.[1173,1174]

These changes infuriated many in the academic and science communities. "Without this research, they aren't going to Mars, period," Simon Ostrach warned at the time. A National Research Council report stated that NASA was attempting to do too much with too little.[1175] Hundreds of researchers at universities and research labs who had spent years developing NASA-sponsored science programs saw them summarily canceled. Griffin justified these decisions on the basis that he was just trying to implement congressionally mandated priorities for the Agency. Again NASA's budget volatility and decreasing research opportunities caused some younger researchers to look elsewhere to perform their work.[1176,1177]

Image 378: *Associate Administrator for Aeronautics Lisa Porter is briefed on Glenn's turbomachinery accomplishments during a visit in April 2006 (GRC–2006–C–00765).*

❖ ❖ ❖ ❖ ❖ ❖

It was in this context that Administrator Griffin appointed Woodrow Whitlow to replace retiring Director Julian Earls in October 2005. Whitlow had begun his career as a research scientist at Langley in 1979, then served in several high-level positions at Langley and headquarters before being named Glenn's Director of Research and Technology in 1998. He served as Deputy Director at Kennedy Space Center from 2003 until he returned to Glenn as director in 2005.[1178]

In early 2006 Center Director Whitlow began traveling to headquarters on a weekly basis to stake out a role for Glenn in the Constellation Program. Prior to his retirement, Earls had assembled a team of former Glenn managers, including Larry Ross and Lonnie Reid, to analyze the center's ability to manage new space programs. Their report, which emerged in 2006, concluded that Glenn's lack of a space program office and capable project managers were the primary reasons for the dearth of projects in recent years. It also expressed concern that the continual budget reductions, lack of new projects, and insufficient support were eroding the center's technical competency. Nonetheless the authors considered Glenn to be technically superior to the Johnson Space Center or Marshall.[1179,1180] They

concluded that the center's experience placed it in a better position to manage large programs than it had been before taking on the Centaur or space station power programs.[1181]

Whitlow and Deputy Director Rich Christiansen immediately began to reorganize the Glenn staff. This process included the creation of the Space Flight Systems Directorate to manage the center's space systems development. They also merged the center's systems engineers and technicians into the new Engineering Directorate and consolidated the aeronautics management into the Research and Technology Directorate.[1182,1183]

In the midst of these activities, Griffin decided that NASA should perform more of the Constellation work in-house. He introduced the "10 healthy centers" motto and assigned key exploration, science, and aeronautics roles for each center. In addition he successfully instructed top management personnel to focus on improving work assignments to minimize the "uncovered" staffing issue. Griffin also asked that the centers attempt to use civil servants at other centers before contracting work

Image 379: *Center Director Woodrow Whitlow and Deputy Director Rich Christiansen (GRC–2005–C–01706).*

*Image 380: Display in the Research Analysis Center, which housed Glenn's Constellation staff. The display featured historical photographs and artifacts such as a Gemini capsule and Mercury escape tower (GRC–2010–C–0206).*

*Image 381: Ares I–X engineering team emulates their Apollo predecessors during a review meeting in July 2007 (GRC–2007–C–01580).*

out. Issues with the number of NASA employees and incorrect work assignments were largely eradicated in 2007 and 2008.[1184]

Glenn immensely benefited from Griffin's decisions. On 12 May 2006 headquarters assigned Glenn the responsibility for the development, budget, staffing, and contracting for the *Orion* service module.[1185] Glenn established an Orion project office under Bryan Smith, and Jim Free managed the service module work. The center also was responsible for the adapter that joined the *Orion* stage to the Ares booster and was asked to provide the Johnson crew module development team with guidance on seals, controls, and combustion.[1186] Glenn's fortunes continued to improve in 2007 as staff reductions ceased, and two new Constellation programs were secured—*Orion* vibration testing in the Space Power Facility (SPF) and the development of a lunar lander and rover. Glenn had been awarded over $1 billion in Constellation work, and Whitlow announced that the efforts would keep the staff occupied for the next decade.[1187,1188]

## Glenn *Orion* Module Work

Despite the concern over Glenn's science and aeronautic areas, employees quickly embraced and adapted to the new VSE work. Many were invigorated by these new, large assignments, and people worked with an enthusiasm that had been absent during the previous decade of decline. They worked countless hours to meet the deadlines imposed by human spaceflight, where schedule was a higher priority than in the research fields. New intercenter collaborations emerged during this period that endure to this day.[1189]

NASA selected Lockheed Martin to construct the *Orion* vehicle. Glenn was responsible for designing, building, and testing the vehicle's propulsion system, which consisted of a main thruster surrounded by eight smaller backup engines and 16 attitude-control thrusters. Unlike previous spacecraft designs, a single propellant source supplied all of the thrusters. In addition, NASA sponsored a study into the applicability of using a liquid-oxygen/liquid-methane propellant system.[1190]

*Image 382: Testing of a liquid-oxygen/liquid-methane thruster for CEV (GRC–2009–C–01936).*

Liquid methane is denser than liquid hydrogen and, therefore, requires smaller tanks. It also burns well with liquid oxygen and theoretically could be manufactured from natural elements in the martian atmosphere.[1191] In 2005 Glenn sponsored the development of two different methane engines. Just as the effort was getting under way, Orion program managers decided to replace the methane and oxygen combination with hypergolic fuels similar to those used for the shuttle thrusters.[1192]

Glenn decided to continue the oxygen/methane research for possible future implementation. The center had investigated liquid methane propulsion systems at the Plum Brook Rocket Systems Area in the 1960s. The current studies were conducted at a new set of small test facilities that replaced those destroyed by the airport runway expansion. The tests included hot firing of the liquid-oxygen/liquid-methane thruster in simulated altitude conditions at the Altitude Combustion Stand (which included one of the engine test stands from the former Rocket Engine Test Facility).[1193] Other researchers used the Small Multipurpose Research Facility to test long-duration insulation systems for methane tanks in simulated lunar conditions.[1194] Researchers successfully demonstrated the operation of the engine, thrusters, and propellant feed system. The technology is available for future use.[1195]

Glenn managed the development of the *Orion* service module, which included its engines, control thrusters, power system, and cooling system. Glenn also contributed to the development of various subsystems for the Ares I launch vehicle, and developed the thrust control

*Image 383: Stacking of two Ares I–X segments in the Fabrication Shop, which was renamed the "Ares Manufacturing Facility" (GRC–2008–C–00421).*

system, electrical power, the payload shroud for Ares V. In addition, the center worked on technology for the lunar lander, lunar rover, and spacesuits;[1196] analyzed coatings to discharge static energy from the *Orion* solar panels; worked on the radiators used to cool the capsule interior; and studied advanced seals.[1197] An acoustic reverberation facility was added to the SPF's high-bay area for future *Orion* vibration testing.

Glenn researchers were also very interested in lunar soil—both its effects on spacecraft and the tools to gather it for use in space. The Simulated Lunar Operations facility was created in the Engine Research Building. It included a 60-foot-long box filled with a sand/clay mixture that resembled lunar soil. Glenn worked with the Canadian Space Agency to develop computer models and physical tools to mine the lunar soil.[1198]

❖ ❖ ❖ ❖ ❖ ❖

The Ares launch vehicles, like any new rockets, required developmental flights to verify their performance before any actual missions were attempted. Engineers were concerned that Ares's single engine might not be able to control the tall, slender rocket. To minimize expenses, these test launches usually carry models that simulate the mass and shape of the upper stages or payloads. NASA assigned Glenn the responsibility for assembling mass models of the second stage, service module, and adapter for the Ares I–X development launch. The launch was the first of five planned flight tests. NASA had usually contracted commercial manufacturers to create flight hardware, but with NASA's new effort to increase in-house work, Glenn decided to build these models in its own Fabrication Shop.[1199,1200]

In June 2007 Glenn began manufacturing the 11 cylindrical steel sections for the simulated *Orion* stage. After each section was completed, it was moved from the Fabrication Shop to the PSF. Technicians stacked the segments on top of one another in the facility's clean room to verify their integrity.[1201] Then, in October 2008 Glenn employees trucked the large segments to Wellsville, Ohio, to begin their 12-day journey down the Ohio and Mississippi rivers, out into the Gulf of Mexico, and over to Kennedy. There they were assembled and added to the Ares booster.[1202]

Members of Glenn's Ares I–X launch team were at their stations at 4 a.m. on 28 October 2009 at Kennedy. An approaching storm forced the team to repeatedly recalculate for the winds and threatened to cancel the launch altogether. At 11:30 a.m. Ares I–X lifted off, and the control room, which had grown quiet during the last 4 minutes of the countdown, erupted in cheers.

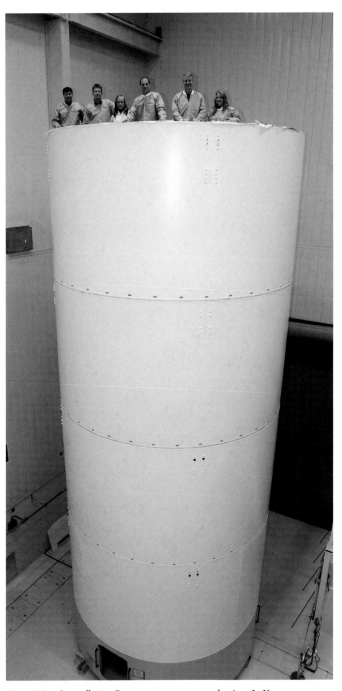

*Image 384: Constellation Program managers tour the Ares I–X segment assembly in the PSF (GRC–2008–C–00836).*

*Image 385: Glenn employees view the Ares I–X launch on 28 October 2009 from the Visitor Center (GRC–2009–C–03936).*

Six minutes later the Ares booster parachuted into the sea for later recovery while Glenn's upper-stage simulator plummeted to the ocean floor as intended.[1203] The successful launch not only demonstrated the flight dynamics, control, and stage separation of Ares I, but Glenn's ability to construct and deliver spaceflight hardware on schedule.

### Resetting the Space Program

Despite the success of the Ares I–X development flight, Constellation's days were numbered. In January 2009 President Barack Obama took office and accepted Administrator Griffin's resignation. Griffin had come under fire from many for his funding of Constellation at the expense of science and aeronautics and clashed with the President's NASA transition team. President Obama selected former shuttle astronaut Charles Bolden as the new Administrator.[1204] The president also created an advisory panel led by Norm Augustine to review the now five-year old Constellation Program. The vision had again become blurred. The VSE's original incremental mission approach had been supplanted by a general push toward developing a system to send humans to Mars. Most of NASA's energy was now directed at space transportation

systems, not the activities that actually would be done on the Moon or Mars. Some claimed that Constellation was no longer supporting the VSE; instead, the VSE was being used to advance Constellation.[1205]

It would have been difficult to manage such a large endeavor with proper funding, but NASA's flat budget made it nearly impossible. In October 2009 Augustine's panel forecasted that the Moon landing schedule would slip by over 10 years, into the 2030s, unless the project received an additional $3 billion annually; and the new Ares 1 launch vehicle would not be ready by the time that the shuttle fleet was supposed to be retired. This meant that NASA would have to divert Constellation funds toward the additional shuttle flights that were needed to complete the ISS. The panel offered several options, including substituting a mission to an asteroid for the Moon landing and reconfiguring the Ares vehicles.[1206] Constellation's prospects did not appear bright, and once again, a sense of gloom settled over the program personnel.

The official announcement came on 1 February 2010. Based on the panel's recommendations, President Obama decided to cancel the Ares 1 rocket, purchase

Image 386: *MPCV pamphlet.*

flights to the ISS from the Russians, and contract private firms to provide access to the ISS in the future. The specific goal of sending humans to the Moon or Mars was replaced by an effort to develop technologies for an undetermined destination. Congress eventually retained the Ares V heavy lift booster and *Orion*, renaming them the Space Launch System (SLS) and the *Multi-Purpose Crew Vehicle* (MPCV), respectively. NASA transferred all management responsibilities for the MPCV to Johnson, but Glenn would continue work on the service module, albeit in a smaller capacity.[1207,1208]

Despite the loss of some funding, the reorientation of the space program realigned Glenn with its traditional role of the research and development of advanced technology. Where Constellation tried to build on the technology from the 1960s, NASA now initiated a new effort to develop innovative technologies that would expedite the exploration process. Headquarters assigned Glenn the responsibility for the Exploration Technology Development and Demonstration Office and the Space Technology Research Grants Program Office. The former sought to develop new technologies for space exploration, such as solar electric propulsion, nuclear surface power, nanotechnology, cryogenic fluid management, and modular power systems. The latter office would establish a grant program for universities to develop new technologies that could be applied to human space missions.[1209] NASA also increased its aeropropulsion and alternative energy research—two of Glenn's fortes.[1210]

Image 387: *The acoustics chamber in the SPF facility to be used for vehicle vibrational and structural dynamics testing (GRC–2011–C–04392).*

*Image 388: NASA's new S–3 Viking cruising along the Lake Erie shoreline in March 2010 (GRC–2010–C–01394).*

*Image 389: View from the cockpit of Glenn's S–3 Viking (GRC–2010–C–03192).*

*After nearly being eliminated in the late 1990s, Glenn's Flight Operations group reemerged in the 2000s. The group supported several of the Aviation Safety Program efforts, continued icing flight research with the Twin Otter and solar cell calibration efforts with the Learjet. In March 2004 Flight Operations acquired a new tool for icing research— a Lockheed S–3 Viking. The S–3 was a former Navy antisubmarine aircraft that surpassed the Twin Otter in range, speed, and power. The S–3 underwent a two-year transformation from military fighter to research aircraft. Its first mission was an icing research study near Puerto Rico in fall 2008.[1211]*

## Aviation Safety

Despite the deep cuts in NASA's aeronautics program, Glenn researchers were able to make several key contributions during the 2000s—particularly in aviation safety. As the aviation industry continued to grow in the 1990s, several of its safety standards began to gradually erode. The percentage of aircraft crashes remained low, but the increased number of flights meant that overall figure was rising. In 1997 President Bill Clinton formed a commission on aircraft safety that called for a reduction of fatal accidents by 80 percent over the next 10 years. Glenn, Langley, and Ames partnered with the Federal Aviation Administration, the Department of Defense, and industry to develop new technologies that industry could use to improve safety on airliner and general aviation aircraft.[1212]

The $500-million program focused on preventing accidents, reducing accident damage, and improving aircraft-monitoring systems. Each one of these topics covered a wide array of research areas.[1213] The system monitoring included the creation of controls and diagnostics to detect and remedy the engine problems that often caused pilot errors.[1214] The accident prevention efforts addressed weather hazards, aircraft-monitoring systems, and development of synthetic vision technology that would improve a pilot's ability to see in inclement weather and darkness. The accident mitigation efforts included fire prevention and improved crash survivability.[1215] Glenn's legacy of flight safety research, which had begun with the icing research and crash fire programs of the 1940s and 1950s, continued into the 1970s with engine control and flight simulation technologies.

Glenn managed two of the Aviation Safety Program's efforts—Weather Information Communications (WINCOMM) and a high-altitude ice crystal research program. The WINCOMM program sought to accelerate the transmission of weather and turbulence information between aircraft and ground-based stations and from aircraft to aircraft. In the early 2000s Glenn researchers analyzed and tested four different first-generation commercial systems that transmitted

*Image 390: Don Campbell is briefed on the Honeywell Weather Information Network, which was one of the off-the-shelf technologies tested during the first phase of WINCOMM (GRC–1999–C–02205).*

## Glenn Shares in Collier Trophy Award

On May 28, the National Aeronautics Association presented the 2008 Robert J. Collier Trophy to the Commercial Aviation Safety Team (CAST), an industry and government partnership that was established in 1997 with the goal of reducing the U.S. commercial aviation fatal accident rate by 80 percent in 10 years.

**Mary Reveley**, a researcher in Glenn's Multidisciplinary Design, Analysis and

*Reveley receives Collier Trophy for her role in the award-winning CAST.*

Optimization Branch, was one of eight key CAST members who received the coveted trophy. The trophy is presented annually for the greatest achievement in American aeronautics or astronautics, with respect to improving the performance, efficiency and safety of air or space vehicles demonstrated during the preceding year. The

award is named for Robert J. Collier, a prominent publisher, patriot, sportsman and aviator.

CAST represents thousands of people in public agencies and private industry "who have worked diligently since 1997 to produce the safest commercial aviation system in the world," according to the award nomination submitted by the Air Transport Association. Reveley represented NASA's Aviation Safety Program serving as a member of the CAST Joint Implementation Measurement and Data Analysis Team. Additionally, three researchers in Glenn's Icing Branch —**Tom Ratvasky, Andy Reehorst** and **Tom Bond** (currently with the Federal Aviation Administration)—served on the CAST Loss of Control Joint Safety Assessment Team.

### Apollo 11 at the VC

On Saturday, July 18, Glenn's Visitor Center will celebrate the 40th Anniversary of Apollo 11. Visitors will learn about NASA's new spacecraft that will return humans to the moon and to Mars. Visitors can also tour the Glenn facilities that contributed to the Apollo missions. Admission will be first-come, first-served. No registration required. Contact 216–433–9653 for more information.

*Image 391: Mary Reveley and three colleagues were among a team that received the prestigious Collier Trophy in 2009 recognizing the Aviation Safety Program's success (Aerospace Frontiers, July 2009).*[1216]

data from the ground to the air. The researchers then expanded the size, quality, and speed of the weather data transmission and tested those systems. In 2005 Glenn flight-tested WINCOMM on the Learjet, finding that the data transmission rates were 20 to 100 times greater than with the system currently in use.[1217]

NASA extended the Aviation Safety Program to include new high-altitude engine icing investigations, electro-optical sensor studies, and analysis of the effect of lightning strikes on composite materials. The systemwide safety research sought to develop new software to anticipate and resolve safety issues, to

identify the causes of risks and predict failure, the interaction of pilots with flight equipment and air traffic controls, and the lifespan of hardware and software. The vehicle systems safety project included efforts to evaluate the health of aircraft components and systems, study the loss of aircraft control, and develop methods for assessing and improving a pilot's ability to recover from failures or disturbances.[1218]

❖ ❖ ❖ ❖ ❖

Glenn undertook a new type of icing study involving ice crystals that form at high altitudes above the weather. Typical engine icing occurs with the accretion of supercooled liquid droplets that freeze on engine components and disrupt airflow or break off and damage the engine. The new studies focused on ice crystals that inexplicably build up inside the hot engine, causing surge, stall, or flameout. Most instances of core icing occurred over 22,000 feet, which is the upper threshold of supercooled liquid clouds.[1219]

Several aeronautics companies formed a working group and built new test facilities to explore the issue. NASA began conducting its own effort, which included flight research, altitude engine testing, and computer simulations.[1220]

Glenn researchers found that ice crystal clouds can form near thunderstorms in which the wet air quickly rises. Despite previous research to the contrary, Glenn researchers discovered that the crystals melt on high-temperature surfaces, then form ice when those surfaces get colder. Glenn installed a new ice simulation system in a Propulsion Systems Laboratory (PSL) altitude test cell to study this phenomenon. When, in 2013, Glenn researchers created ice crystals in simulated altitudes for the first time, the PSL was the nation's only facility with this capability.[1221,1222]

*Image 392: PSL's No. 3 test chamber was modified for high-altitude engine icing tests (GRC–2012–C–04153).*

*Image 393: The Dr. Edward R. Sharp Alcove of Honor. In 2007 the pick and shovel used for the groundbreaking events at Lewis Field in 1941 and Plum Brook in the 1956 were found in a storage barn (GRC–2008–C–01791).*

## The Changing Face of Glenn

Since the groundbreaking in 1941 the Glenn campus has been almost constantly in flux. Technology advanced so quickly that some of the original facilities were soon out of date. The center modified several of these facilities and built newer ones to address future developments. Support buildings and offices were modified and upgraded, as well. By the mid-1950s the entire 200-acre property was occupied. With the lack of space, the center was forced to transform several of its facilities to meet the needs of the space program in the 1960s. The center's decreased budget and programmatic shifts in the 1970s resulted in the closure of several of its legendary facilities. Glenn began a new wave of physical change in the 2000s.

In 2003, for the first time in its history, NASA Headquarters allocated funds for the demolition of unused facilities and asked its centers to submit structures for consideration. The annual upkeep of the unused sites was expensive, so Glenn proposed the removal of nine buildings, including the Altitude Wind Tunnel and PSL No. 1 and 2. The two facilities had played significant roles in the advancement of the nation's propulsion technology, but they had not been used for testing since the 1970s. They were demolished in spring and summer 2009. In 2010 Glenn also removed the Jet Propulsion Static Laboratory and the Plum Brook Rocket Systems Area.

Other recent physical changes to the campus include the relocation of the Visitor Center to the Great Lakes Science Center (where Glenn's work was seen by at least five times more people and school groups each year than at the site at Glenn), the demolition of the Guerin House and its subsequent replacement with a more modern meeting facility, improvement of the main gate structures, and the addition of two

*Image 394: PSL wasteland in July 2009. In just over two months, the wrecking crew leveled a facility that had been part of the center's landscape for nearly 60 years (GRC–2009–C–02018).*

*Image 395: The Altitude Wind Tunnel shell is cut down in February 2009 (GRC–2009–C–00752).*

*Image 396: The Apollo capsule used for the Skylab 3 mission is transferred from Glenn to the Great Lakes Science Center in June 2010 (GRC–2010–C–02647).*

*Image 397: The Mission Integration Center (MIC) (GRC–2014–C–02988).*

new meeting centers. Perhaps the most significant addition has been the new MIC office building near the center of the campus. The MIC, located near the center of the campus, houses hundreds of engineers and managers working side-by-side on the center's main aerospace projects. The structure incorporates an array of design elements to conserve energy.

## Space Research and Technology

NASA established the Human Research Program in 2005 to develop improved processes and technologies to protect humans on the VSE's long-term space missions. Johnson managed the wide-ranging effort, which included everything from a better quality diet and improved psychological health to mitigation of environmental risks like solar radiation.[1223] Glenn has been able to use its expertise in areas such as microgravity research, computational modeling, and instrumentation to support the Human Research Program. The center's biomedical engineering work had begun in earnest a decade before.

In 1994 Administrator Dan Goldin urged Glenn to use its microgravity science expertise to support NASA's life science program and Cleveland's expanding medical community. In response, the center created a taskforce to facilitate technology transfer to this community. The local medical industry responded with requests for new materials, instrumentation, software, and telemedicine projects, as well as for items for specific medical applications such as lasers for cataract treatment and artificial heart actuators. For a variety of reasons, it was more difficult to establish partnerships with hospitals and research institutions than with industry at this time, and NASA terminated Lewis's work in this area during the 1995 Zero Base Review.[1224]

The center was able to connect with the local medical and research institutions several years later. In May 2002 NASA created the John Glenn Biomedical Engineering Consortium, a partnership between Glenn, Case Western Reserve University, the Cleveland Clinic Foundation, University Hospitals of Cleveland, and the National Center for Space Exploration Research. Seven three-year collaborative projects were undertaken to mitigate health and safety risks to the human space crews through research in fluids physics and sensor technology. Examples included tools to measure and prevent bone loss,

*Image 398: Glenn Biomedical Consortium agreement signing event on 7 June 2002 at Glenn's Zero Gravity Research Facility. From left to right: Don Campbell, Mary Kicza, and Howard Ross from Glenn; Huntington Willard (University Hospitals), and Bill Sanford (Bioenterprise Corporation) (GRC–2002–C–01425).*

*Image 399: The vertical treadmill is demonstrated in Glenn's Exercise Countermeasures Lab. The treadmill, referred to as the "Standalone Zero Gravity Locomotion Simulator," allows researchers to study the effects of low gravity on astronaut exercise (GRC–2010–C–03733).*

implantable medicine release systems, and instruments to detect cardiac, radiation, and metabolism irregularities.[1225] These studies led to Glenn's work with NASA's Human Research Program in 2005.

Glenn has contributed to two primary areas of the Human Research Program—Human Health Countermeasures and Exploration Medical Capabilities. The former seeks to identify and mitigate a range of physiological problems caused by spaceflight. Glenn's focus has been on the development of advanced exercise equipment to prevent bone and muscle atrophy. The challenge has been to design this equipment to fit into a comparatively small spacecraft. Researchers have created computational programs to analyze the effect of these devices on bone and muscle. Glenn has also been developing computer models to help determine the cause of visual impairment caused by microgravity.[1226]

The Exploration Medical Capabilities effort develops technology to forecast and prevent health risks associated with inflight medical conditions. These include a system to generate saline from existing resources, an ultrasound that can identify internal health conditions such as bone fractures, a portable medical oxygen concentrator, and a medical suction device that can operate in microgravity. Glenn also manages a computer model that uses medical data from previous missions to predict future risks. The center's contributions to the Human Research Program continue today.[1227]

❖ ❖ ❖ ❖ ❖ ❖

The ISS began operating in November 2000, with six-person crews typically serving four- to five-month terms.[1228] The largest space structure ever constructed, the ISS provides researchers with ample space and power to perform extended studies.[1229] Glenn researchers took advantage of this new tool to further

their study of the behavior of fluids and combustion in a microgravity environment. Although the shuttle had provided the opportunity to perform microgravity experiments that were significantly larger and longer than those on Earth, the two-week shuttle missions did not provide much time to modify an experiment on the basis of early results. The ISS not only provided an unlimited amount of time for the experiments but room for larger setups.[1230]

Glenn's most significant contribution to the station has been the Fluids and Combustion Facility (FCF), one of six permanent ISS experiments. The FCF consists of two modular, adaptable racks that house multiple experiments. Glenn researchers designed these racks—the Combustion Integration Rack (CIR) and the Fluids Integration Rack (FIR)—in the mid-1990s, basing the design on the Combustion Module-1 shuttle experiment rack. After several years of ground testing, the racks were installed on the station in 2008 and 2009.[1231] The FCF enables researchers to carry out experiments remotely from Glenn with minor assistance from the astronauts.[1232]

*Image 400: The FCF's CIR undergoing testing in the Acoustic Testing Laboratory during April 2005 (GRC–2005–C–00559).*

*Image 401: Astronaut Nicole Scott installs the FIR on 22 October 2009 (NASA ISS021E011440).*

Since beginning operation in 2009, Glenn researchers have used the CIR to conduct three long-term liquid fuel combustion experiments involving thousands of tests.[1233] Researchers have used the FIR to investigate heat transfer and the characteristics of colloid materials in space.[1234]

❖ ❖ ❖ ❖ ❖ ❖

As the Constellation project was transforming in 2010, a Glenn team raced to complete a new space communications program in time to catch a ride on one of the few remaining shuttle flights to the ISS. Despite the success of the Advanced Communications Technology Satellite (ACTS) satellite in the 1990s, NASA management downsized Glenn's space communications program and instructed the group to shift its focus from potential commercial technology to applications for NASA. Glenn researcher Richard Reinhart developed the Space Communications and Navigation (SCaN) communication device, which employs computer codes instead of electronics to transmit signals.

SCaN technology can be used for communications, networking, and navigation both in space and on Earth.

The key to SCaN's technology was its use of software to generate waveforms, which allows repairs or upgrades to be performed remotely through software upgrades.[1235] Researchers integrated three different software-defined radios and an antenna-pointing system into a test package to be installed on the ISS. Each of these radios operated on a different bandwidth— including the Ka-band spectrum that had first been explored by ACTS.[1236]

Glenn, which initiated the program in 2006, was responsible for developing, building, and testing the technology. Over the course of just five years, the SCaN team converted the theoretical concepts into working technology and tested it at length. Glenn tested the systems extensively in the center's vacuum chambers and shake facilities.

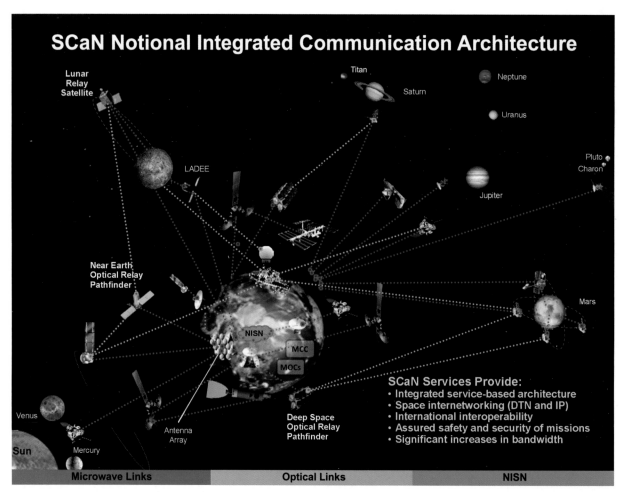

*Image 402: SCaN's architecture and services.*

Image 403: The SCaN testbed is prepared for shipment to Japan for its launch (GRC–2012–C–00698).

The program suffered some setbacks, however, and missed the opportunity to secure a place on the remaining shuttle launches. NASA made an alternative arrangement to have a Japanese rocket transport the SCaN hardware to the ISS in July 2012. Researchers control the SCaN testbed from the Glenn Telescience Support Center that was developed to monitor microgravity experiments on the shuttle.[1237] SCaN is being used to demonstrate software-designed radio systems; conduct communications, navigation, and networking experiments; and test capabilities for future missions.

## New Day

Days after the Constellation Program was canceled in 2010, Woodrow Whitlow accepted the Associate Administrator for Mission Support Directorate position at headquarters. Administrator Bolden selected then Deputy Director Ramon Lugo to fill Glenn's Center Director vacancy and promoted Jim Free from Director of Space Flight Systems to Deputy Director. Lugo had spent most of his career in the construction and launch vehicle areas at Kennedy before transferring to Glenn in 2007.[1238]

Glenn continued its Constellation assignments for several months while Congress reviewed President's Obama's plans. As NASA adjusted to its new mission, Congress refused to pass President Obama's budget for 2011. This began a series of continuing resolutions and threatened government shutdowns. Despite the rancor, Congress passed a three-year NASA authorization bill that stemmed impending layoffs.[1239]

Federal budget arguments continued into 2011, and a Senate proposal to eliminate NASA's aeronautics work was blocked.[1240] In April 2011 Congress approved a budget that maintained 2010's overall spending levels. The center ended up receiving $100 million more than it had in 2010, mostly

geared toward the new Exploration Technology Development and Demonstration Program. Seven new space technology projects were initiated, including a new solar-electric propulsion demonstration and an in-space cryogenic fluid management program.[1241]

*Image 404: Ray Lugo (right) and Jih Fen Lei (behind Lugo), Director of the Research and Development, take Ohio Representative Dennis Kucinich (left) on a tour of the PSF (GRC–2011–C–03296).*

*Image 405: Astronaut Michael Foreman speaks at a gala gathering of Ohio astronauts in 2008. Foreman flew on STS–123 and STS–129. He was one of several astronauts who served on one-year details leading Glenn's External Programs Office (GRC–2008–C–02464).*

*Image 406: Jim Free speaks to the media during a visit to the EPL by Administrator Charlie Bolden and local congressional representatives Marcy Kaptur and Sherrod Brown in January 2014 (GRC–2014–C–00269).*

Despite the seeming good news, the center's general operations budget was severely decreased. Forty-five members of the operations contract staff were immediately terminated, and center services were reduced.[1242]

In August 2011 Congress passed a bill that mandated indiscriminate cuts across federal agencies in 2013 if expenditures were not reduced. Glenn took steps in 2012 to prepare for this by reducing its spending through cuts in the contract staff, consolidation of its management, and reduction of its facilities maintenance.[1243] Lugo also brought in consultants to carry out a technical review of the center. The reviews concluded that "Glenn has far too many lines of research and technology development and needs to focus its efforts to have the ability to impact the Agency and the nation in a credible way."

Lugo and his management team began searching for ways to focus the center's recently expanded role in research and technology and improve efficiency in applying the smaller number of staff to the many projects.[1244] They initiated a reorganization process, but in mid-November 2012 Lugo announced his impending retirement.

On 3 January 2013, Administrator Bolden appointed Jim Free as Glenn's Center Director. The 44-year-old Free was the first Glenn Director to have grown up in the Cleveland area. After graduating from Miami University in Ohio, he had the unique experience of earning his graduate degree in Space Systems Engineering at the Delft University in the Netherlands. Free began his NASA career in 1990 as a propulsion and systems engineer at Goddard Space Flight Center.[1245]

Free joined Glenn in 1999 as the Fluids and Combustion Facility liaison for the ISS. He went on to lead the NEXT electric propulsion effort and to manage the *Prometheus* spacecraft. Free transferred to Johnson in 2008 where he served as the Orion Test and Verification Manager and the Orion Service Module Manager.

He returned to Glenn in September 2009 as Director of Space Flight Systems, in which he managed all of the center's space activities, including *Orion*, the shuttle, and the ISS. Free became Lugo's Deputy Director in November 2010.[1246]

As Deputy Director, Free understood that the center would have to live with the 2010 budget reductions for the foreseeable future. Glenn began taking steps to operate institutionally in that restrictive scenario.[1247] As part of this effort, management began planning a reorganization to focus on the center's five identified core competencies—space power, aeronautics and space propulsion systems, advanced materials, physical sciences and biomedical technology, and advanced communications technology.

Free solicited input from the center staff as he continued to pursue this new organization after becoming Director. He was supported by Deputy Director Greg Robinson (recently replaced by Janet Kavandi)

and Associate Director Janet Watkins. Free also guided the center through a series of congressional budgetary threats including the "fiscal cliff," sequestration, and ultimately a three-week government shutdown in October 2013.

### Casting a Keen Eye Back

On 25 September 2015 the Glenn Research Center inducted the first class into its new Hall of Fame. The ceremony was the culmination of a series of events at the center to mark the NACA centenary. All nine individuals had begun their careers during the NACA era.

"I find it striking," commented NASA Chief Historian Bill Barry, "that ... as we pursue our many demanding goals and we look toward the future, ...we're not just rushing into the future headlong. ...we are casting a keen eye back on our accomplishments, relevant role models and leaders from the past that built this legacy that we continue to build on. ...And what an incredible legacy at Glenn!"[1248]

*Image 407: First class inducted into the Glenn Hall of Fame (PS–01671–3–1015).*

*Image 408: Simon Ostrach, Elaine Siegel, and Robert Siegel at the 2015 Glenn Hall of Fame Induction Ceremony held in the new MIC (GRC–2015–C–06102).*

Glenn's inaugural Hall of Fame class was diverse. It featured not only seminal leaders, like Ray Sharp, Abe Silverstein, and Bruce Lundin, but aviation safety expert Irv Pinkel, computer programmer Annie Easley, graphic artist Jim Modarelli, and a trio of experts nominated as "The Giants of Heat Transfer"—Simon Ostrach, Robert Siegel, and Robert Deissler.[1249]

The event was among the first to officially recognize the importance of individuals to the center's success. During his remarks, Center Director Jim Free noted, "We can build all the buildings we want, and we can have great test facilities, but what matters is the people who come to work here every day—and the incredible intellect and spirit and caring and devotion that these folks had."[1250]

Ostrach and Siegel were present, as were families of deceased inductees, joined by former colleagues and many current employees. For the hour-long ceremony, daily concerns and problems faded into the background, and the audience was filled with pride and admiration of the people, the center, and all they had accomplished for our nation.

❖ ❖ ❖ ❖ ❖ ❖

Glenn's relentless pursuit of the future has endured through massive technological transformations like jet propulsion, spaceflight, and computerization; through changes in leadership, staffing fluctuations, and decreasing budgets; and through an evolution from military research to managing contracts and collaborating with private corporations. Ultimately Glenn's research has produced a persistent flow of new insights and technologies that have improved human activities on Earth, in the air, and in space.

The center's state-of-the-art facilities and equipment expedited many of these advancements, but ultimately, it was generations of talented and motivated people who made the successes reality. Research and development is not easy; there are always wrong turns and roadblocks. Thus, success requires not only intellect and skills, but patience and fortitude. As exemplified by the 2015 Hall of Fame inductees, employees across the center have exhibited these qualities for 75 years. It is on their shoulders that the legacy of the NASA Glenn Research Center rests.

## Endnotes for Chapter 10

1089. NASA, "The Columbia Accident Investigation Board (CAIB)" (17 September 2009), *http://history.nasa.gov/columbia/CAIB.html* (accessed 1 April 2015).

1090. George W. Bush, "The Vision for Space Exploration," 14 February 2004 (Washington, DC: NASA NP–2004–01–334–HQ, 2004).

1091. "Work Assignments for Moon Mission: Ares I and Orion," 5 June 2006, *http://www.nasa.gov/centers/glenn/moonandmars/work_assignments.html* (accessed 8 December 2015).

1092. Howard Ross to Robert Arrighi, "Changing Missions," 20 November 2015, NASA Glenn History Collection, Cleveland, OH.

1093. Tori Woods, "New Assignments, New Responsibilities Proposed for Glenn" (15 June 2010), *http://www.nasa.gov/centers/glenn/exploration/new_glenn.html* (accessed 3 August 2015).

1094. Sean O'Keefe, "Pioneering the Future" address, *NASA Facts* (12 April 2002), *http://www.hq.nasa.gov/office/codez/plans/Pioneer.pdf*

1095. W. Henry Lambright, "Leadership and Change at NASA: Sean O'Keefe as Administrator," Public Administration Review, March/April 2008.

1096. "NASA's Implementation Plan for International Space Station Continuing Flight" (28 October 2003), *http://www.nasa.gov/pdf/53067main_station_imp_plan.pdf* (accessed 1 April 2015).

1097. Ross to Arrighi.

1098. Ann Over, "Combustion Module-2 Achieved Scientific Success on Shuttle Mission STS-107" *Research & Technology 2003* (Washington, DC: NASA/TM—2004-212729, 2004), *http://ntrs.nasa.gov*

1099. "STS-107 To Carry Glenn-Developed Experiments," *Aerospace Frontiers* (July 2002).

1100. Doreen Zudell, "STS-107 Crew Train on Glenn Hardware," *Aerospace Frontiers* (March 2001).

1101. A. Abbud-Madrid, J. T. McKinnon, F. Amon, and S. Gokoglu, "Suppression of Premixed Flames by Water Mist in Microgravity: Findings From the MIST Experiment on STS-107," Proceedings of the Halon Options Technical Working Conference (May 2004), *http://www.nist.gov/el/fire_research/upload/R0401173.pdf* (accessed 8 December 2015).

1102. Ross to Arrighi.

1103. Over, "Combustion Module-2 Achieved Scientific Success," *http://ntrs.nasa.gov*

1104. Stephen J. Altemus, Jon N. Cowart, and Warren H. Woodworth, "STS-107 Reconstruction Report" (Washington, DC: NASA NSTS–60501, 30 June 2003), *http://www.nasa.gov/externalflash/CAIB/docs/STS-107_Columbia_Reconstruction_Report_NSTS-60501_30Jun03.pdf* (accessed 8 December 2015).

1105. Brian M. Mayeaux, et al., "Materials Analysis: A Key to Unlocking the Mystery of the Columbia Tragedy," *Journal of the Minerals, Metals & Materials Society* (February 2004).

1106. "Critical Contributions to the Space Shuttle Return-to-Flight Effort Made by Glenn's Ballistic Impact Team," *Research & Technology 2005* (Washington, DC: NASA/TM—2006-214016, 2006), *http://ntrs.nasa.gov*

1107. Jan Wittry, "Shooting for Safety: Ballistics Team Helps Return Shuttle to Flight" (5 July 2005), *http://www.nasa.gov/returntoflight/crew/Ballistics_RTF_Feature.html* (accessed 1 April 2015).

1108. Doreen Zudell, "Glenn Team Made Big Impact on Shuttle Investigation," *Aerospace Frontiers* (November 2003).

1109. Mayeaux, "Materials Analysis."

1110. "Critical Contributions," *http://ntrs.nasa.gov*

1111. Wittry, "Shooting for Safety," *http://www.nasa.gov/returntoflight/crew/Ballistics_RTF_Feature.html*

1112. Zudell, "Glenn Team Made Big Impact."

1113. Jayanta Panda, Fred Martin, and Daniel Sutliff, *Estimation of the Unsteady Aerodynamic Load on Space Shuttle External Tank Protuberances From a Component Wind Tunnel Test* (Washington, DC: NASA/TM—2008-215155, 2008), *http://ntrs.nasa.gov*

1114. "External Tank Protuberance Air Load (PAL) Ramps," *NASA Facts* (Washington, DC: NASA FS–2005–04–08–MSFC, Pub 8–40393, April 2005).

1115. Wayne Hale, et al., Wings in Orbit: *Scientific and Engineering Legacies of the Space Shuttle, 1971–2010* (Washington, DC: NASA SP–2010–3409, 2010), *http: ntrs.nasa.gov*

1116. Tariq Malik, "NASA Nixes Foam Ramp for Next Space Shuttle Flight," Space.com (15 December 2005), *http://www.space.com/1871-nasa-nixes-foam-ramp-space-shuttle-flight.html* (accessed 1 April 2015).

1117. NASA, "The Columbia Accident Investigation Board (CAIB)," *http://history.nasa.gov/columbia/CAIB.html*

1118. Jenise Veris, "Glenn Employees Reflect on RTF Efforts, *Aerospace Frontiers* (August 2005). Special Notice, NASA Glenn History Collection, Directors Collection, Cleveland, OH.

1119. Mrityunjay Singh and Tara Shpargel, "Glenn Refractory Adhesive for Bonding and Exterior Repair (GRABER) Developed for Repairing Shuttle Damage" (Washington, DC: NASA/TM—2005-213419, 2005), *http://ntrs.nasa.gov*

1120. James Zakrajsek, "Return to Flight: Shuttle Actuators," (May 2005), *http://www.nasa.gov/centers/glenn/pdf/90864main_M-1587_actuators.pdf* (accessed 8 December 2015).

1121. Frank Sietzen and Keith L. Cowing, *New Moon Rising: The Making of America's New Space Vision and the Remaking of NASA* (Burlington, Ontario: Apogee Books, 2004).

1122. Sietzen, *New Moon Rising*.

1123. House Science Committee Hearing Charter, "The Future of Human Space Flight" (16 October 2003) *http://www.spaceref.com/news/viewsr.html?pid=10658* (accessed 8 December 2015).

1124. Michael D. Griffin, "Hearing on the Future of Human Space Flight Committee on Science" testimony (16 October 2003), *http://www.spaceref.com/news/viewsr.html?pid=10683* (accessed 8 December 2015).

1125. Wesley Huntress, testimony to Full Committee Hearing, "Future of NASA" (29 October 2003), *http://history.nasa.gov/columbia/Troxell/Columbia%20Web%20Site/Documents/Congress/Senate/OCTOBE~1/huntress_testimony.html* (accessed 8 December 2015).

1126. Lambright, "Leadership and Change at NASA."

1127. Sietzen, *New Moon Rising*.

1128. Lambright, "Leadership and Change at NASA."

1129. Sietzen, *New Moon Rising*.

1130. Bush, "The Vision for Space Exploration."

1131. "Dr. Julian M. Earls Selected To Lead NASA Glenn Research Center," 2008, NASA Headquarters

1132. Graham Warwick, "Playing With Fire," *Flight International* (3 March 2003).

1133. Randall Taylor, *Prometheus Project Final Report* (Washington, DC: NASA 982–R120461, 1 October 2005).

1134. John E. Foster, Tom Haag, and Michael Patterson, *The High Power Electric Propulsion (HiPEP) Ion Thruster* (Washington, DC: NASA/TM—2004-213194, 2004), *http://ntrs.nasa.gov*

1135. "Glenn Successfully Tests Ion Engine," *Aerospace Frontiers* (January 2004).

1136. "Electric Propulsion Technology Developed for Prometheus 1," *Research & Technology 2005* (Washington, DC: NASA/TM—2006-214016, 2006), *http://ntrs.nasa.gov*

1137. Lee Mason, "Dynamic Energy Conversion: Vital Technology for Space Nuclear Power," *Journal of Aerospace Engineering* 26 (2013): 352–360.

1138. Steven Oleson, et al., *Mission Advantages of NEXT: NASA's Evolutionary Xenon Thruster* (Washington, DC: NASA/TM—2002-211892, 2002), *http://ntrs.nasa.gov*

1139. Rohit Shastry, Daniel A. Herman, and George C. Soulas, *Status of NASA's Evolutionary Xenon Thruster (NEXT) Long-Duration Test as of 50,000 h and 900 kg Throughput* (Washington, DC: NASA GRC–E–DAA–TN11548, 6 October 2013).

1140. "Aerojet Rocketdyne Receives Contract to Continue Development of NASA's Evolutionary Xenon Gridded Ion Thruster System" (Sacramento, CA: Aerojet Rocketdyne, 6 April 2015) *http://www.rocket.com/article/aerojet-rocketdyne-receives-contract-continue-development-nasa%E2%80%99s-evolutionary-xenon-gridded* (accessed 2 December 2015).

1141. Dan Herman interview, Cleveland, OH, by Lindsey Frick, "What's the Difference Between Gridded Ion Thrusters and Hall Effect Thrusters" (21 August 2013), *http://www.engineeringtv.com/video/Whats-the-Difference-Between-Gr; Whats-the-Difference-Videos-Ser* (accessed 23 July 2015).

1142. John M. Sankovic, John A. Hamley, and Thomas W. Haag, *Performance Evaluation of the Russian SPT-100 Thruster at NASA LeRC* (Washington, DC: NASA TM–106401, 1993), http://ntrs.nasa.gov

1143. John M. Sankovic, Leonard H. Caveny, and Peter R. Lynn, The BMDO Russian Hall Electric Thruster Technology (RHETT) Program: From Laboratory to Orbit (Reston, VA: AIAA 97–2917, 1997).

1144. John M. Sankovic, *NASA Technology Investments in Electric Propulsion: New Directions in the New Millennium* (Washington, DC: NASA/TM—2002-210609, 2002). *http://ntrs.nasa.gov*

1145. I. G. Mikellides, G. A. Jongeward, B. M. Gardner, I. Katz, M. J. Mandell, and V. A. Davis, *A Hall-Effect Thruster Plume and Spacecraft Interactions Modeling Package* (Pasadena, CA: Conference Paper IEPC–01–251, 15 October 2001). *http://erps.spacegrant.org.uploads/*

images/images/iepc_articledownload_1988-2007/
2001index/251_2.pdf (accessed 2 December 2015).

1146. NASA Marshall Space Flight Center, "Discovery Program," *http://discovery.nasa.gov/missions.cfml* (accessed 4 December 2015).

1147. Hani Kamhawi, et al., *Overview of Hall Thruster Activities at NASA Glenn Research Center* (Wiesbaden, Germany: Conference Paper IEPC–2011–339, 11 September 2011).

1148. "New Hall Thruster Passes Performance Testing," *Aerospace Frontiers* (October 2005).

1149. Jacobson, NASA's 2004 Hall Thruster Program.

1150. "Tank 5: Tops in Electrical Propulsion Testing," *Aerospace Frontiers* (January 2002).

1151. Kamhawi, *Overview of Hall Thruster Activities.*

1152. Ross to Arrighi.

1153. Ross to Arrighi.

1154. Paul Spudis, "The Vision for Space Exploration: A Brief History (Part 2)," Spudis Lunar Resources Blog (26 October 2012), *http://www.spudislunarre-sources.com/blog/ the-vision-for-space-exploration-a-brief-history-part-2* (accessed 18 December 2015).

1155. Jennifer L. Rhatigan, Jeffrey M. Hanley, and Mark S. Geyer, "Formulations of NASA's Constellation Program," *Constellation Program Lessons Learned, Volume I* (Washington, DC: NASA SP–2011–6127–VOL–1, 2011), *http://ntrs.nasa.gov*

1156. "President Bush Offers New Vision for NASA," *Aerospace Frontiers* (2 February 2004).

1157. Bush, "The Vision for Space Exploration."

1158. Lambright, "Leadership and Change at NASA."

1159. Paul Spudis, "The Vision for Space Exploration: A Brief History (Part 3)," Spudis Lunar Resources Blog (2 November 2012), *http://www.spudislunarresources.com/blog/ the-vision-for-space-exploration-a-brief-history-part-3* (accessed 18 December 2015).

1160. Becky Gaylord, "Glenn Could Get Work on Shuttle Replacement But Management Needs Improvement, Reports Say," *Cleveland Plain Dealer* (12 February 2006).

1161. Doreen Zudell, "Glenn Brings CEV Concept to Life Through Full-Scale Model," *Aerospace Frontiers* (February 2006).

1162. "Constellation Program: America's Spacecraft for a New Generation of Explorers: The Orion Crew Exploration Vehicle," *NASA Facts* (Washington, DC: NASA FS–2006–08–022–JSC, 2006).

1163. Erik Seedhouse, *Lunar Outpost: The Challenges of Establishing a Human Settlement on the Moon* (Berlin, Germany: Springer Science + Business Media, 2009).

1164. Daniel Morgan, *Future of NASA: Space Policy Issues Facing Congress* (Washington, DC: Congressional Research Service, 14 January 2010).

1165. Bart Gordon, *Hearing Before the Committee on Science, House of Representatives, One Hundred Ninth Congress, First Session* (Washington, DC: NASA Earth Science, 28 April 2005).

1166. Becky Gaylord and Stephen Koff, "NASA Glenn to Lose 700 Jobs as Cleveland Holds Its Breath," *Cleveland Plain Dealer* (8 February 2005).

1167. Gaylord, "NASA Glenn to Lose 300 Jobs by the End of 2007," *Cleveland Plain Dealer* (7 February 2006).

1168. Gaylord, "NASA Glenn Bleeding Jobs; Prospects Are Poor," *Cleveland Plain Dealer* (21 October 2005).

1169. Gaylord, "NASA Glenn to Lose 300 Jobs."

1170. Gaylord, "NASA Glenn to Lose 300 Jobs."

1171. Doreen Zudell, "Glenn's Evolving Role Within the Vision for Space Exploration," *Aerospace Frontiers* (March 2005).

1172. Dhanireddy R. Reddy, "Seventy Years of Aeropropulsion Research at NASA Glenn Research Center," *Journal of Aerospace Engineering* (April 2013).

1173. Gaylord, "NASA Glenn Bleeding Jobs."

1174. Zudell, "Glenn's Evolving Role."

1175. John Mangels, "NASA's Science Programs in Jeopardy Missions To Put Humans in Space Lead to Deep Cuts," *Cleveland Plain Dealer* (28 May 2006).

1176. Larry Wheeler, "Scientists, Researchers Feel Pain of NASA Budget Cuts," *USA Today* (3 April 2006).

1177. Mangels, "NASA's Science Programs."

1178. Morrie Goodman, "NASA Administrator Names Woodrow Whitlow Associate Administrator for Mission Support," NASA press release 10–032, 3 February 2010, Press Release Collection, Cleveland, OH.

1179. Julian Earls interview, Cleveland, OH, by Rebecca Wright, 22 February 2006, NASA Glenn History Collection, Oral History Collection, Cleveland, OH.

1180. Becky Gaylord, "NASA Glenn Sets Its Sights on Space Travel—Director Works to Get New Office Off Ground," *Cleveland Plain Dealer* (20 April 2006).

1181. Gaylord, "Glenn Could Get Work on Shuttle Replacement."

1182. Woodrow Whitlow to Glenn Civil Service Employees, "Reorganization of the NASA Glenn Research Center-Update," 5 January 2007, NASA Glenn History Collection, Directors Collection, Cleveland, OH.

1183. Doreen Zudell, "Dr. Whitlow Shares First Step in Restructuring Plan," *Aerospace Frontiers* (June 2006).

1184. Ross to Arrighi.

1185. Zudell, "Dr. Whitlow Shares First Step."

1186. Doreen Zudell, "Glenn Leads, Supports CEV Efforts," *Aerospace Frontiers* (July 2006).

1187. Grant Segall, "Cleveland's Shot at the Moon Lunar Project Ensures Local NASA Jobs for a Decade," *Cleveland Plain Dealer* (31 October 2007).

1188. Grant Segall, "NASA Denies Flaws, Backs Ares Program for Flight to Station," *Cleveland Plain Dealer* (25 June 2008).

1189. Ross to Arrighi.

1190. *NASA's Exploration Systems Architecture Study* (Washington, DC: NASA/TM—2005-214062, 2005), *http://ntrs.nasa.gov*

1191. Robert Braeunig, "Rocket Propellants," Rocket & Space Technology (2008), *http://www.braeunig.us/space/propel.htm* (accessed 23 July 2015).

1192. Mark Klem, *Liquid Oxygen/Liquid Methane Propulsion and Cryogenic Advanced Development* (Paris, France: Conference Paper IAC–11–C4.1.5, 2011).

1193. Michael L. Meyer, et al., *Testing of a Liquid Oxygen/Liquid Methane Reaction Control Thruster in a New Altitude Rocket Engine Test Facility* (Washington, DC: NASA/TM—2012-217643, 2012), *http://ntrs.nasa.gov*

1194. Helmut H. Bamberger, R. Craig Robinson, and John Jurns, *Liquid Methane Conditioning Capabilities Developed at the NASA Glenn Research Center's Small Multi-Purpose Research Facility (SMiRF) for Accelerated Lunar Surface Storage Thermal Testing* (Washington, DC: NASA/CR—2011-216745, 2011), *http://ntrs.nasa.gov*

1195. Klem, *Liquid Oxygen/Liquid Methane.*

1196. "Introduction," *Research & Technology 2007* (Washington, DC: NASA/TM—2008-215054, 2008), *http://ntrs.nasa.gov*

1197. John Mangels, "Lost in Space: NASA Glenn Work in Limbo Shift in Agency Mission Alters Project Priorities," *Cleveland Plain Dealer* (26 March 2010).

1198. "Glenn Tests Planetary Excavation Tools and Vehicles," *Aerospace Frontiers* (February 2011).

1199. "Constellation Program: Ares I-X Flight Test Vehicle," *NASA Facts* (Washington, DC: NASA FS-2008-04-142-LaRC, 2008).

1200. Doreen Zudell, "Glenn's In-House Talents Utilized for Ares I-1 Flight Test Vehicle," *Aerospace Frontiers* (December 2006).

1201. "Ares I-X" Press Kit, October 2009, NASA Glenn History Collection, Constellation Collection, Cleveland, OH.

1202. "Ares I-X Segments Leave Center, Bound for Space Flight," *Aerospace Frontiers* (November 2008).

1203. Tori Woods, "Glenn Helps Ares I-X Soar" (17 November 2009), *http://www.nasa.gov/centers/glenn/moonandmars/grc_helps_aresIX.html#* (accessed 6 August 2015).

1204. "Charles F. Bolden, Jr., NASA Administrator (July 17, 2009 – present)" (July 2013), *http://www.nasa.gov/content/charles-f-bolden-jr-nasa-administrator-july-17-2009-present/* (accessed 2 April 2015).

1205. Paul Spudis, "The Vision for Space Exploration: A Brief History (Part 4)," Spudis Lunar Resources Blog (8 November 2012), *http://www.spudislunarresources.com/blog/the-vision-for-space-exploration-a-brief-history-part-4* (accessed 18 December 2015).

1206. Review of Human Spaceflight Plans Committee, "Seeking a Human Spaceflight Program Worth of a Great Nation" (October 2009), *http://www.nasa.gov/pdf/396093main_HSF_Cmte_FinalReport.pdf* (accessed 3 April 2015).

1207. Office of Inspector General, *Status of NASA's Development of the Multi-Purpose Crew Vehicle* (Washington, DC: Report No. IG–13–022, 15 August 2013).

1208. Marcia Smith, "NASA's Project Constellation and the Future of Human Spaceflight," SpacePolicy Online.com Fact Sheet (23 February 2011), *http://www.spacepolicyonline.com/images/stories constellation_fact_sheet_jan_2011.pdf* (accessed 2 April 2015).

1209. Woods, "New Assignments," *http://www.nasa.gov/centers/glenn/exploration/new_glenn.html*

1210. Tom Breckenridge, "New Duties Cast NASA Glenn as Leader, Chief Says," *Cleveland Plain Dealer* (10 April 2010).

1211. Roger Bloom, "Aviation Safety Research is Alive at NASA," *Airline Pilot* (November/December 2001).

1212. Jaiwon Shin, "The NASA Aviation Safety Program: Overview" (Washington, DC: NASA/TM—2000-209810, 2000), *http://ntrs.nasa.gov*

1213. Shin, "The NASA Aviation Safety Program."

1214. *Achieving the Extraordinary: NASA Glenn Research Center at Lewis Field* (Cleveland, OH: B–1202, July 2006).

1215. *NASA Aviation Safety Program: Initiative Will Reduce Aviation Fatalities* (Washington, DC: NASA FS-2000-02-47-LaRC, 2000).

1216. "NASA Shares Collier Trophy Award for Aviation Safety Technologies," Langley Research Center press release, 29 May 2009.

1217. NASA Glenn Research Center, "WINCOMM (Weather Information Communications)" (31 December 2013), *https://acast.grc.nasa.gov/projects-completed/wincomm/* (accessed 30 May 2014).

1218. "Aviation Safety Program," *NASA Facts* (Washington, DC: NF–2011–04–535–HQ, 2011).

1219. Guy Norris and Michael Mecham, "Icing Thrust Loss Mystifies Researchers," *Aviation Week* (23 April 2012).

1220. Harold Addy and Joseph Veres, *An Overview of NASA Engine Ice-Crystal Icing Research* (Washington, DC: NASA/TM—2011-217254, 2011), *http://ntrs.nasa.gov*

1221. Thomas A. Griffin, Dennis J. Dicki, and Paul J. Lizanich, *PSL Icing Facility Upgrade Overview* (Reston, VA: AIAA 2014–2896, 2014).

1222. Nancy Kilkenny, "Frigid Heat: How Ice Can Menace a Hot Engine" (14 May 2013), *http://www.nasa.gov/centers/glenn/aeronautics/frigid_heat.html* (accessed 6 August 2015).

1223. Timothy Gushanas, "About HRP," 14 July 2015, *https://www.nasa.gov/hrp/about* (accessed 21 December 2015).

1224. "NASA Cleveland Biomedical Assets," Section VI, 1996, NASA Glenn History Collection, Cleveland, OH.

1225. "The John Glenn Biomedical Engineering Consortium Helping Astronauts, Healing People on Earth," NASA B–1036–01, May 2002, NASA Glenn History Collection, Directors Collection, Cleveland, OH.

1226. Lealem Mulugeta, "Digital Astronaut Project," OpenNASA (25 May 2012), *http://www.opennasa.org/digital-astronaut-project.html* (accessed 18 December 2015).

1227. NASA, Exploration Medical Capability" (6 February 2013), *https://www.nasa.gov/exploration/humanresearch/elements/research_info_element-exmc.html* (accessed 18 December 2015).

1228. "About the Space Station: Facts and Figures" (2015), *http://www.nasa.gov/mission_pages/station/main/onthestation/facts_and_figures.html* (accessed 1 May 2015).

1229. "Unlocking Mysteries in Microgravity," *NASA Facts* (Washington, DC: FS–1999–07–007–GRC, July 1999).

1230. Lauren M. Sharp, Daniel L. Dietrich, and Brian J. Motil, "Microgravity Fluids and Combustion Research at NASA Glenn Research Center," *Journal of Aerospace Engineering* 26 (April 2013): 439–450.

1231. Tim Reckhart, "ISS Fluids & Combustion Facility" (20 April 2015), *https://spaceflightsystems.grc.nasa.gov/sopo/ihho/psrp/fcf/* (accessed 20 April 2015).

1232. "Unlocking Mysteries."

1233. Tim Reckart, "ISS Fluids and Combustion Facility" (26 June 2014), *https://spaceflightsystems.grc.nasa.gov/sopo/ihho/psrp/fcf/print/* (accessed 17 July 2015).

1234. Tim Reckart, "FIR" (27 June 2014), *https://spaceflightsystems.grc.nasa.gov/sopo/ihho/psrp/fcf/cir/print/* (accessed 17 July 2015).

1235. Space Operations Project Office, "SCaN Testbed," *SpaceNews* (25 June 2014), *http:/spaceflightsystems.grc.nasa.gov/SOPO/SCO/SCaNTestbed/* (accessed 1 February 2015).

1236. John Mangels, "NASA Glenn-Designed Space Hardware Will Test New Communication Technologies," *Cleveland Plain Dealer* (11 February 2012).

1237. Mangels, "NASA Glenn-Designed Space Hardware."

1238. "Ramon Lugo III" (August 2010), *https://www.nasa.gov/centers/glenn/about/bios/lugo_bio.html* (accessed 1 April 2015).

1239. Jason Rhian, "President Signs NASA 2010 Authorization Act," Universe Today (11 October 2010), *http://www.universetoday.com/75522/president-signs-nasa-2010-authorization-act/* (accessed 1 April 2015).

1240. Sabrina Eaton, "NASA Glenn Anxiously Awaits News in Budget Plan," *Cleveland Plain Dealer* (14 February 2011).

1241. Sabrina Eaton, "NASA Glenn Would Get More Funding, But Social Programs Hit," *Cleveland Plain Dealer* (15 February 2011).

1242. Richard Flaisig memorandum for the record, "LTID Service Changes Due to Staffing Reductions," 8 April 2011, NASA Glenn History Collection, Directors Collection, Cleveland, OH.

1243. Doreen Zudell, "Lugo Outlines Areas Impacted by Budget," *Aerospace Frontiers* (September 2012).

1244. Ramon Lugo, "Positioning Glenn to Be Competitive," *Aerospace Frontiers* (April 2012).

1245. Chuck Soder, "NASA Glenn Research Center Director Ramon Lugo III To Exit Post at Year's End," *Crain's Cleveland Business* (16 November 2012).

1246. "Biography of James M. Free" (3 January 2013), *https://www.nasa.gov/centers/glenn/about/bios/ free_bio.html* (accessed 1 April 2015).

1247. Leone, "Profile: James Free."

1248. Bill Berry, "Comments at the 2015 Glenn Hall of Fame Induction Ceremony," 25 September 2015, NASA Glenn History Collection, Cleveland, OH.

1249. "NASA Glenn Inaugurates Hall of Fame to Honor Exemplary Employees" (16 September 2015), *http://www.nasa.gov/feature/nasa-glenn-inaugurates-hall-of-fame-to-honor-exemplary-employees* (accessed 18 December 2015).

1250. Jim Free, "Address at the 2015 Glenn Hall of Fame Induction Ceremony," 25 September 2015, NASA Glenn History Collection, Directors Collection, Cleveland, OH.

*Image 409: Glenn's S-3 Viking (GRC–2013–C–02281).*

# The Future

What a journey that was through the history and achievements of NASA Lewis and NASA Glenn!

Today Glenn remains deeply connected with its roots in aircraft propulsion and spacecraft propulsion while continuing to lead the nation in increasing the capabilities of our space missions and taking us farther than humans have ever gone. Our advancements in materials, electric propulsion, space communications, and microgravity science will propel NASA, the country, and the world to well beyond the Earth.

We stand here today, poised for the future, based on the great minds that came before us during these 75 years. We excel because of the people that embody the spirit of NASA and because of our incredible tools and facilities.

Thank you for sharing in the experience of our first 75 years through the words and images of this book. The NASA Glenn Research Center will always be working to Dream Big!

*Jim Free*
Director, Glenn Research Center

Image 410: William "Eb" Gough poses on the lab's McDonnell F2H-2B Banshee in February 1958 in front of the hangar. Gough had served as the lab's chief pilot since 1945. In seven months the NACA would transition into NASA. Gough left shortly thereafter as Lewis phased out its aeronautics work for nearly a decade (GRC–1958–C–47086).

# Acknowledgments

$\mathcal{T}$his publication would not have been possible without the determined advocacy of Anne Mills with center management and the NASA Headquarters History Office. Anne's support has been extraordinary throughout this effort. I want to thank Jim Free for backing this effort and taking time to contribute the Foreword and Concluding Remarks. Many thanks to Howard Ross for his input on the growth of Lewis's microgravity program in the late 1980s and into its apex in the 1990s. The efforts of Robyn Gordon and Richard Flaisig in this process are also appreciated.

I also want to thank fellow Alcyon Technical Services employees on the Technical Information, Administrative, and Logistics Services 2 (TIALS 2) contract for their indispensable contributions. Donald Reams, Caroline Rist, and Jaime Scibelli provided oversight for the project. Nancy O'Bryan has done an exceptional job editing the text and organizing the endnotes. Kelly Shankland was responsible for the excellent layout of the pages and cover. Lorraine Feher performed the daunting task of formatting the citations, proofreading the text, and checking the index. Deborah Demaline provided access to important documents prior to their shipment to the National Archives. Marvin Smith, Mark Grills, and Bridget Caswell scanned and catalogued many of the images. In addition, Teresa Hornbuckle provided images from the Langley Photograph Collection, and Colin Fries at Headquarters assisted with portions of the research.

Thanks to Bill Barry and Steve Garber at the NASA Headquarters History Office for supporting this project and coordinating the peer reviews. I am especially appreciative to Howard Ross, Thomas St. Onge, Larry Ross, and the anonymous peer reviewers whose comments helped transform this work into a more cohesive history. Larry Viterna, Simon Ostrach, Erwin Zaretsky, Michael Patterson, and David Jacobson graciously reviewed sections of the book and offered key insights.

Finally, I want to thank my wife Sarah. Completion of this project could not have been possible without her love and encouragement.

Image 411: Technicians prepare a spherical liquid-hydrogen tank in June 1960 to study the fluid's behavior in low gravity during a flight on the center's AJ–2 Savage aircraft (GRC–1960–C–53861).

# Bibliographic Essay

The materials used to create this publication can be broken into five categories—publications, reports, newspapers, interviews, and primary documents. The majority of these materials come from the Glenn History Collection.

Numerous secondary resources from the NASA History Series and external journals were consulted for contextual information. Not surprisingly Virginia Dawson's *Engines and Innovation* and George Gray's *Frontiers of Flight* were used extensively. A 2013 special issue of the *Journal of Aerospace Engineering* dedicated to NASA Glenn was particularly useful. Researchers from across the center wrote brief histories of their particular field for the issue. Other key publications were John Sloop's *Liquid Hydrogen as Propulsion Fuel*, Alex Roland's *Model Research*, Dawson and Mark Bowles's *Taming Liquid Hydrogen*, and the 2011 paper "Role of NASA in Photovoltaic and Wind Energy" by Sheila Bailey, Larry Viterna, and K. R. Rao.

Although no interviews were conducted specifically for this publication, transcripts from 45 previous interviews in the Glenn History Collection provided many key details and insights. In addition, transcripts of the technical talks given by researchers during the NACA Inspections were extremely valuable in adding context to the technical reports in the Aeronautics and Space Database's massive archive. The History Collection, which includes the entire run of center newspapers from 1942 to the present, provided many of the details not available in other sources.

The Glenn History Collection also includes past correspondence from the Director's Office, the personal papers of a number of key employees, program files, and other documents that provided the basis for much of this publication. The collection contains research materials from previous book projects, including *Engines and Innovation, Liquid Hydrogen* as *Propulsion Fuel, Revolutionary Atmosphere, Taming Liquid Hydrogen*, and *Pursuit of Power*. In addition I was able to review a new batch of directors' materials that were essential to the final three chapters.

Of course, the Glenn Imaging Technology Center's extensive photograph collection was essential to this book, not just in providing the hundreds of photographs, but also in helping to date activities and identify individuals.

Image 412: Assembly of experimental wind turbine parts in the Engine Research Building during March 1979 (GRC–1979–C–01034)

# Glenn History Resources

## Publications
(Available at: http://history.nasa.gov/series95.html)

Virginia Dawson, *Engines and Innovation* (Washington, DC: NASA SP–4306, 1991).

William Leary, *We Freeze to Please: A History of NASA's Icing Research Tunnel and the Quest for Flight Safety* (Washington, DC: NASA SP–2002–4226, 2002).

Mark Bowles and Robert Arrighi, *NASA's Nuclear Frontier: The Plum Brook Reactor Facility* (Washington, DC: NASA SP–2004–4533, 2004).

Virginia Dawson and Mark Bowles, *Taming Liquid Hydrogen: The Centaur Upper Stage Rocket 1958–2002* (Washington, DC: NASA SP 2004–4230, 2004).

Virginia P. Dawson and Mark D. Bowles, editors, *Realizing the Dream of Flight: Biographical Essays in Honor of Centennial of Flight* (Washington, DC: NASA SP–2005–4112, 2005).

Virginia Dawson, *Ideas Into Hardware: A History of the Rocket Engine Test Facility at the NASA Glenn Research Center* (Washington, DC: NASA, 2004).

Mark Bowles, *Science in Flux: NASA's Nuclear Program at Plum Brook Station 1955–2005* (Washington, DC: NASA SP–2006–4317, 2006).

Robert Arrighi, *Revolutionary Atmosphere: The Story of the Altitude Wind Tunnel and Space Power Chambers* (Washington, DC: NASA SP–2010–4319, 2010).

Robert Arrighi, *Pursuit of Power NASA's Propulsion Systems Laboratory No. 1 and 2* (Washington, DC: NASA SP–2012–4548, 2012).

## Video Documentaries
(Available through the history office)

Of Ashes and Atoms—An Emmy-nominated full-length documentary on the Plum Brook Reactor Facility

Fueling Space Exploration—A documentary on the Rocket Engine Test Facility (RETF)

A Tunnel Through Time: The History of NASA's Altitude Wind Tunnel—A history of the Altitude Wind Tunnel and Space Power Chambers

## Websites and Multimedia

NASA Glenn History Office—A website gateway to the history of Glenn Research Center and Plum Brook Station. *http://grchistory.grc.nasa.gov/*

Rocket Engine Test Facility—A website designed to preserve the legacy of the RETF. Includes photographs and videos of RETF and interactive lessons on rocket engine testing. *http://retf.grc.nasa.gov/*

Altitude Wind Tunnel and Space Power Chambers—An interactive look at what was one of NASA's most significant wind tunnels. Includes images, videos, reports, timelines, and materials for students. *http://awt.grc.nasa.gov/*

Propulsion Systems Laboratory—A website dedicated to PSL's history and facility. Includes related documents and reports, an interactive layout, and students' section. *http://pslhistory.grc.nasa.gov/*

Lessons of a Widowmaker (and Other Aircraft)—An interactive history of NASA's research aircraft, focusing primarily on former NASA Glenn vehicles. Includes images, histories, and specifications for 17 aircraft, as well as educational materials and videos. *http://www.nasa.gov/externalflash/aero/*

B1 and B3 Test Stands—A website providing the history and physical description of the B1 and B2 Test Stands at Plum Brook Station. Includes photographs, documents, timelines, and the 100-plus page Historic American Engineering Record (HAER) Report. *http://pbhistoryb1b3.grc.nasa.gov/*

Glenn—The Early Years—A photographic retrospective of the first 40 years of operation at the center (1941–79). Drawn from a large collection of archived photographs, this gallery of images along with in-depth descriptions offers a unique look into the research, people, and activities that have generated numerous technological advancements over the years. Available for both the Web and iPad *http://www.nasa.gov/centers/glenn/multimedia/grchistory.html*

The NACA Inspections—A site serving as a repository for images, presentations, media materials, correspondence, and other documents from the NACA's post-war Inspections. The materials provide a snapshot of the state of the NACA, its aeronautical research, and its major personalities during this era. *http://grchistory.grc.nasa.gov/inspections/*

Glenn's NACA-Era Aircraft—A website providing images, histories, and specifications for 11 aircraft used at Glenn between 1943 and 1958. *http://www.nasa.gov/externalflash/NACA/*

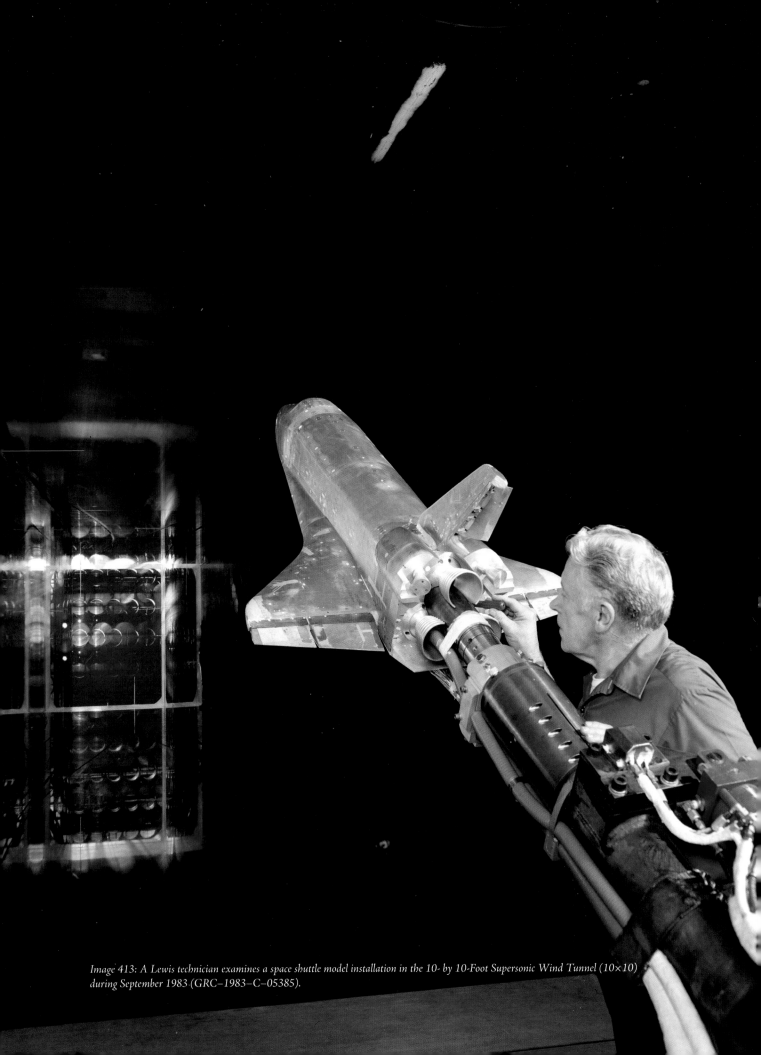

Image 413: A Lewis technician examines a space shuttle model installation in the 10- by 10-Foot Supersonic Wind Tunnel (10×10) during September 1983 (GRC–1983–C–05385).

# List of Images

Image 31:     Westward view of the Steam Plant and general AERL construction area in 1942 (GRC–2007–C–02309).

Image 32:     Ray Sharp at his desk in 1942 (GRC–2015–C–06568).

Image 33:     NACA contract with the Emerson Company.

Image 34.     In May 1942 the Prop House became the first operating facility at the AERL. It contained four test cells designed to study large reciprocating engines. Researchers tested the performance of fuels, turbochargers, water-injection, and cooling systems here during World War II. The facility was also used to investigate a captured German V–I buzz bomb during the war (GRC–1942–C–01134).

Image 35.     The laboratory established its own fire department while the lab was being constructed in the early 1940s. The group, which was based at the Utilities Building, not only responded to emergencies but conducted safety inspections, checked fuel storage areas, and supervised evacuation drills and training. In addition, they frequently assisted local fire departments and responded to accidents at the adjacent Cleveland Municipal Airport (GRC–1943–C–04291).

Image 36:     Drafting staff members at work in the temporary hangar offices during 1942 (GRC–2015–C–06545).

Image 37:     Temporary offices constructed inside the hangar to house the architectural and drafting personnel as well the machine shops (GRC–2015–C–06557).

Image 38:     Program for the initiation of research of the AERL.

Image 39:     AERL mechanics work on an engine installation on one of the Prop House's test stands in 1943 (GRC–1943–C–03349).

Image 40:     Receptionist Mary Louise Gosney enjoys the new Administration Building in July 1943. She started at the lab in November 1941 and spent an entire year in the hangar. She also served as the lab's clearance officer and would later head the Administrative Services Division (GRC–1943–C–01842).

Image 41:     *Wing Tips* kept the staff updated on the progress of the construction.

Image 42:     Construction of the AWT in April 1943. The facility would be up and running in eight months (GRC–2008–C–00817).

Image 43:     Cartoon showing AWT's ability to simulate altitude conditions (*Wing Tips*).

Image 44:     Aerial view of the AERL in June 1945 (GRC–1945–C–10493).

Image 45:     General Henry Arnold addressing Aircraft Engine Research Laboratory personnel in the hangar on 9 November 1944. Arnold was only at the lab for a few hours, but he managed to take a comprehensive tour that included the Jet Propulsion Static Laboratory, a turbojet engine run in the Altitude Wind Tunnel, and the testing of a carburetor for the B–29 Superfortress in the Engine Research Building (GRC–1944–C–07493).

Image 46:     A Fuels and Lubrication Division researcher at work in August 1943 (GRC–1943–C–02124).

Image 47:     AERL recruiting pamphlet.

Image 48:     AERL staff members wish Ray Sharp happy holidays in December 1945 (GRC–1945–C–13948).

Image 49:     Addison Rothrock's speech to AERL staff in December 1942.

Image 50:     P–39 Mustang fuselage being tested in the IRT in October 1944 (GRC–1944–C–7062).

Image 51:     View from the control room during a June 1944 engine cooling investigation in an Engine Research Building test cell (GRC–1944–C–05498).

Image 52:     PBOW staff gather for an October 1941 ceremony in front of the PBOW Administration Building (GRC–2015–C–06563).

Image 53:     One of three acid-producing facilities at the PBOW. The acid was used to manufacture TNT and DNT, c1941 (GRC–2015–C–06540).

Image 54:     PBOW fire station located near Taylor and Columbus avenues, c1941 (GRC–2015–C–06564).

Image 55:     *PBOW News* feature showing safety ceremony on a stage constructed with empty powder crates in May 1943.

Image 56:     "Service Stars" was a regular wartime column in the *Wing Tips* newsletter.

Image 57:     Women working alongside male colleagues in the AERL's Fabrication Shop (GRC–1944–C–05380).

Image 58:    Zella Morowitz worked in the AERL design office at Langley prior to her transfer to Cleveland in 1941, where she served as Ray Sharp's secretary for six years (GRC–2015–C–06555).

Image 59:    Memphis Belle crew Robert Hanson, Vincent Evans, and Charles Leighton; AERL Manager Raymond Sharp; Robert Morgan; William Holliday of the Cleveland Chamber of Commerce; Army Liaison Officer Colonel Edwin Page; Airport Commissioner John Berry; Cecil Scott; John Quinlan; and James Verinis. Kneeling are Harold Loch, Casimer Nastal, and Charles Winchell (GRC–1943–C–01870).

Image 60:    Researchers work with test setups in a Fuels and Lubrication Building lab room during March 1943 (GRC–1943–C–01370).

Image 61:    Bisson with physicist and mathematician Lucien C. Malavard (GRC–1949–C–24300).

Image 62:    Reprint of "Fuels Talk" by Bisson.

Image 63:    A representative from the Allison Engine Company instructs AERL mechanics on the operation of a basic Allison powerplant. The staff was taught how to completely disassemble and reassemble the engine components and systems (GRC–1943–C–03045).

Image 64:    An AERL researcher demonstrates the improved fuel injection system for the R–3350 engine at a tour stop in the Engine Research Building in June 1945 (GRC–1945–C–10678).

Image 65:    Wright R–3350 installed in the AWT test section on 4 July 1944 (GRC–1944–C–05554).

Image 66:    *NACA Wartime Reports* were classified until after the war.

Image 67:    A B–29 bomber on display in the hangar during June 1945 (GRC–1944–C–10587).

Image 68:    P–38 Lightning fuselage in the IRT during March 1945 (GRC–1945–C–8832).

Image 69:    On 13 September 1944 an AERL technician prepares for the initial test run in the IRT (GRC–1944–C–06552).

Image 70:    A P–38J Lightning in front of the blower on the hangar apron that was utilized as a crude rain-simulating device. The blower was used extensively during the 1940s to supplement wind tunnel and flight research data (GRC–1945–C–09650).

Image 71:    The Jet Static Propulsion Laboratory as it nears completion in August 1943. The secret facility was officially called the Supercharger Laboratory to disguise its true nature (GRC–2015–06544).

Image 72:    The secret test of the Bell YP–59A Airacomet in the spring of 1944 was the first investigation in the new AWT. The Airacomet, which was powered by two General Electric I–A centrifugal turbojets, was the first U.S. jet aircraft (GRC–1944–C–04830).

Image 73:    Abe Silverstein, Head of the AWT, discusses the tunnel's research during the war, concentrating on the several General Electric and Westinghouse jet engines that were studied (GRC–1945–C–10661).

Image 74:    Lockheed's YP–80A, powered by two General Electric I–40 turbojets, in the AWT test section in March 1945. The P–80 was the first U.S. aircraft to fly faster than 500 mph (GRC–1945–C–09576).

Image 75:    AWT wartime test schedule.

Image 76:    On 8 May 1945 the staff awoke to news that Germany had surrendered. A mid-morning ceremony was held at the Administration Building, but work at the lab continued on (GRC–1945–C–09905).

Image 77:    Final issue of the *Plum Brook News* from August 1945.

Image 78:    An AERL metallurgist examines a supercharger in January 1944 (GRC–1944–C–03814).

Image 79:    Mechanics lower an inlet duct for a Westinghouse J40 engine into the Altitude Wind Tunnel's 20-foot-diameter test section (GRC–1951–C–28463).

Image 80:    A mechanic inspects a General Electric I–40 turbojet engine. The lab had begun investigating jet engines during the war, but the "big switch" to jet propulsion began in October 1945 (GRC–1946–C–15674).

Image 81:    Kathryn "Nicki" Crawford demonstrates that there are sufficient coins in the bucket to match her weight in 1946. The group of mechanics contributed the money to celebrate Crawford's upcoming marriage to their colleague Bill Harrison, who had recently returned from the Army Air Corps. The Harrisons spent the next 66 years together (GRC–2015–C–06814).

Image 82:    1949 advertisement seeking to sell the Plum Brook Ordnance Works.

Image 83:    Interior of 1 of the 99 Plum Brook bunkers that were used to store crates of trinitrotoluene (TNT) and dinitrotoluene (DNT) during the war, c1941 (GRC–2015–C–06565).

Image 84:    The lab's new management team—Addison Rothrock (left) and Raymond Sharp (center)—with NACA Director of Research George Lewis (GRC–1945–C–12029).

Image 85:    *Wing Tips* article about AERL's reorganization.

Image 86:    NACA Secretary John Victory (left) and Ray Sharp (right) lead General Dwight Eisenhower on a tour of the Cleveland lab on 11 April 1946. The former supreme commander of Allied Expeditionary Forces in Europe was visiting several U.S. cities at the time (GRC–1946–C–14688).

Image 87:    Page from a photo album of visitors to the Administration Building in the postwar years (GRC–1951–C–27147).

Image 88:    Tour stop schedule for the 1947 Inspection.

Image 89:    George W. Lewis (GRC–2015–C–06556).

Image 90:    Myrtle Lewis with her sons George, Jr., and Harvey during an October 1951 visit to the laboratory (GRC–1951–C–28570).

Image 91:    A mechanic examinines compressor blades on a General Electric J47 engine (GRC–1949–C–22850).

Image 92:    A failure of a Westinghouse J34 engine in the AWT test section (GRC–1950–C–26294).

Image 93:    A mechanic with a fire extinguisher watches the firing of twin afterburners (GRC–1949–C–23744).

Image 94:    Mechanics install a turbojet engine in a Four Burner Area test cell (GRC–1950–C–25120).

Image 95:    Aircraft mechanics work on an early jet aircraft in the hangar (GRC–1946–C–14739).

Image 96:    Page from a compiled apprentice roster.

Image 97:    Some apprentices take a break from their studies to pose for a photograph. Only 150 of the 2,000 hours of annual training were spent in the classroom (GRC–1956–C–43227).

Image 98:    A Consolidated B–24D Liberator (left), Boeing B–29 Superfortress (background), and Lockheed RA–29 Hudson (foreground) parked inside the hangar. A P–47G Thunderbolt and P–63A King Cobra are visible in the background (GRC–1944–C–05413).

Image 99:    Lewis pilot Howard Lilly poses with his P–63 King Cobra, which he flew in the 1946 National Air Races (GRC–2015–C–06813).

Image 100:   AERL pilots during the final days of World War II: from left to right, Joseph Vensel, Howard Lilly, William Swann, and Joseph Walker. William "Eb" Gough joined the group months after this photograph. Vensel, a veteran pilot from Langley, was the Chief of Flight Operations and a voice of reason at the laboratory. In April 1947 Vensel was transferred to lead the new Muroc Flight Tests Unit in California until 1966 (GRC–1945–C–11397).

Image 101:   A flight research member examines instrumentation in the B–24D during a 1945 icing flight (GRC–1945–C–10377).

Image 102:   NACA memo authorizing icing flight tests of jet engines.

Image 103:   Abe Silverstein measures ice buildup on the Westinghouse J34 engine (GRC–1948–C–20836).

Image 104:   The XB–25E Mitchell searches for icing clouds in January 1947. The aircraft dubbed "Flamin Mamie," includes nose art depicting a fiery woman chasing off icing researchers (GRC–1947–C–17763).

Image 105:   The wooded picnic grounds as it appeared in August 1945. The area was improved in the ensuing years (GRC–1945–C–12065).

Image 106:   Harold Mergler with his differential analyzer (GRC–1951–C–27875).

Image 107:   A computer at work in one of the three offices on the second story of the 8- by 6-Foot Supersonic Wind Tunnel (8×6) office building. The largest room housed approximately 35 women with advanced mathematical skills (GRC–1954–C–35057).

Image 108:   The Farm House as it appeared shortly after the NACA took over in 1941 (GRC–2011–C–00345).

Image 109:   The Administrative Services Building after the modifications (GRC–1946–C–15355).

Image 110:   The Administrative Services Building after it was moved behind the Administration Building (GRC–1967–C–01234).

Image 111:   Harold Friedman with an 8-inch-diameter ramjet model (GRC–1949–C–23083).

Image 112:   Construction of the lab's first supersonic tunnel. Eventually the building would house three small supersonic tunnels (GRC–1945–C–10764).

Image 113:   A technician operates a Schlieren camera to view the airflow dynamics inside the 24- by 24-inch test section of one of the Stack Tunnels (GRC–1949–C–24977).

Image 114:   John Evvard with a missile model in February 1957 (GRC–1957–C–44223).

Image 115:   A B–29 bomber that was modified to serve as a ramjet testbed for Lewis researchers. The experimental ramjet was lowered from the bomb bay and fired (GRC–1948–C–21990).

Image 116:   A North American XF–82 Twin Mustang prepares for flight with a ramjet missile under its right wing (GRC–1949–C–23330).

Image 117:   The seven-stage axial compressor that powers the 8×6. The compressor was driven by three electric motors with a total output of 87,000 horsepower, resulting in airspeeds from Mach 0.36 to 2.0 (GRC–1949–C–23277).

Image 118:   A researcher inspects a 16-inch-diameter ramjet engine in the 8×6 test section. Researchers studied the ramjet's performance at different speeds and varying angles of attack. The engine performed well, and the findings correlated with nonfueled studies in the smaller wind tunnels (GRC–1950–C–25776).

Image 119:   A researcher prepares a jet-assisted take off (JATO) rocket for a combustion study at the Rocket Lab (GRC–1945–C–10724).

Image 120:   Firing of a nitric acid aniline JATO rocket at the Rocket Lab in March 1946. The Rocket Lab was expanded over the next 10 years and eventually included its own hydrogen liquefier (GRC–1946–C–14478).

Image 121:   John Sloop demonstrates a small rocket setup in Cell 4 of the Rocket Lab (GRC–1947–C–19769).

Image 122:   Proceedings from the NACA Conference on Fuels.

Image 123:   A ramjet installation in the 8×6 in May 1949 (GRC–1949–C–23522).

Image 124:   A 5,000-pound-thrust rocket engine is fired from the Rocket Lab's Cell 22 in January 1955. The series of tests proved to be Lewis's first successful liquid-hydrogen/liquid-oxygen runs (GRC–1955–C–37428).

Image 125:   Technicians install an experimental hypersonic test missile on the NACA's McDonnell F2H–2B Banshee in August 1957. Lewis pilots launched the missiles over the Atlantic Ocean at Wallops Island (GRC–2015–C–06812).

Image 126:   Sharp and Silverstein share a moment in 1958 (GRC–2015–C–06570).

Image 127:   Memo announcing Silverstein's promotion.

Image 128:   Lewis researchers Harold Mirels, Franklin Moore, Stephen Maslen, and Simon Ostrach in September 1987 celebrating Maslen's induction into the National Academy of Engineering (GRC–2015–C–06552).

Image 129:   Robert Deissler receives an NACA Exceptional Service Award from NACA Director Hugh Dryden in October 1957. Deissler was cited for "achieving significant scientific results in the solution of fluid flow and heat-transfer problems associated with aircraft nuclear propulsion" (GRC–1957–C–46286).

Image 130:   *Thermal Radiation Heat Transfer.* Bob Siegel and John Howell started putting together notes for in-house classes to teach fellow employees about heat transfer in the early 1960s. The researchers fleshed out the information and published it in 1968 as NASA SP–164. Siegel and Howell updated the Special Publication (SP) three times, then published it as a textbook in the mid-1970s. The textbook has become the standard heat transfer textbook. It has been translated into numerous languages and was recently issued in its sixth edition.

Image 131:   Frank Rom was one of Lewis's chief nuclear propulsion researchers. He designed nuclear aircraft, pursued tungsten-based reactors for the nuclear rocket program, and helped design the Plum Brook Reactor Facility (GRC–1957–C–43739).

Image 132:   A c1956 roster of participants in Lewis's in-house "nuclear school" and the branches from which they came.

Image 133:   The General Electric-designed cyclotron in the extended basement of the Materials and Stresses Building (GRC–1957–C–45988).

Image 134:   Ray Sharp and Congressman Albert Baumhart break ground for the Plum Brook Reactor Facility in September 1956. The pick and shovel were the same as those used for the AERL ground-breaking in January 1941 (GRC–1956–C–43033).

Image 164: Nuclear propulsion display at the Parade of Progress Event at the Cleveland Public Auditorium in August 1964. There is a model of a nuclear spacecraft in the foreground and a Plum Brook Reactor Facility display behind (GRC–1964–C–71686).

Image 165: Lewis technicians examine a Centaur rocket in the Space Power Chambers shop (GRC–1964–C–71100).

Image 166: Bruce Lundin (left) and Walter Olson in May 1956 (GRC–1956–C–42155).

Image 167: Lundin's paper on the NACA's role in space, as marked up by Silverstein.

Image 168: During the last few months of the NACA's existence, its leadership made a final tour of its three research laboratories. The group arrived in Cleveland on 24 June 1958. At one of the stops Lewis mechanic Leonard Tesar demonstrated the machining of a 20,000-pound-thrust rocket engine for the group in the Fabrication Shop. From left to right, Associate Director Eugene Manganiello, researcher Edward Baehr, NACA Chairman James Doolittle, NACA Executive Secretary John Victory, NACA Committee member Frederick Crawford, Tesar, Lewis Director Ray Sharp, and mechanic Curtis Strawn (GRC–1958–C–48117).

Image 169: The official NASA seal.

Image 170: Jim Modarelli (GRC–1956–C–43683).

Image 171: The NASA logo, often referred to as the "meatball."

Image 172: Silverstein represents the new space agency on CBS's "Face the Nation" television program on 8 March 1959 (GRC–2015–C–06538).

Image 173: *Orbit* announcement of Lewis transfers to NASA Headquarters. During the transition to NASA, *Wing Tips* was designed and renamed *Orbit*. The name was changed the *Lewis News* in February 1964.

Image 174: Gale Butler examines the *Mercury* capsule's retrograde rockets prior to a test run inside the AWT (GRC–1960–C–53146).

Image 175: Technicians in the Fabrication Shop align the *Mercury* capsule afterbody with its pressure chamber in May 1959 (GRC–1959–C–50759).

Image 176: Lewis technicians and engineers prepare the Big Joe capsule for launch from Cape Canaveral (GRC–2009–C–02180).

Image 177: A mock-up *Mercury* capsule and escape tower rockets mounted in the AWT for testing in July 1960 (GRC–1960–C–53287).

Image 178: Lewis pilot Joe Algranti explains the MASTIF operation to Alan Shepard in February 1960. Shepard was the first astronaut to operate the MASTIF (GRC–1960–C–52706).

Image 179: The MASTIF was erected in the wide end of the AWT, where the nitrogen thrusters generated a series of loud hisses as they were fired (GRC–1959–C–51723).

Image 180: Displays at the November 1962 Space Science Fair at the Cleveland Public Auditorium (GRC–1962–C–62704).

Image 181: Lewis staff with one of the Spacemobile vehicles in October 1964 (GRC–C–1964–72829).

Image 182: Gene Manganiello (right) welcomes Wernher von Braun to Lewis in December 1959 (GRC–1959–C–52148).

Image 183: Recommendations of the Silverstein Committee regarding the use of liquid hydrogen in the Saturn upper stages.

Image 184: Saturn model installation in the 8×6 in September 1960 (GRC–1960–C–54466).

Image 185: Silverstein holds a 3 November 1961 press conference announcing additional recruiting efforts Over the previous months, the center had hired 135 new staff members, interviewed over 700 prospects, and had over 300 applications on file (GRC–1961–C–58359).

Image 186: Interior of the 20-foot-diameter vacuum tank in the Electric Propulsion Laboratory. The circular covers on the floor sealed the displacement pumps located beneath the chamber (GRC–1961–C–57748).

Image 187: The Guerin House (GRC–1964–C–72264).

Image 188: The J Site crew on the "portable" rig on 13 August 1960 before the first test at Plum Brook Station (GRC–2015–C–06550).

Image 189:    Initially, NASA let the PBOW structures stand. As more and more acres were acquired, however, workers began to destroy a large number of the buildings. In 1961 a local company was hired to raze all unusable structures and to dismantle three acid plants. However, a number of the nonmanufacturing structures were retained (GRC–2015–C–06565).

Image 190:    B Complex with the B–1 and B–3 test stands (GRC–1965–C–03012).

Image 191:    Three bulkheads were placed inside the AWT to create the SPC. The largest is seen here being inserted approximately where the wind tunnel fan was located (NASA C–1961–58551).

Image 192:    Interior of the SPC's 51-foot-diameter high-altitude test area inside the former AWT (GRC–1963–C–67001).

Image 193:    Lewis report from 1962 describing the history of electric propulsion and requirements for long-duration interplanetary missions.

Image 194:    *Space Electric Rocket Test I (SERT–I)* spacecraft and thrusters tested in EPL's Tank No. 3 in June 1964 (GRC–1964–C–70258).

Image 195:    Kaufman with his electron bombardment thruster in the early 1960s (GRC–2001–C–01603).

Image 196:    Technicians prepare the SERT I spacecraft in an EPL cleanroom in February 1964 (GRC–1964–C–68553).

Image 197:    An unfueled Kiwi B–1–B reactor and its Aerojet Mark IX turbopump being prepared for installation in the B–3 test stand (GRC–1967–P–01289).

Image 198:    The B–1 and B–3 test stands, 135 and 210 feet tall, could test different components of high-energy rocket engines under flight conditions (GRC–1964–C–01310).

Image 199:    Ron Roskilly demonstrates the testing of a hydrogen turbopump (right) at the Rocket Systems Area's A Site (left) (GRC–1962–C–61077).

Image 200:    Bob Siegel with a test rig for high-speed filming of liquid behavior in microgravity in the 8×6 tunnel (GRC–1960–C–54149).

Image 201:    Fred Haise, Lewis pilot and future *Apollo* astronaut, monitors the cameras and instrumentation for the experimental liquid-hydrogen container in the bomb bay of the AJ–2 aircraft (GRC–1960–C–54979).

Image 202:    An Aerobee rocket being prepared for launch in January 1961 (GRC–1961–C–55686).

Image 203:    *Orbit* article on the recovery of an Aerobee telemetry unit from the Atlantic Ocean.

Image 204:    A NASA AJ–2 Savage makes a pass for cameramen at the Cleveland Municipal Airport in November 1960. The AJ–2 was a Navy-carrier-based bomber in the 1950s (GRC–1960–C–54979).

Image 205:    Memo authorizing the transfer of Centaur to Lewis in 1962.

Image 206:    Atlas booster being hoisted into the Dynamics Stand at Plum Brook Station. The Atlas and Centaur were tested individually, as a pair, and with a simulated *Surveyor* payload (GRC–1963–P–01700).

Image 207:    A Centaur stage is lowered through the dome and into the SPC vacuum tank (GRC–1964–C–68846).

Image 208:    AC–2 on a launch pad at Cape Canaveral in November 1963. It was the first successful launch of a liquid-hydrogen rocket. The Centaur upper stage from the AC–2 launch remains in orbit today (GRC–2015–C–06539).

Image 209:    Researchers prepare a Centaur-Surveyor nose cone shroud for a separation test in the SPC vacuum tank (GRC–1964–C–71091).

Image 210:    Lewis's new DEB.

Image 211:    Lewis launch team monitors the Thor-Agena launch of an Orbiting Geophysical Observatory satellite in 1965 (GRC–2015–C–60541).

Image 212:    Bill Harrison films a test analyzing the effect of a lander's jets on simulated Moon dust. This experimental tank was located in the 8×6 complex (GRC–1960–C–53768).

Image 213:    The new Space Power Facility opened in late 1969. It was one of three world-class facilities brought online at Plum Brook Station (GRC–1969–C–03156).

Image 214:    A test capsule is suspended over the mouth of the Zero Gravity Research Facility (Zero-G) prior to the first drop test on 6 June 1966 (GRC–1966–C–02290).

Image 215:    Zero-G drop preparations in September 1966 (GRC–1966–C–03685).

Image 241:    Damage to the PSL Equipment Building (GRC–1971–C–01422).

Image 242:    Wrecked F–8 with Pinkel, seen to the right, who led the ensuing investigation (GRC–1969–C–02422).

Image 243:    A Centaur D1–A rocket is readied for a test firing inside the B–2 vacuum chamber. The test chamber, 55 feet high by 33 feet in diameter, can handle rockets up to 22 feet long (GRC–1969–C–02596).

Image 244:    The shroud for Skylab installed inside the SPF vacuum chamber for a jettison test. The shroud enclosed the multiple docking adapter, the top of the airlock, and the *Apollo* telescope mount. Problems with the ejection system were found during two tests in winter 1970. The issues were remedied, and the shroud was successfully jettisoned at a simulated 330,000-foot altitude in June 1971 (GRC–1969–C–03690).

Image 245:    Irving Pinkel's retirement mat signed by the Silversteins, Bruce Lundin, Bernie Lubarsky, and others (NASA Pinkel Collection).

Image 246:    Silverstein bids Christine Truax farewell as she retires as head of the Computing Section in July 1967. She had joined the NACA in the early 1940s (GRC–2015–C–06573).

Image 247:    Silverstein badge for *Apollo 11* launch.

Image 248:    Lewis's front entrance reflects the center's new focus (GRC–1980–C–04980).

Image 249:    A 5,000-pound-thrust rocket engine is fired at the Rocket Engine Test Facility as part of a thermal fatigue investigation in September 1975 (GRC–1975–C–03125).

Image 250:    Lundin memo announcing the termination of the nuclear program.

Image 251:    The staff gathers in the control room on 5 January 1973 as the Plum Brook Reactor is shut down one final time (GRC–2003–C–00847).

Image 252:    Silverstein and Lundin talk at a 1968 reception for Lundin at the Guerin House (GRC–2015–C–06553).

Image 253:    NASA's F–106B Delta Dart was acquired as the chase plane for the center's first F–106B. After that program ended, the chase plane was equipped with air-sampling and ocean-scanning equipment and performed remote sensing throughout the 1970s. The ocean-scanning equipment was stored in the nose section of the F–106B (GRC–1979–C–02423).

Image 254:    A NASA OV–10 aircraft participates in the Project Icewarn program during March 1973 (GRC–1973–C–00948).

Image 255:    Two-way ship traffic through Neebish Channel, Michigan, in January 1976 (GRC–1976–C–00365).

Image 256:    A J–58 engine in a new test chamber in the Propulsion Systems Laboratory (PSL) with a swirl-can combustor on display (GRC–1973–C–03376).

Image 257:    Lundin with Gerald Soffen, Cleveland native and Chief Scientist for the *Viking* missions, at an ALERT event in the Developmental Engineering Building auditorium in February 1977 (GRC–1977–C–02740).

Image 258:    Acting Director Bernard Lubarsky takes questions from staff members during a 13 April 1978 AWARENESS forum (GRC–1978–C–01278).

Image 259:    Display for the Big Boost From Rockets presentation at the 1973 Inspection, which discussed the applications of rocket technology for everyday life (GRC–1973–C–3372).

Image 260:    Bruce Lundin watches as NASA Administrator James Fletcher and ERDA Administrator Robert Seamans start the Plum Brook wind turbine for the first time (GRC–1975–C–03866).

Image 261:    Lewis engineers set up a Mod–0A–2 wind turbine in Culebra, Puerto Rico (GRC–1978–C–02389).

Image 262:    Wind energy program pamphlet.

Image 263:    Installation of 2–MW wind turbine in Goldendale, Washington, in November 1980 (GRC–1980–C–05886).

Image 264:    The Boeing 7.2-MW wind turbine began operation in Oahu in July 1987 (GRC–1987–C–05991).

Image 265:    Robert Ragsdale briefs Senator Howard Metzenbaum on the Solar Simulation Laboratory during a visit to Lewis on 13 February 1974 (GRC–1974–C–00599).

Image 266:    Residents of Schuchuli, Arizona, attend the dedication ceremony for the solar village project (GRC–1978–C–05107).

Image 267:    Medical personnel treat patients in the cyclotron area (GRC–1978–C–04709).

Image 268:    American Motors Pacer vehicle modified to run on twenty 6-volt batteries (GRC–1977–C–02096).

Image 305: Atlas-Centaur 1 shroud jettison test at the newly reactivated Space Power Facility (SPF) for the upcoming Combined Release and Radiation Effects Satellite (CRRES) launch. It was the first major hardware test at the SPF in over 15 years (GRC–1990–C–00216).

Image 306: Shuttle/Centaur Program report.

Image 307: Centaur G on a work stand being prepared for the Ulysses mission at the Vertical Processing Center on 6 February 1986 (NASA KSC–86PC–0088).

Image 308: Artist's rendering of the ACTS deployment from *Discovery* (GRC–1987–C–01695).

Image 309: ACTS antennas mounted on the 8- by 6-Foot (8×6) Supersonic Wind Tunnel (GRC–1996–C–03369).

Image 310: OAI (GRC–1991–C–01824).

Image 311: John Klineberg (GRC–1987–C–75023).

Image 312: Plum Brook status report.

Image 313: Interior of the B–2 test chamber with the lid removed in April 1987 (GRC–1987–C–02665).

Image 314: Ross instituted a regular column in the *Lewis News*.

Image 315: Larry Ross (right) meets with Bob Angus in the Administration Building during August 1993 (GRC–1993–C–06422).

Image 316: Ross as a Centaur test engineer in the early 1960s (GRC–2008–C–01424).

Image 317: Conceptual drawing of a lunar-oxygen-augmented nuclear thermal rocket (NTR) (GRC–1997–C–00827).

Image 318: NASA Technical Memorandum about NTR.

Image 319: William Klein examines settings for the slush hydrogen test rig in the K Site vacuum chamber during June 1991 (GRC–1991–C–07458).

Image 320: 13-foot-diameter propellant tank installed in K Site's 25-foot-diameter vacuum chamber (GRC–1967–C–03315).

Image 321: Historian Virginia Dawson conducting research in 1984 for her history of the center, *Engines and Innovation* (GRC–1984–C–03885).

Image 322: Time capsule installed in front of the Administration Building in 1991 (GRC–1991–C–08985).

Image 323: Fluids and Combustion Facility during testing at the Structural Dynamics Laboratory (GRC–2004–C–01827).

Image 324: Technician analyzes a NASA Solar Electric Power Technology Application Readiness (NSTAR) thruster in the Electric Propulsion Laboratory (EPL) during December 1993 (GRC–1993–08855).

Image 325: Dan Goldin began his professional career in 1962 working on electric propulsion systems at Lewis. He left NASA for TRW in 1967 and worked his way up to be the firm's vice president (GRC–1962–C–60944).

Image 326: Artist's rendering of Space Station Freedom's Alpha design (GRC–1994–C–00566).

Image 327: Goldin and Campbell converse in the Lewis hangar in August 1998 (GRC–1998–C–01638).

Image 328: Space Experiments Division's internal review for NASA's Zero Base Review.

Image 329: ISS solar array testing in the Space Power Facility (SPF) (GRC–2000–C–00279).

Image 330: Testing for the Mir Cooperative Solar Array (GRC–1995–C–00994).

Image 331: Titan-Centaur launch of the *Cassini*/Huygens spacecraft on 15 October 1997 (NASA KSC–97PC–1543).

Image 332: Twin peaks on the martian surface as seen by Sojourner (NASA).

Image 333: *Pathfinder* airbag drop in the SPF (GRC–1995–C–01614).

Image 334: *Lewis News* article about Lewis's experiments on *Pathfinder*.

Image 335: NSTAR thruster test (GRC–2015–C–06537).

Image 336: A researcher prepares the NSTAR thruster for a 2000-hour wear test in the EPL during November 1994 (GRC–1994–C–05234).

Image 337: Flight Operations veterans Kurt Blankenship and Bill Rieke flying the center's Learjet (GRC–2001–C–003108).

Image 338: Aircraft consolidation plan.

the development of the nation's aerospace technology. The symposium was one of a series of activities held across the nation in 2003 to commemorate the 100th anniversary of the Wright Brothers' initial flight in 1903 (GRC-2003-C-02005).

Image 369: NASA's Prometheus Project Final Report.

Image 370: A rectangular HiPEP thruster being removed from the EPL vacuum tank in June 2005 after the duration test (GRC-2005-C-01061).

Image 371: Testing of a prototype NEXT in the EPL during May 2006 (GRC-2006-C-01260).

Image 372: Ivanovich Anatoli Vassine poses with a Russian T-160E Hall thruster being tested at the center in November 1997 (GRC-1997-C-04095).

Image 373: HiVAC thruster installed in tank 5 of the EPL during May 2013 (GRC-2013-C-03742).

Image 374: Glenn economic impact document.

Image 375: Michael Griffin visits Glenn on 16 May 2005, one month after being confirmed as NASA Administrator (GRC-2005-C-00704).

Image 376: While NASA was still defining the Constellation mission and parameters, Glenn sought ways to participate in the program. In 2005 Glenn teamed with Marshall to design the *Orion* service module mock-up. The Glenn Fabrication Shop then created an 18-foot-diameter full-scale model with a 180° cutaway highlighting the module's internal components. Ever since, it has been on display next to the Administration Building (GRC-2007-C-00566).

Image 377: Model of the Ares V vehicle built behind the Administration Building (GRC-2009-C-01301).

Image 378: Associate Administrator for Aeronautics Lisa Porter is briefed on Glenn's turbomachinery accomplishments during a visit in April 2006 (GRC-2006-C-00765).

Image 379: Center Director Woodrow Whitlow and Deputy Director Rich Christiansen (GRC-2005-C-01706).

Image 380: Display in the Research Analysis Center, which housed Glenn's Constellation staff. The display featured historical photographs and artifacts such as a *Gemini* capsule and *Mercury* escape tower (GRC-2010-C-0206).

Image 381: Ares I-X engineering team emulates their Apollo predecessors during a review meeting in July 2007 (GRC-2007-C-01580).

Image 382: Testing of liquid oxygen/liquid-methane thruster for *CEV* (GRC-2009-C-01936).

Image 383: Stacking of two Ares I-X segments in the Fabrication Shop, which was renamed the "Ares Manufacturing Facility" (GRC-2008-C-00421).

Image 384: Constellation Program managers tour the Ares I-X segment assembly in the PSF (GRC-2008-C-00836).

Image 385: Glenn employees view the Ares I-X launch on 28 October 2009 from the Visitor Center (GRC-2009-C-03936).

Image 386: *MPCV* pamphlet.

Image 387: The acoustics chamber in the SPF facility to be used for vehicle vibrational and structural dynamics testing (GRC-2011-C-04392).

Image 388: NASA's new S-3 Viking cruising along the Lake Erie shoreline in March 2010 (GRC-2010-C-01394).

Image 389: View from the cockpit of Glenn's S-3 Viking (GRC-2010-C-03192).

Image 390: Don Campbell is briefed on the Honeywell Weather Information Network, which was one of the off-the-shelf technologies tested during the first phase of WINCOMM (GRC-1999-C-02205).

Image 391: Mary Reveley and three colleagues were among a team that received the prestigious Collier Trophy in 2009 recognizing the Aviation Safety Program's success (*Aerospace Frontiers*, July 2009).

Image 392: PSL's No. 3 test chamber was modified for high-altitude engine icing tests (GRC-2012-C-04153).

Image 393: The Dr. Edward R. Sharp Alcove of Honor. In 2007 the pick and shovel used for the groundbreaking events at Lewis Field in 1941 and Plum Brook in the 1956 were found in a storage barn (GRC-2008-C-01791).

*Image 414: Maureen Umstead with a Sonic Boom Model in the Abe Silverstein 10- by 10-Foot Supersonic Wind Tunnel (10×10) (GRC–2005–C–00926).*

# Acronym List

| | |
|---|---|
| 8×6 | 8- by 6-Foot Supersonic Wind Tunnel |
| 9×15 | 9- by 15-Foot Low-Speed Wind Tunnel |
| 10×10 | Abe Silverstein 10- by 10-Foot Supersonic Wind Tunnel |
| AC | Atlas-Centaur |
| ACEE | Aircraft Energy Efficiency Program |
| ACTS | Advanced Communications Technology Satellite |
| AEC | Atomic Energy Commission |
| AEDC | Arnold Engineering Development Center |
| AERL | Aircraft Engine Research Laboratory |
| AIAA | American Institute of Aeronautics and Astronautics |
| ALERT | Alerting Lewis Employees on Relevant Topics |
| AMG | AM General |
| ANP | Aircraft Nuclear Propulsion |
| ASE | Automotive Stirling Engine |
| ASRDI | Aerospace Safety Research and Data Institute |
| AST | Advanced Subsonic Technology |
| ATP | Advanced Turboprop |
| ATS | Applications Technology Satellite |
| AWARENESS | Acquainting Wage Board, Administrative and Research Employees with New Endeavors of Special Significance |
| AWT | Altitude Wind Tunnel |
| B–1 | High Energy Rocket Engine Research Facility |
| B–2 | Space Propulsion Research Facility |
| B–3 | Nuclear Rocket Dynamics and Control Facility |
| CADDE | Computer Automated Digital Encoder |
| CAIB | Columbia Accident Investigation Board |
| CEV | Crew Exploration Vehicle |
| CIR | Combustion Integration Rack |
| CM–2 | Combustion Module-2 |
| COLD–SAT | Cryogenic On-Orbit Liquid Depot Storage, Acquisition, and Transfer Satellite |
| CPC | IBM Card Programmed Electronic Calculator |
| CPU | central processing unit |
| CRRES | Combined Release and Radiation Effects Satellite |
| CSD | Computer Services Division |
| CTS | Communications Technology Satellite |
| DEB | Development Engineering Building |
| DNT | Dinitrotoluene |
| DOE | Department of Energy |
| DOS | disk operating system |
| ECI | Engine Component Improvement |
| ECL | Energy Conversion Laboratory |
| E$^3$ | Energy Efficient Engine |
| EIDI | Electro-Impulse Deicing System |
| EPA | Environmental Protection Agency |
| EPDM | Electric Propulsion Demonstration Module |
| EPL | Electric Propulsion Laboratory |
| EPRB | Electric Propulsion Research Building |

| | |
|---|---|
| ERDA | Energy Research and Development Administration |
| ESA | European Space Agency |
| FAA | Federal Aviation Administration |
| FADEC | Full Authority Digital Engine Control |
| FCF | Fluids and Combustion Facility |
| FIR | Fluids Integration Rack |
| FLTSATCOM | Fleet Satellite Communications |
| GAP | General Aviation Propulsion |
| GASP | Global Air Sampling Program |
| GEnx | next-generation General Electric engine |
| HiPEP | High Power Electric Propulsion |
| HISS | Helicopter Icing Spray System |
| HiVAC | High Voltage Hall Accelerator |
| HSR | High Speed Research |
| HTF | Hypersonic Tunnel Facility |
| IRT | Icing Research Tunnel |
| ISS | International Space Station |
| JATO | jet-assisted take off |
| JPL | Jet Propulsion Laboratory |
| K Site | Cryogenic Tank Storage Site |
| LDI | lean direct injection |
| LeSAC | Lewis Social Activities Committee |
| LEWICE | LEWis ICE accretion program |
| LIMS | Lewis Information Management System |
| LINK | Lewis Information Network |
| MASTIF | Multi-Axis Space Test Inertia Facility |
| MIC | Mission Integration Center |
| *MPCV* | *Multi-Purpose Crew Vehicle* |
| MSL | Microgravity Science Laboratory |
| MTI | Mechanical Technology Incorporated |
| NACA | National Advisory Committee for Aeronautics |
| NASA | National Aeronautics and Space Administration |
| NASP | National Aerospace Plane |
| NEPA | Nuclear Energy for the Propulsion of Aircraft |
| NERVA | Nuclear Engine for Rocket Vehicle Application |
| NEXT | NASA's Evolutionary Xenon Thruster |
| NOAA | National Oceanic and Atmospheric Administration |
| $NO_x$ | nitrogen oxide |
| NSF | National Science Foundation |
| NSTAR | NASA Solar Electric Power Technology Application Readiness |
| NTR | nuclear thermal rockets |
| OAI | Ohio Aerospace Institute |
| OAO | Orbiting Astronomical Observatory |
| OART | Office of Advanced Research and Technology |
| OAST | Office of Aeronautics and Space Technology |
| OMB | Office of Management and Budget |
| PBOW | Plum Brook Ordnance Works |
| PBRF | Plum Brook Reactor Facility |
| PSF | Power Systems Facility |
| PSL | Propulsion Systems Laboratory |

| | |
|---|---|
| QCSEE | Quiet Clean Short Haul Experimental Engine |
| RAC | Research Analysis Center |
| RAM | random access memory |
| RCC | reinforced carbon-carbon |
| RETF | Rocket Engine Test Facility |
| RQL | Rich-burn/Quick-mix/Lean-burn |
| SAMS | Space Acceleration Measurement System |
| SCaN | Space Communications and Navigation |
| SEI | Space Exploration Initiative |
| *SERT–I* | *Space Electric Rocket Test* |
| *SERT–II* | *Space Electric Rocket Test II* |
| S–IB | Saturn IB |
| S–IC | Saturn IC (Saturn V second stage) |
| SLAR | Side Looking Airborne Radar |
| SLS | Space Launch System |
| SNAP | Systems for Nuclear Auxiliary Power |
| SNPO | Space Nuclear Propulsion Office |
| SOHO | Solar Heliospheric Observatory |
| SPC | Space Power Chambers |
| SPF | Space Power Facility |
| SPHINX | Space Plasma, High Voltage Interaction Experiment |
| SST | supersonic transport vehicle |
| STDCE | Surface-Tension-Driven Convection Experiment |
| STEX | Space Technology Experiments |
| STG | Space Task Group |
| STS | space transportation system |
| TALON | Technology for Advanced LOw NOx |
| TAPS | twin-annular pre-mixing swirler |
| TIALS | Technical Information, Administrative, and Logistics Services |
| TNT | trinitrotoluene |
| UDF | Unducted Fan |
| UEET | Ultra-Efficient Engine Technology |
| U.S. | United States |
| USML–1 | U.S. Microgravity Laboratory 1 |
| USML–2 | U.S. Microgravity Laboratory 2 |
| VIP | very important person |
| VSE | Vision for Space Exploration |
| VSP | Vehicle Systems Program |
| V/STOL | vertical or short-takeoff and landing |
| WAA | War Assets Administration |
| WINCOMM | Weather Information Communications |
| Zero-G | Zero Gravity Research Facility |

*Image 415: Delivery of the European Service Module for the Orion spacecraft to the Space Power Facility in November 2015. The module, built by the European Space Agency, will undergo a battery of tests at the facility in 2016 (GRC–2015–C–07166).*

# Index

*Bold numbers indicate that the topic is depicted in an image.*

Lewis Information Network (LINK) 220
See also Employees, Positions; Engineering Research Associates; International Business
Machines (IBM); and Cray, Inc.
COMSAT Corporation 175, 217
Conferences 172
Aircraft Noise Reduction Conference 167
Conference for Power Industry **172**
Conference on Aerospace Related Technology for Industry in Commerce **172**
Conference on Fuels **83**
Flight Propulsion Conference 114, **115**, 116, 124, 142
International Workshop on Aircraft Icing 235
Propfan Acoustics Workshop **213**
Rotor Icing Consortium 236
Space Power Conference **178**
Wind Energy Workshop 200
See also Inspections
Congress, U.S. 17, 124, 272, 310, 317, 330
Acts
Authorization Act of 1974 200
Clean Air Act 197
National Aeronautics and Space Act of 1958 126
Unitary Plan Wind Tunnel Act 107
Veterans Administration—Housing and Urban Development Appropriation 286
Establishment of AERL 11, 12, 24
Establishment of NASA 5, 123, 125
Establishment of the NACA 4
Federal downsizing 261–263, 330–332
Funding and support 115, 133, 137, 193, 194, 199, 222, 235, 243, 261–264, 275, 277, 310, 330
Hearings 206
Centaur 152
NASA Appropriations 273
Space station 241, 262
Human spaceflight 297, 303
Ohio Delegates 96, 222, 249, 286, 317, 330, **331**
Consolidated Aircraft Corporation
B–24D Liberator 44, **71**, **72**, 73, 74, 97
PBY Catalina 72
Constellation Program 5, 303, 304, **309**, 310
Glenn contributions 311, **312**, 313, 314, 315, 317
Program restructure 311, 316, 328, 330
See also Ares rocket, *Crew Exploration Vehicle (CEV)*, and Vision for Space Exploration (VSE)
Construction (Glenn) **8**, **10**, 15, 20, 21, 23, 24, **26**, 27
Dedication 24, **26**
Design 12, 16, **26**, 27
Groundbreaking 14, 253, **321**, 322
Site selection 13, 21
Convair Corporation
B–58 Hustler 107, 116
F–106B Delta Dart **165**, 166, **195**, 196

# L

# M

North American Aviation Rocketdyne
>    AJ–2  135, **148, 149**, 171, 279, 345
Northrop Corporation, see Northrup Grumman Corporation
Northrup Grumman Corporation
>    Gulfstream I (NASA 5)  271
>    Gulfstream II  **234**, 235
>    P–61 Black Widow  65
>    T–38 Talon  180
>    V/STOL engines  208
Nozzles (aircraft engine)  67, 97, 197, 274
>    Chevron  275, 276, **277**
>    Noise reduction  103, 166
>    Variable-area nozzles  68
Nozzles (rocket engines)  83, 146, 216
>    Base heating  135
>    Regenerative cooling  115, 145, 251
Nuclear Energy for the Propulsion of Aircraft (NEPA)  94, 95, 135, 145
Nuclear Engine for Rocket Vehicle Application (NERVA)  146, 147, 250
>    Lewis management  123, 146
>    Lewis testing  146, 169
Nuclear propulsion  91, 93, 94, 142, 145, 155
>    Nuclear-powered aircraft  92, 94, 135, 145
>        See also Nuclear Energy for Propulsion of Aircraft (NEPA) and Aircraft Nuclear Propulsion (ANP)
>    Nuclear-powered rocket  **122**, 145, 146, 249
>        Nuclear thermal rockets  **249, 250**
>        See also Nuclear Engine for Rocket Vehicle Application (NERVA) and Project Rover
Nuclear Regulatory Commission, see Atomic Energy Commission (AEC)
Nuclear Rocket Dynamics and Control Facility (B–3)  139, **140**, 146
>    Testing  **145, 146**, 164, 214

# O

Oak Ridge National Laboratory  94
Oakar, Mary Rose  222, 249
Obama, Barak  297, 316
Oberth, Herman  142
Ocean scanning  **195**, 196
Office of Advanced Research and Technology, see Office of Aeronautics and Space Technology
Office of Aeronautics and Space Technology (OAST)  137, 138, 194, 199
Office of Management and Budget (OMB)  246
Ohio  70
>    Ohio River  315
>    See also Cleveland and Great Lakes
Ohio Aerospace Institute (OAI)  **245**, 246
O'Keefe, Sean  **297**, 298, 301, 303, 304, 308
Olson, Sandra  283
Olson, Walter T.  44, 82, 83, 134, 142, 194, 233
>    NACA role in space  115, 124
>    Technology utilization  172, 199

# S

# V

# W